international AIR POWER REVIEW

AIRtime Publishing

United States of America • United Kingdom

international AIR POWER REVIEW

Published quarterly by AIRtime Publishing Inc.
120 East Avenue, Norwalk, CT 06851
Tel (203) 838-7979 • Fax (203) 838-7344

© 2004 AIRtime Publishing Inc.
F/A-18 and F-100 cutaways © Mike Badrocke
Photos and other illustrations are the copyright of their respective owners

Softbound Edition ISSN 1473-9917 / ISBN 1-880588-60-9
Hardcover Deluxe Casebound Edition ISBN 1-880588-61-7

Publisher
Mel Williams

Editor
David Donald   e-mail: airpower@btinternet.com

Assistant Editor
David Willis

Sub Editor
Karen Leverington

US Desk
Tom Kaminski

Russia/CIS Desk
Piotr Butowski, Zaur Eylanbekov   e-mail: zaur@airtimepublishing.com

Europe and Rest of World Desk
John Fricker, Jon Lake

Correspondents
Australia: Nigel Pittaway
Belgium: Dirk Lamarque
Brazil: Claudio Lucchesi
Bulgaria: Alexander Mladenov
Canada: Jeff Rankin-Lowe
France: Henri-Pierre Grolleau
Greece: Konstantinos Dimitropoulos
India: Pushpindar Singh
Israel: Shlomo Aloni
Italy: Luigino Caliaro
Japan: Yoshitomo Aoki
Netherlands: Tieme Festner
Romania: Danut Vlad
Spain: Salvador Mafé Huertas
USA: Rick Burgess, Brad Elward, Mark Farmer (North Pacific region), Peter Mersky, Bill Sweetman

Artists
Mike Badrocke, Chris Davey, Juanita Franzi, Keith Fretwell, John Weal, Iain Wyllie

Designer
Zaur Eylanbekov

Controller
Linda DeAngelis

Origination by Universal Graphics, Singapore
Printed in Singapore by KHL Printing

All rights reserved. No part of this publication may be copied, reproduced, stored electronically or transmitted in any manner or in any form whatsoever without the written permission of the publishers and copyright holders.

International Air Power Review is published quarterly in two editions (Softbound and Deluxe Casebound) and is available by subscription or as single volumes. Please see details opposite.

Acknowledgments
We wish to thank the following for their kind help with the preparation of this issue:

Larry Davis
Robert F. Dorr
Peter Liander
Terry Panopalis

The authors of the Escuadrilla Aeronaval de Exploración article wish to thank Capitán de Corbeta Alejandro L. Angarola, Capitán de Corbeta J.M. Pernuzzi and the squadron as a whole.

The editors welcome photographs for possible publication but can accept no responsibility for loss or damage to unsolicited material.

Contact and Ordering Information (hours: 9am-5pm EST, Mon-Fri)
addresses, telephone and fax numbers

International Air Power Review, P.O. Box 5074, Westport, CT 06881, USA
 Tel (203) 838-7979 • Fax (203) 838-7344
 Toll free within USA and Canada: 1 800 359-3003
 Toll free from Australia (13-15 hours ahead): 0011 800 7573-7573
 Toll free from New Zealand (17 hours ahead): 00 800 7573-7573
 Toll free from Japan (14 hours ahead): 001 800 7573-7573

International Air Power Review, Postbus 3946, 4800 DX Breda, Holland
 Toll free (to our US East Coast office) from the United Kingdom, Belgium, Denmark, France, Germany, Holland, Ireland, Italy, Luxembourg, Norway, Portugal, Sweden and Switzerland (5-6 hours ahead): 00 800 7573-7573
 Toll free from Finland (6 hours ahead): 990 800 7573-7573

website
 www.airtimepublishing.com
e-mails
 airpower@airtimepublishing.com
 inquiries@airtimepublishing.com

Subscription & Back Volume Rates

One-year subscription (4 quarterly volumes), inclusive of ship. & hdlg./ post. & pack.:

Softbound Edition
USA $59.95, UK £48, Europe EUR 88, Canada Cdn $99, Rest of World US $79 (surface) or US $99 (air)

Deluxe Casebound Edition
USA $79.95, UK £68, Europe EUR 120, Canada Cdn $132, Rest of World US $99 (surface) or US $119 (air)

Two-year subscription (8 quarterly volumes), inclusive of ship. & hdlg./ post. & pack.:

Softbound Edition
USA $112, UK £92, Europe EUR 169, Canada Cdn $187, Rest of World US $148 (surface) or US $188 (air)

Deluxe Casebound Edition
USA $149, UK £130, Europe EUR 232, Canada Cdn $246, Rest of World US $187 (surface) or US $227 (air)

Single/back volumes by mail (each):

Softbound Edition
US $16, UK £10.95, Europe EUR 18.50, Cdn $25.50
(plus ship. & hdlg./post. & pack.)

Deluxe Casebound Edition
US $20, UK £13.50, Europe EUR 22, Cdn $31
(plus ship. & hdlg./post. & pack.)

Prices are subject to change without notice. Canadian residents please add GST. Connecticut residents please add sales tax.

Ship. & Hdlg./ Post. & Pack. Rates
(for back volume and non-subscription orders)

	USA	UK	Europe	Canada	ROW (surface)	ROW (air)
1 item	$4.50	£4	EUR 8	Cdn $7.50	US $8	US $16
2 items	$6.50	£6	EUR 11.50	Cdn $11	US $12	US $27
3 items	$8.50	£8	EUR 14.50	Cdn $14	US $16	US $36
4 items	$10	£10	EUR 17.50	Cdn $16.50	US $19	US $46
5 items	$11.50	£12	EUR 20.50	Cdn $19	US $23	US $52
6 or more	$13	£13	EUR 23.50	Cdn $21.50	US $25	US $59

**Volume Eleven
Winter 2003/2004**

CONTENTS

AIR POWER INTELLIGENCE
Programme Update 4
Project Development 4
Upgrades and Modifications 6
Procurement and Deliveries 8
Air Arm Review 13
*John Fricker and Tom Kaminski;
additional material by David Donald, Nigel Pittaway
Jos Schoofs, Holger Stüben and Jens Schymura, and
Roberto Yañez*

DEBRIEF
Austria's Black Hawks 16
Erich Strobl
F-16C Block 30/JDAM 17
David Donald
Canadair CT-133 Silver Star 18
Stefan Degraef and Edwin Borremans
F-16C/D Block 42 re-engining 20
Tom Wolfe

TECHNICAL BRIEFING
Erieye: Sweden's Eyes on the World 22
Robert Hewson

PHOTO-FEATURE
Austria's Drakens:
Last of the Dragons 32
Luigino Caliaro

FOCUS AIRCRAFT
Boeing F/A-18E/F Super Hornet 36
Bigger and smarter than its predecessor,
the Super Hornet is entering service at
a considerable rate to replace elderly
Tomcats and first-generation Hornets.
Brad Elward

VARIANT FILE
US Military King Airs: Part 1
Beech/Raytheon U-21 and C-12 74
Tom Kaminski

COMBAT COLOURS
Escuadrilla Aeronaval de Exploración 94
Santiago Rivas and Juan Carlos Cicalesi

AIR COMBAT
Marine Corsairs in Korea 104
Warren Thompson

WARPLANE CLASSIC
North American F-100 Super Sabre 120
Jon Lake

TYPE ANALYSIS
Heinkel He 177 Greif 158
Dr Alfred Price

INDEX 174

MAJOR FEATURES planned for VOLUME TWELVE
Focus Aircraft: Bell AH-1 Cobra, **Warplane Classic:** Avro Lancaster, **Variant Briefing:** US Military King Airs Part 2,
Air Power Analysis: Russia Part 1, **Special Report:** Last of the Rhinos, **Photo Feature:** Swiss air force,
Special Feature: RAAF Mirage III, **Pioneers & Prototypes:** Convair F2Y Sea Dart

Air Power Intelligence

Programme Update

F-16E/F Block 60 for the UAE

The perennial Lockheed Martin F-16 entered a new era on 6 December 2003 with the first flight of the Block 60 aircraft developed for the United Arab Emirates. Steve Barter, LM's F-16 chief test pilot, took aircraft 3001/N161LM aloft at Fort Worth for a 50-minute maiden flight. To mark the considerable advances of this version, it has been designated F-16E (single-seater) and F-16F (two-seater). In the 1980s these designations were provisionally to have been assigned to the single- and two-seat versions of the F-16XL cranked-arrow wing aircraft if it had been adopted by the USAF.

The UAE Air Force has ordered 55 F-16Es and 25 F-16Fs, with first deliveries slated for April 2004 and all to be complete by 2007. The aircraft features the 'big spine' first seen on Israeli F-16Ds, and also has overwing conformal fuel tanks as fitted to Greece's latest Block 52+ aircraft and the F-16I. Power is provided by the General Electric F110-GE-132 engine, which provides a considerable thrust increase – to 32,500 lb (144.62 kN) with afterburning.

Internally, however, the Block 60 represents a new generation of F-16, with a major revision to the structure, flight control system and avionics. At the heart of the aircraft's increased capabilities is the Northrop Grumman APG-80 Agile Beam Radar, which uses active electronically-steered antenna technology. Fitment of the radar requires the removal of the nose-mounted pitot probe. The radar is augmented by an Integrated FLIR Targeting System (IFTS), also by Northrop Grumman. This comprises a steerable navigation FLIR turret on the nose, and a streamlined targeting pod under the intake. A completely revised Falcon Edge EW system is incorporated, including jammers and towed decoys. The cockpit is a thoroughly modern work-station, each position being provided with three 5 x 7-in (12.7 x 17.8-cm) colour

For its early test work in the US the first F-16F Block 60 (known as 'RF01') has a temporary US civil registration applied. The fixtures on the rear fuselage are mounts for a spin recovery chute which will be fitted when aerodynamic envelope trials are performed.

Project Development

China

New fighter flies
On 25 August 2003 Chengdu flew the first prototype of its JF-17 Thunder fighter. This project has been under development for many years, and was originally known as the FC-1 Super 7. The aircraft is being developed by Chengdu primarily for the Pakistan Air Force, which requires up to 150 – mainly to replace the Mirage III/5 fleet. The long gestation of the aircraft is attributed to the lack of suitable Western avionics, several manufacturers having been refused export licenses. The engine is the Klimov RD-93, but no radar has been chosen. The FIAR Grifo is a likely candidate.

Italy

767 Tanker Transport
Boeing has begun modifying the first 767-200ER into 767 Tanker Transport configuration at its Wichita Development and Modification Center in Kansas. Destined for service with the Italian Air Force, the aircraft will be equipped with a number of modifications that include the installation of the advanced remote aerial refuelling operator station, refuelling boom, wing pods and centreline hose-and-drogue refuelling equipment, refuelling receptacle, freighter/passenger interior and cargo door. Additional military, navigation and communications equipment will also be installed.

Russia

Yak-130 production
The first production Yak-130 was completed in the autumn of 2003 at the Sokol factory at Nizhny Novgorod. The first aircraft is expected to be delivered to the training school at Krasnodar in 2004.

South Korea

T-50 LIFT takes flight
On 29 August 2003 KAI (Korean Aircraft Industries) flew the first example of the Lead-In Fighter Trainer version of the T-50 Golden Eagle. The T-50 LIFT differs from the standard trainer by being fitted with a 20-mm cannon and an APG-67(V)4 radar. Israel is to evaluate the T-50 LIFT as a possible TA-4 replacement.

United States

Firescout systems demonstrated
Northrop Grumman recently conducted a demonstration that verified the compatibility of its RQ-8A Fire Scout vertical take-off and landing tactical unmanned air vehicle (VTUAV) with shipboard equipment. Operating from Naval Base Ventura County/Point Mugu, the air vehicle responded to commands received from a Tactical Control System (TCS) aboard the USS *Denver* (LPD 9) which was operating at sea. The ship was equipped with the Tactical Control System (TCS). During the mission the air vehicle launched from Point Mugu, flew for more than an hour in the vicinity of the ship, and was recovered safely ashore. In addition to demonstrating the TCS, the flight test also verified the

Typhoons for Spain

A ceremony held at EADS/CASA Getafe plant on 9 October 2003 marked the official entry into Spanish Air Force service of the first two Ala 11 EF-2000 Typhoons. The new aircraft, registered CE16-01/11-01 and CE16-02/11-02, were formally accepted by the Spanish AF Chief of Staff, General González-Gallarza in the presence of King Juan Carlos I. Both aircraft will be initially operated from Getafe AB by a detachment of Ala 11 pilots and ground crew personnel, who are receiving training on the aircraft's operation and maintenance from EADS/CASA test pilots and technicians. Also present at the ceremony was the third two-seater for the wing, which was shortly to begin test flying. After receipt of its six two-seaters planned for the first half of 2004, 113 Escuadrón will return to Morón AB to act as the type OCU for Spanish Typhoon pilots. The first single-seaters will also go to Morón-based Ala 11. Ala 14 at Albacete will be the second unit to receive the new aircraft from 2007 onwards.

Roberto Yañez

The first two EdA two-seat Typhoons are seen at Getafe during the entry-to-service ceremony, with the first nearest the camera. At the far end of the line-up is the German-built prototype DA.1, which is temporarily assigned to EADS-CASA for development work. All three took part in a flypast during the ceremony.

Air Power Intelligence

multi-function displays, and a 25° x 25° wide-angle HUD.

Among the new features of the Block 60 is a pre-loadable digital terrain database, which allows the aircraft to undertake automatic low-level terrain-following flight. It can also use the APG-80 for terrain-following. Automatic terrain avoidance and deep-stall/disorientation recovery functions are also included.

The first three F-16Fs will be fully instrumented for testing. As the Block 60 represents such an advance over previous aircraft, a sizeable trials campaign is required, especially as the flight control system is much revised. The test instrumentation is packaged into the gun bay, so that the gun can be fitted when trials work is concluded. The CFTs and F110-GE-132 engine have already been flight-rated on the F-16 Block 50.

Initial production will be to an interim Standard 0, which will allow Block 60 training to commence, initially at Fort Worth. In 2003 UAEAF pilots began conversion to the F-16 Block 50 in the US in preparation for Block 60 training. Standard 1 represents the final hardware configuration, while Standards 2 and 3 are software updates, the last of which is the full contracted-for capability, scheduled to be achieved in 2008.

F-16I for Israel

On 14 November 2003 the Israel Defence Force formally accepted its first F-16I aircraft. Under the Peace Marble V contract the IDF/AF is buying 102 F-16Is, raising its F-16 purchases to 362. Named Soufa (storm) in IDF/AF service, the two-seat F-16I was developed specifically to fulfil an Israeli long-range interdiction requirement and features a big spine for additional avionics, conformal fuel tanks and a Northrop Grumman APG-68(V)9 radar with synthetic aperture ground-mapping function.

For the ceremony the first aircraft was displayed (above) carrying JDAMs and Rafael Python 4 missiles under the wings, with AMRAAMs on the wingtips. The F-16I is powered by the Pratt & Whitney F100-PW-229, which also equips the IDF/AF's F-15I Ra'am fleet. The aircraft on display featured LANTIRN nav/targeting pods, but it can also use the Litening II. The cockpit is fitted with full-colour displays and it can support the helmet-mounted display systems which are in common use with the IDF/AF. The first F-16I flew at Fort Worth on 23 December 2003.

Tactical Common Data Link's (TCDL) ability to downlink data from the UAV's sensors, as well as the latest version of the Unmanned Common Automatic Recovery System (UCARS-V2). UCARS-V2 is the Navy's primary recovery and control system for UAVs operating aboard ships and provides precise position information to the air vehicle during shipboard take-offs and landings. It supports a near all-weather, day and night capability for UAV operations.

Another series of test flights demonstrated the compatibility of a General Atomics synthetic aperture radar/moving target indicator (SAR/MTI) system with the RQ-8A Fire Scout VTUAV system and its ground station. The tests verified the system's operation in conjunction with the RQ-8A's existing payload of electro-optical/infrared (EO/IR) sensors and a laser designator/rangefinder. A subsequent series of tests is planned to evaluate the Northrop Grumman tactical UAV synthetic aperture radar (TUAVR).

J-UCAS team formed

Lockheed Martin has joined Northrop Grumman's joint unmanned combat air system (J-UCAS) team and the two companies will work together to design, develop and produce a UCAV system for the US Navy and Air Force. The Department of Defense will establish a Joint Systems Management Office to manage the new J-UCAS programme.

Global Hawk tests new sensor

The USAF's RQ-4A Global Hawk unmanned aerial vehicle completed the first of five planned flight tests from the German Navy's base at Nordholz, Germany, on 21 October 2003. For these tests, which were intended to demonstrate interoperability between USAF and German Ministry of Defense unmanned aerial vehicle systems, the Global Hawk was equipped with a German-designed electronic-intelligence (ELINT) sensor payload. The UAV arrived in Germany from Edwards AFB, California, on 15 October 2003. Prior to the first flight, the German sensor was installed in the air vehicle and engine runs and taxi tests were conducted.

KC-130J enters OPEVAL

The USMC's KC-130J tanker/transport recently entered into operational evaluation (OPEVAL) with the Navy's VX-1 at NAS Patuxent River, Maryland. The squadron and its contingent of Marines will put a pair of KC-130Js through a series of operational scenarios over a three-month period. The programme will test the aircraft's ability to refuel fixed and rotary-wing aircraft inflight, conduct rapid ground refuelling of aircraft, ground vehicles and fuel dumps, perform aerial delivery of equipment and personnel, conduct low-level operations, as well as operate from "temporary landing zones, non-standard runways and airfields." The aircraft will fly approximately 270 hours during the OPEVAL, which is being conducted under varied conditions from multiple locations including NAF El Centro, California. The 'Hercs' deployed to El Centro in late September 2003 for the initial phase of testing.

UH-1Y and AH-1Z approved

The Defense Acquisition Board (DAB) recently approved the H-1 Upgrade Program to begin Low-Rate Initial Production (LRIP). This milestone gives the Navy and Marine Corps approval to proceed with the remanufacture of six UH-1N Huey and three AH-1W Super Cobra helicopters to the UH-1Y and AH-1Z configuration during FY04. A second LRIP lot, also comprising six UH-1Ys and three AH-1Zs, will be included in the FY05 defence budget. As of October 2003, the H-1 Upgrade integrated test team had completed more than 1,300 flight test hours with five aircraft, comprising three AH-1Z and two UH-1Y test aircraft. With the exception of a single NAH-1Z, all of these aircraft are production representative. The NAH-1Z will eventually be transferred to NAWS China Lake, California, where it will be subjected to destructive live fire testing. The remaining test aircraft will support the upcoming Operational Evaluation (OPEVAL). The successful conclusion of OPEVAL will result in the DAB's approval of full rate production.

Tanker deal progress

The Senate Armed Services Committee has reached a compromise that, if accepted, will allow the USAF to lease 20 Boeing KC-767A tankers while acquiring 80 further examples through a standard purchase. The new plan will reportedly reduce the cost of the project by as much as $4 billion over the original plan to lease 100 aircraft from Boeing. Although the project still faces several hurdles, this is a positive step in the USAF's attempt to begin replacing its elderly fleet of KC-135 tankers. Under this plan Boeing will still meet the delivery schedule set forth in the original tanker lease proposal, which called for delivery of the first four aircraft to the USAF in 2005. The initial squadron of 20 aircraft will be stationed at Fairchild AFB, Washington. Three other congressional panels that had previously approved the 100 aircraft leasing plan must approve the proposal.

'Super-blimp' for warning?

The Department of Defense's Missile Defense Agency (MDA) has selected Lockheed Martin to develop a high-altitude remotely-operated blimp that will monitor the US borders, supplementing ground-based radar and satellites. The contractor has received a $40 million contract to develop the helium-filled blimp, which will be

In late August the USAF selected Boeing over Lockheed Martin to continue development of its Small Diameter Bomb, which will become the principle armament of USAF aircraft in the future. Weighing in the 250-lb (113-kg) class, the SDB measures around 5 ft 11 in (1.8 m) long and is 7.5 in (19 cm) in diameter. Its small dimensions allow a greater number of weapons to be carried. It uses GPS guidance and has fold-out wings to increase its range in some launch scenarios to over 40 miles (64 km). Production at Boeing's St Charles, Missouri, plant is expected to reach 24,000 bombs and 2,000 carriages over the next 10 years, and service entry is slated for October 2005, initially on the F-15E. Here two test bombs are fitted to the rotary launcher of a B-2.

Air Power Intelligence

Raptor 03 lands at Edwards in October 2003. The aircraft wears a large Raptor fin marking, and below the cockpit has missile silhouettes recording 15 AIM-9 Sidewinder launches and seven of the AIM-120 AMRAAM.

Raptor 11 is one of the aircraft undergoing IOT&E at Nellis AFB with the 53rd Wing. It is seen here at Edwards making a precautionary landing after the port side weapon bay doors failed to close.

F/A-22 Raptor

A significant event occurred at Tyndall AFB, Florida, on 26 September 2003 when the first F/A-22A was delivered to the 325th Fighter Wing's 43rd Fighter Squadron. The unit is the first operational unit to fly the Raptor and will serve as the initial training squadron. In related news, Air Combat Command announced that the 27th Fighter Squadron (FS) 'Fighting Eagles' at Langley AFB, Virginia, will be the first of three squadrons assigned to the 1st Fighter Wing (FW) to convert to the F/A-22A Raptor.

152.4 m (500 ft) long, with a diameter of 48.8 m (160 ft) and a volume of 147248 m³ (5.2 million cu ft), making it about 25-times larger than the average advertising blimp in use today. The solar-powered blimp would fly at an altitude of 19812 m (65,000 ft), have a 1814-kg (4,000-lb) payload and will be able to remain airborne for a month. From its stationary position over the earth the blimp will be capable of providing surveillance over an area that is 1287 km (800 miles) in diameter. Ten such blimps could provide complete coverage of the east and west coasts and the southern borders, and a prototype could be flying as early as 2006. The contract includes a $50 million option under which Lockheed Martin Maritime Systems & Sensors will build one blimp. The contractor had already conducted preliminary design work under a $2 million contract from the Missile Defense Agency.

Aerial Common Sensor platform
Northrop Grumman has announced that it has selected Gulfstream Aerospace as a partner in its bid to build the US Army's Aerial Common Sensor (ACS) intelligence, surveillance, and reconnaissance (ISR) system. The company's Gulfstream 450 will serve as the aerial platform for ACS. Based upon the earlier Gulfstream III/IV and current G400 platforms, the G450 was unveiled on 7 October 2003. It shares many features with the current G400 but is 0.30 m (12 in) longer and is powered by an improved Rolls-Royce Tay 611-8C turbofan engine, each of which produces 61.61 kN (13,850 lb). Eight versions of the earlier Gulfstream III and IV variants are already in service with the US Air Force, Army, Navy and Marine Corps under the mission design series designation C-20. The larger G500 is under development for a variety of special mission roles with Israel.

Upgrades and Modifications

Brazil

AMX upgrade plans
In late 2003 Brazil's air force was expected to finally sign a contract covering the upgrade of 53 AMX International A-1s. The upgrade covers a new cockpit with three multi-function displays, NVG compatibility, HOTAS (hands on throttle and stick) controls and a wide-angle head-up display. New avionics include the SCP-01 Scipio radar and a datalink.

Finland

F-18 mid-life update
Finland has announced that it is to upgrade its fleet of Boeing F/A-18 Hornets, and has requested details. The required modifications include JHMCS (Joint Helmet-Mounted Cueing System), TAMMAC (tactical aircraft moving map capability), APX-111 IFF transponders, and launchers compatible with the AIM-9X missile.

India

Re-engined Cheetah
HAL has re-engined a Cheetah (a licence-built version of the Aérospatiale SA 315 Lama) with the Turbomeca TM333-2M2 engine which powers HAL's Dhruv (ALH). The aircraft has also received a number of other modifications, such as new rotors and cockpit instrumentation, and has been christened the Cheetal. Trials have been conducted in the high mountain region.

Pakistan

Upgrades for Fighting Falcons
The US Department of Defense has agreed to provide refurbishment and upgrade of Pakistan's F-16A/B fleet as part of a $341 million defence upgrade. The project will return the Pakistan Air Force's Fighting Falcon

Based at Edwards AFB with the 412th TW, this F-15B carries the ATIMS (Airborne Turret Infrared Measurement System) pods. This multi-sensor data collection system is used to test the effectiveness of countermeasures and to test infra-red missile lock-on ranges. The articulated pod heads can contain missile seekers and IR measuring devices, and the pods can be mounted to face forward or aft. Control is effected from the F-15's back seat, which is fitted with a video display. The F-15/ATIMS is closely involved in F/A-22 Raptor tests.

Latest HUG phase completed

In a ceremony at Williamtown on 27 August 2003, Boeing Australia handed back the final RAAF Hornet Upgrade (HUG) Phase 2.1 aircraft. HUG 2.1 comprised the fitment of Raytheon's AN/APQ-73 radar, an Enhanced Interference Blanking Unit, KY-100 Secure Voice Encryption capability, and the installation of a crash data recorder. Phases 2.2 and 2.3, currently under development, will include the procurement of a Joint Mounted Helmet Cueing System (an example of which has recently been purchased from Boeing for trial), ALE-47 CMDS, Link 16 Datalink, EWSP and further software upgrades.

Concurrently, HUG 3.1 is now underway, with the first aircraft dispatched to Canada for structural survey as part of the International Follow On Structural (IFOS) upgrade. This will determine the make-up of a minor structural upgrade designed to rectify a range of known problems with early generation Hornets. Once HUG 3.1 is defined, Boeing Australia will cycle the RAAF fleet through its Williamtown facility.

HUG 3.2 is a major structural upgrade that would replace the centre fuselage barrels with new-build items, but it is hoped that the NACC type (New Air Combat Capability = F-35) will have entered service before this work needs to be done. The RAAF is keeping a close eye on the Hornet's fatigue index and, if HUG 3.2 proves unnecessary, the entire project will draw to a close in 2007.

No. 75 Squadron recently deployed to Qatar with HUG 2.1 aircraft, flying Combat Air Patrol and Close Air Support sorties as part of the war against Iraq. The RAAF was very pleased with the upgraded Hornet's performance, despite acknowledging that the aircraft was deficient in ECM capability across the whole spectrum (a deficiency to be corrected in HUG 2.3). The Commanding Officer of the RAAF's Air Combat Group, Air Commodore John Quaife, noted that integrated operations with both US and UK forces were achieved without problem.

Nigel Pittaway

Left: During the RAAF's contribution to Operation Iraqi Freedom – known as Operation Falconer – No. 75 Squadron HUG 2.1 Hornets flew fighter and attack missions. For the latter the aircraft routinely carried GBU-12 500-lb (227-kg) laser-guided bombs.

Carrying the 'Grumpy Monkey' insignia of No. 77 Squadron on its spine, a HUG 2.1 aircraft displays its ability to carry four AIM-120 AMRAAMs. The AIM-9s on the wingtips will give way to ASRAAMs.

fleet to 40 aircraft and likely involve the sale of 11 Belgian aircraft to Pakistan. Pakistan also plans to acquire an unidentified airborne early warning and control system (AWACS) aircraft. Both projects still require the approval of the US Congress.

Russia

Su-24MK upgrade

A key part of the Russian air force's modernisation is the upgrade of its Su-24MK 'Fencer-D' fleet, thought to number around 400. A number of aircraft have been upgraded for tests at the GLITs test centre at Akhtubinsk. Both design bureau Sukhoi and the air force's upgrade specialist Gefest i T originally offered competing upgrades, but they have now been combined. At the centre of the programme is an overhauled nav system with GPS, new cockpit displays and enhanced weapons capability, adding the Kh-31P (AS-17 'Krypton') and Kh-59M (AS-18 'Kazoo') weapons to the already impressive Su-24 repertoire. Conformal fuel tanks have also been developed to extend range.

Saudi Arabia

Sentry upgrades

In late 2003 Boeing completed the last installation of upgraded IFF equipment and a new mission computer, and related hardware, on the Royal Saudi Air Force's five Boeing E-3A AWACS aircraft, bringing them into line with the current USAF standard. Further upgrade work is planned, mirroring advances made in the USAF fleet. This will focus on enhanced communications and navigation systems for greater combat effectiveness.

United Kingdom

Reworked Pumas

To augment its hard-worked Puma fleet, the RAF has acquired six SA 330Hs from the South African Air Force. In late 2003 they were undergoing a major overhaul and upgrade work at Eurocopter Romania to bring them up to RAF standard. After a test flight in Romania, the aircraft will be shipped to AgustaWestland's Yeovil factory for final modifications.

United States

Advanced Hawkeye approved

Northrop Grumman has been awarded a $1.9 billion system development and demonstration (SDD) contract associated with the E-2 Advanced Hawkeye. The new model will offer greatly improved threat-detection capabilities over land and water. It will have greater range and precision than existing systems and will provide the basis for the Navy's theatre air missile defence function. It will also be equipped with new communications systems that will enable it to integrate information and surveillance data, and provide forward control and communications capabilities. The Advanced Hawkeye will also feature a new 'glass' cockpit that will allow either the pilot or co-pilot to participate as a fourth mission system operator. It will also be equipped with terrain avoidance systems and global air traffic management system enhancements. Under SDD the contractor will modify two E-2C Hawkeye 2000 aircraft to the new E-2 AHE configuration at its St Augustine, Florida, facility. Operational testing will begin in 2007 and the new aircraft will achieve initial operational capability in 2011.

Final BMUP Orion delivered

The US Navy recently took delivery of the tenth P-3C to be updated under the Block Modification Upgrade Program (BMUP). The BMUP was designed to provide P-3C Update II and II.5 aircraft with the same capabilities as the Update III variant. As part of the BMUP a number of pieces of so-called 'legacy' equipment was replaced by new equipment, including the AN/USQ-78B acoustic processor, AN/ASQ-227 digital tactical computer, flat-panel displays, touch-screen entry panels, enhanced on-line weapons controls and an upgraded electronic support measures system. Although the initial two examples were completed by Lockheed Martin at its Greenville, South Carolina, facility, the remaining eight examples were updated by L3 Communications Integrated Systems in Greenville, Texas.

Tankers with relay station

The USAF has announced plans to equip its KC-135 tanker aircraft with a removable communications relay station. The roll-on beyond-line-of-sight enhancement (ROBE) will enable the Stratotanker to serve as a data relay platform while conducting its air refuelling mission. The ROBE, which is approximately the same size as the galley already in the tanker, is strapped to the floor of the aircraft like any other pallet and will allow line-of-sight/beyond-line-of-sight communication between battle staff and personnel en route to or in a theatre of operations.

AFSOC flight tests new tanker

As part of a plan to increase the number of tankers capable of supporting its long-range helicopter missions, Air Force Special Operations Command has begun flight-testing a new refuelling system for the MC-130H Combat Talon II. The Talon II fleet is not currently configured for tanker operations and a fleet of 50

VX-20 at Patuxent River has completed testing of the E-2C with new-generation Hamilton Sundstrand NP-2000 propellers. Full-rate production has been approved for the E-2 fleet, with C-2As to follow.

Air Power Intelligence

Sikorsky's first UH-60M made its first flight at West Palm Beach, Florida, on 17 September 2003. The M features uprated T700-701D engines and improved wide-chord rotor blades, among other improvements. 1,217 of the planned 1,680-helicopter acquisition will be converted from existing UH-60A/Ls, but given new serials.

With RSK MiG pilot Yevgeniy Gorbunov at the controls, BVVS MiG-29 '24' takes off for a post-modification flight on 21 August 2003. Three Bulgarian MiGs were returned to airworthy condition in short order by the RSK MiG/TEREM team. The longer term BVVS MiG-29 upgrade, with Thales avionics, has been delayed, with first flight not expected until mid-2004 at the earliest.

MC-130E Combat Talons and MC-130P Combat Shadow handles inflight refuelling of AFSOC's helicopters. The removable system, which was developed by Flight Refuelling Ltd, offers greater capacity and improved reliability and maintainability over the existing system installed on the command's tankers. AFSOC intends to modify all 22 MC-130Hs as tankers by late 2006. Boeing will carry out a $197 million modification programme to incorporate the equipment.

Updated Harrier delivered

Naval Air Depot (NADEP) Cherry Point in North Carolina recently returned the first fully upgraded two-seat TAV-8B trainer, modified under the Harrier Trainer Upgrade Program (TUP), to VMAT-203. As part of the TUP the depot is equipping 14 TAV-8Bs with major modifications that provide the trainer with new wiring, enhanced night vision capabilities and the more powerful F402-RR-408 engine. All of these systems are already incorporated on the operational single seat AV-8B. Two TAV-8Bs that had previously received partial upgrades are included in the total. The last example will be returned to service during 2007.

B-2A releases new weapons

The USAF's 419th Flight Test Squadron and the Global Power Bomber Combined Test Force at Edwards AFB, California, recently conducted a series of nine test flights that culminated in a B-2A releasing 80 inert JDAM separation test vehicles on a single drop that lasted just 22 seconds. Each of the weapons was assigned to attack separate individual targets. The primary objective of the tests is to integrate the Smart Bomb Release Assembly (SBRA) and the GBU-38 227-kg (500-lb) version of the JDAM.

A B-2A released a pair of live 2268-kg (5,000-lb) GBU-28B/B series weapons over the Utah Testing and Training Range on 14 August 2003. The event marked the first successful drop of the newly updated weapon and the last dedicated flight for the developmental test and evaluation phase of the programme.

The GPS/laser-guided weapon is an enhanced version of the GBU-28A/B, which was designed specifically for the B-2A. Testing of the new variant began in March 2003 with the release of an inert GBU-28B/B at NAWS China Lake, California. Unlike the integration of the GBU-38 JDAM, no physical modifications were required for the integration of the weapon. Additional tests will be conducted by the Operational Test and Evaluation team at Whiteman AFB, Missouri, before the weapon is declared operational in late 2004.

PROCUREMENT AND DELIVERIES

Australia

Super Seasprite delivered

The Royal Australian Navy accepted the first of 11 new SH-2G(A) helicopters for testing, evaluation and training at Naval Air Station Nowra, New South Wales, on 18 October 2003. Armed with Penguin anti-ship missiles and torpedoes, the helicopters will operate from Australia's fleet of eight 'ANZAC'-class frigates. The Seasprites are expected to be fully operational by mid-2005.

Canada

Final Cormorant delivered

The Canadian Air Force recently took delivery of the last of 15 CH-149 search and rescue helicopters. The aircraft was delivered to 8 Wing at Canadian Forces Base Trenton in July 2003 after making intermediate stops in France, England, Norway, Belgium, Scotland, the Faroe Islands, Iceland, Greenland, Iqaluit, Kuujjuaq and Bagotville. The first Cormorant was delivered to CFB Comox in October, 2001 and the aircraft are now flying with four squadrons. Since then, it has assumed helicopter SAR duties at two of four squadrons, replacing the CH-113 Labrador helicopter, which has been in service since 1963. Greenwood-based 413 Squadron is midway though its transition to the new helicopter while 424 Squadron in Trenton is expected to conclude the Cormorant fleet implementation in March 2004.

Colombia

MPA aircraft delivered

The Colombian Navy recently took delivery of a pair of CN-235 maritime patrol aircraft from EADS CASA. The aircraft will mainly be used in support of anti-narcotics missions but will have a secondary search and rescue function. The aircraft arrived at the Naval Air Base in Bogotá in late September 2003, wearing the serial numbers 1260 and 1261, but will operate from Rafael Nuñoz Air Base in Cartagena and Juan Chaco in Buenaventura.

Denmark

C-130Js accepted

The Royal Danish Air Force accepted three new C-130J-30 Hercules tactical transport aircraft during a ceremony held at Lockheed Martin's Marietta, Georgia, facility on 21 October 2003. The 'Hercs' will replace a similar number of earlier C-130Hs that were delivered in 1975 and are currently being modified with equipment that is specific to the Danish requirements. Delivery to 721 Skvadron at Aalborg Air Base is expected in early 2004.

Germany

Dutch Orions for Marineflieger?

In October 2003 the German government formally signalled its interest in acquiring the 10 Lockheed Martin P-3 CUP Orions operated by the Dutch Navy's MARPAT. The aircraft would fulfil a long-standing requirement to replace the German Navy's Br 1150 Atlantic fleet. However, it is possible that the Orions would form part of a multi-national maritime patrol unit involving Belgian, Dutch and German participation. In November, the first completed CUP (Capabilities Upkeep Programme) aircraft was redelivered.

Greece has taken delivery of its first two (from a total of six) CSAR-configured AS 532A2 Cougar Mk 2s. The new aircraft arrived resplendent in a three-tone 'low-viz' camouflage scheme which resembles that used on the Mirage 2000 fleet. They are equipped with stretched cabins, FLIR, weapons pylons and chaff/flare dispensers, and carry the PA (Polemiki Aeroporia) logo on the nose.

Greece

AH-64D and NH90 contracts

On 28 August 2003 the Hellenic Ministry of Defence MoD/DGA (General Delegation for Armaments) signed a contract for the provision of 12 new Boeing AH-64D Apache attack helicopters, with an option for four more, valued at $675 million. Deliveries are scheduled to begin in the first quarter of 2007. The following day another contract was formally signed, this time for for the acquisition of 20 NH 90 helicopters plus 14 options for the Hellenic Army. The NH 90 firm order, worth close to 657 million Euros, includes 16 Tactical Transport variants and 4 Special Operations variants. All helicopters can be converted into the MedEvac variant with the aid of the four ordered roll change medical kits and will be equipped with Rolls-Royce-Turbomeca RTM322 engines. Deliveries are expected to begin in 2006 and to be completed in 2009.

CF-5s for lead-in trainers

A Hellenic Air Force Committee visited Canada during the summer in order to examine first-hand the possibility of purchasing 24 of the Canadian Armed Forces stored CF-5Bs, to cover the HAF's immediate need for an advanced lead-in trainer, to prepare future fighter pilots for the advanced F-16 and Mirage 2000 aircraft. With the HAF F-5 retired since 2000, the Canadian Freedom Fighters seem to provide the next best solution to the HAF short-term needs. The CAF upgraded a total of 46 aircraft in the early 1990s (13 single-seat and 33

Tornados on test

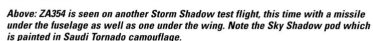

Below: Recent Tornado trials have focused on the MBDA Storm Shadow, fired from the BAE Systems Tornado GR.Mk 1 testbed ZA354 as part of the Package 2 upgrade. Here it carries a test round together with recording pods and a dummy ALARM missile.

Above: ZA354 is seen on another Storm Shadow test flight, this time with a missile under the fuselage as well as one under the wing. Note the Sky Shadow pod which is painted in Saudi Tornado camouflage.

Right: Tornado F.Mk 3 AMRAAM trials are ongoing, mostly by ZE155, BAE System's test vehicle. For recording weapon separation the aircraft carries cameras packaged into a Sky Shadow ECM pod body. It also has a forward-facing camera in a fairing on the leading edge of the fin.

twin-seat), known locally as CF-116A and CF-116D, respectively, to act as lead-in trainers for the CF-188 Hornet. The AUP was carried out by Bristol in two phases and included the addition of a new HUD and Weapons Aiming and Computing System (WACS) with an associated air data computer, improved navaid suite and INS, and structural strengthening of the airframe from 4,000 to 8,000 hours. New avionics were added as part of the second phase. In their final form they comprised a GEC-Marconi HUD/WAC, Litton INS, Magnavox AN/ARC-164 VHF radio, JET stand-by attitude indicator, Conrac AoA sensor, Honeywell radar altimeter and Ferranti video camera, all linked by a MIL-STD 1553B databus. Despite the excessive rework none of the upgraded aircraft entered service (a total of 37 had been completed) because of the sudden 1995 Defence Budget cuts. All 37 were put in long-term storage awaiting potential customers, and it seems they will soon fly again this time wearing the HAF insignia. A final decision on the procurement of these aircraft is expected in the next few months.

India

Trainer news
The Indian government has reached an agreement with BAE Systems covering the purchase of 66 Hawk Mk 115 Advanced Jet Trainers (AJTs) for the Indian Air Force. The deal, which is worth $1.75 billion, includes 24 aircraft to be completed by BAE Systems and 42 further examples that will be manufactured locally under license by Hindustan Aeronautics Limited (HAL). The Adour 871 engines will be built at Rolls-Royce's Patchway facility and a new plant to be established in Bangalore. The deal also includes the training of 25 Indian pilots per year at the Empire Test Pilot School at Boscombe Down.

To bridge an immediate gap in trainer capability, the IAF has procured 27 MiG-21UM two-seat trainers from Kyrgyzstan and the Ukraine. These will help a shortfall in MiG-21 training which will become more acute as upgraded MiG-21UPGs are returned to service.

VIP jets from Brazil
India will purchase five Embraer Legacy executive jets from the Brazilian manufacturer at a cost of $159.3 million. The executive jets will be configured for VIP use and equipped with 14 seats. They will also feature an advanced communications and defensive countermeasures suites. Four of the aircraft will be assigned to the Indian Air Force Headquarters Communication Squadron at Palam Air Base in Delhi, replacing HS.748s currently in service. The final example will be assigned to the Border Security Force (BSF) under the Union Home Ministry.

AWACS developments
Russia and India have agreed to the terms of a contract for three A-50 Mainstay airborne warning and control aircraft (AWACS) at a cost of $1 billion. The A-50, which is based on the Ilyushin Il-76 airlifter, will be equipped with the Israeli Aircraft Industries Phalcon early warning system and powered by new PS-90A engines. The Aircraft Association will be responsible for the certification tests of the A-50/PS-90A, however the Vega Research and Production Association and the Beriev Aircraft Research and Technical Association will be responsible for integrating the systems with the Il-76s, which will be taken from the Russian Air Force inventory.

Indonesia

'Flankers' and 'Hinds' delivered
The Indonesian Air Force took delivery of its initial pair of Sukhoi Su-27 fighters at Iswahyudi Air Base on 20 August 2003. The disassembled aircraft arrived aboard an An-124 military transport and will be reassembled by Russian technicians.

An Antonov An-124 also flew in two Mi-35P 'Hinds' for the Indonesian army, on 15 September. This fulfils the initial Indonesian order, but it is likely that more will be procured if experience with the aircraft proves satisfactory.

Ireland

Learjet for Irish Air Corps
The Irish Minister for Defence recently announced plans for the purchase of a single Bombardier Learjet 45 aircraft for use as a Ministerial air transport. The 7-9 seat aircraft will replace a 23-year old Beech (Raytheon) King Air 200 that is currently serving in this role. Delivery is expected before the end of 2003.

Italy

F-16s arrive at Cervia
On 14 November 2003 the first three ex-USAF F-16s were delivered to the Italian air force's renovated base at Cervia, which is home to the 23° Gruppo, 5° Stormo. Because Cervia was not ready in time, the first F-16s for the AMI initially operated from Trapani, Sicily.

Jordan

More F-16s
Jordan is purchasing a further 17 Lockheed Martin F-16s to equip a second squadron. Under the FMS

Czech AF MiG-21s have begun to adopt a toned down two-tone scheme similar to that applied to the Aero L-159. This 4. ZTL MiG-21UM was seen at Pardubice in August 2003. It was temporarily operating from there while the runway at Caslav was repaired.

This is one of the Mi-24Vs recently delivered by Russia to the Czech air force as part of a debt repayment package. The aircraft were painted in Russia to Czech specifications, but the result was quite different from the existing Czech aircraft.

Air Power Intelligence

contract, 17 aircraft are being received. They are to F-16A/B Block 15 standard, but will be upgraded under the Falcon-Up and Falcon-STAR structural improvement programme and will have modified F100-PW-220E engines.

UH-60s sought
Jordan has requested eight Sikorsky UH-60L Black Hawks through FMS channels. The nation already operates three S-70A Black Hawks which are flown by the Royal Flight.

CASA received
The first of two EADS-CASA C.295s on order for the Royal Jordanian Air Force was delivered to the country in September 2003.

Malaysia

Helicopters ordered
The Malaysian Ministry of Defence recently ordered 11 A109 Light Observation Helicopters (LOH) from AgustaWestland in a deal estimated to be worth $70 million. The helicopters, which will be operated by the Malaysian Army in the reconnaissance, transport and area-suppression roles, will be equipped with 70-mm rockets and 20-mm machine-guns. Agusta will also provide technical support and training when the helicopters are delivered beginning in December 2005.

'Flankers' finalised
On 5 August 2003, during a visit to Kuala Lumpur by Russian President Vladimir Putin, the Royal Malaysian Air Force signed a contract covering 18 Su-30MKM multi-role fighters. Part of the deal will be paid for in palm oil. Training is expected to start in Russia soon, and deliveries are slated for 2006 to 2009.

Mexico

Panthers ordered
The Mexican Navy (Armada de México) has placed an order with Eurocopter for two AS 565 Panther helicopters and taken options on eight additional examples. The helicopters, which will be delivered in 2005, will operate both from land bases and from naval warships, in support of drug interdiction, surveillance, SAR, and troop transport duties. Eurocopter

Following up the article in Volume 10, Upgrading Greece's Fighter Forces, this is the first picture to be released showing the rear cockpit of the Peace Xenia III F-16D Block 52+ aircraft recently delivered to the EPA. The fully night-capable cockpit has two colour head-down MFDs, and a large head-up screen which can display navigation and targeting data from the infra-red systems, and a HUD symbology repeater.

is currently upgrading the Mexican Navy's fleet of BO105 helicopters and the first of these will enter service by the end of 2003.

Pakistan

Helicopters requested
The government of Pakistan has made a request, through the Defense Security Cooperation Agency, to purchase up to 40 Bell 407 helicopters. The aircraft will be equipped with a commercial avionics suite and the package, which could be worth as much as $97 million, includes support equipment, spares, training and technical support. The helicopters would primarily be used in support of anti-terrorist and border security operations.

Philippines

Broncos delivered
The Thai government recently delivered four OV-10C Broncos to the Philippines Air Force at Villamor Air Base in Pasay City. The aircraft, which are the first of eight aircraft to be donated, will be assigned to the 15th Strike Wing at Danielo Atienza Air Base in Cavite. The Broncos were last assigned to the Royal Thai Air Force's 41st Wing/411 Squadron at Chiang Mai Royal Thai Air Force Base.

Poland

First CASA transport
On 15 August 2003 the Polish air force received the first of eight EADS-CASA C.295M transports for service with the 13.PLT (transport squadron) at Krakow-Balice. Delivery of the C.295Ms will allow the retirement of five An-26s from the 13.PLT inventory, although a further five 'Curls' will remain in service until at least 2008. Two of the C.295Ms will be equipped

with inflight-refuelling probes to allow extended-range deployments.

More MiGs?
Having already begun to take delivery of MiG-29s from the German Luftwaffe, the Polish air force has been in talks with RSK MiG over the possible acquisition of new-build 'Fulcrums'. While this would seem to be at odds with Poland's F-16 buy, and will probably not result in any outright purchases, it is possible that more MiGs will be acquired through leasing arrangements.

Russia

Ka-52 deliveries
Reports from Russia suggest that the first batch of 12 Ka-52 Alligator helicopters was delivered between December 2003 and January 2004. The Ka-52 is the side-by-side two-seat derivative of the Ka-50 'Hokum'. In the meantime, air tests have begun of the first prototype Ka-52 fitted with the Phazotron-NIIR Arbalet radar, which offers long-range detection capability against armour and air targets.

Singapore

RSAF to evaluate Strike Eagle
The government of Singapore has added the Boeing F-15T to a short-list of platforms that it will evaluate during the invitation-to-tender phase of the Air Force's Next Fighter Replacement Programme. Based upon the F-15E, the latest variant of the Strike Eagle was one of six types under consideration.

Spain

Tiger chosen
On 5 September 2003 the Spanish government announced that it had selected the Eurocopter Tiger to fulfil its outstanding requirement for an attack helicopter to replace the BO 105. Twenty-four Tigers will be acquired for delivery from 2007. The Spanish helicopters will be to HAD (Hélicoptère d'Appui et de Protection) standard, combining the anti-tank and escort roles. As well as the 30-mm cannon, the Tiger HAD will use a developed MBDA Trigat MR weapon for anti-tank duties and Mistral air-to-air missiles. The HAD also has uprated engines in the form of the Rolls-Royce/MTU MTR390 Enhanced. Spanish industry will play a major part in the development of the HAD, with EADS-CASA and Eurocopter España involved. The latter becomes a full partner in the Tiger programme, and will build rear fuselages. Ahead of the Tiger HAD deliveries, the Spanish army will receive three HAPs from the French order to allow training to begin.

Thailand

Government jet
The RTAF has placed an order for an Airbus ACJ (Airbus Corporate Jet). Based on the A319, the ACJ has a VIP interior for the transport of government officials. The aircraft for the RTAF will have four additional tanks in the fuselage for long-range travel, and is to be powered by CFM International CFM56-5B7 engines. Delivery is set for July 2004.

Helicopters under test at NAS Patuxent River with HX-21 include the Bell AH-1Z 'Zulu' (above) and the MH-60R (right). Two 'Zulus' (this is the first, with redesigned stabiliser) and two UH-1Y 'Yankees' are currently exploring the Cobra/Twin Huey update programme. The MH-60R (BuNo. 166404) is the third low-rate initial production aircraft.

Co-operative Key 2003

Between 1 and 13 September 2003 Bulgaria hosted another Co-operative Key exercise involving Partnership for Peace air arms, as well as full NATO members. In common with previous CK exercises, the scenario focused on multi-national peacekeeping operations and Combat SAR missions. The principal tasking of CK'03 was to find, protect, supply and, where necessary, evacuate a large number of 'refugees'. Paradropping of Special Forces was practised, and the insertion of medical teams and facilities figured prominently. CSAR missions were woven into the wider scenario.

Twenty-one nations sent personnel to the exercise, 15 of which contributed a total of 63 aircraft. Fighters and helicopters operated from Graf Ignatievo, while transports and AWACS flew from Krumovo. The latter function was provided by a single Sentry AEW.Mk 1 from the RAF's No. 23 Squadron. The backbone of the Blue Force fighters was provided by Swedish Gripens, French Mirage 2000s, Romanian Lancers, and MiG-21s and Su-25s from the local air force. Bulgarian L-39s were used as Opposing Force (Opfor) fighters, augmenting F-16s from Turkey, and A-7s and F-4s from Greece. Greek and Turkish aircraft operated from their own bases.

Holger Stüben and Jens Schymura

Above: The largest contingent for CK'03 came from the Swedish Air Force, which sent eight JAS 39As from F17 Blekinge Flygflottilj at Ronneby, together with a single HKP 11 (AB 412) from HkpBat 4. F17's Gripens assumed the SWAFRAP (Swedish Air Force Rapid reaction) commitment on 1 January 2004, this having been relinquished by F21's Viggens on 30 September 2003.

Above: The local air force provided four 3 IAB MiG-21bis for Blue Force's air defences.

Among the miscellany of transport aircraft taking part in CK '03 were two from Moldova, an An-72 'Coaler' (left) and An-2 'Colt' (above). Both were used for paradropping during the exercise.

United States

Predators ordered
General Atomics Aeronautical Systems has been awarded a $41.4 million contract for 19 MQ-1L Predator unmanned aerial vehicles (UAV).

200th Texan II delivered
Raytheon delivered the 200th example of the T-6A trainer aircraft recently when it was accepted by the US Navy. The aircraft was the 21st example for the US Navy and follows 108 delivered to the USAF, 45 to the Hellenic Air Force and 26 to the NATO Flying Training in Canada programme.

Goshawks ordered
Boeing's McDonnell Douglas subsidiary has received a $60.9 million order for five T-45C Goshawk trainers from the US Navy. The FY2003 budget included eight T-45Cs and 15 further examples have been requested in 2004. Those aircraft covered by the latest order aircraft will be delivered during 2006.

Production RQ-4 unveiled
Northrop Grumman unveiled the first production RQ-4A Global Hawk unmanned air vehicle to the USAF at its Antelope Valley Manufacturing Center in Palmdale, California, recently. Known as Global Hawk Air Force 1 – serial 02-2008 – which is the eighth Global Hawk to be built, carried a new operational gray-and-white paint scheme. The first seven air vehicles were built under the advanced concept technology demonstration (ACTD) phase of the programme. Once flight-testing has been completed at Edwards AFB, California, the vehicle will be delivered to the 9th Reconnaissance Wing at Beale AFB near Sacramento, California. Since its first flight in 1998, Global Hawk has logged more than 3,000 flight hours, more than half of them during operational missions that were conducted in support of Operations Enduring Freedom in 2001 and Iraqi Freedom in 2003.

MC2A aircraft ordered
The USAF has awarded Boeing a $126 million contract covering the purchase of a single 767-400ER aircraft that will be modified as the prototype for the E-10A multi-sensor command and control aircraft (MC2A) program. Once delivered in 2005 the aircraft will be equipped with advanced integrated air and ground surveillance and targeting under development by a team led by Northrop Grumman. Northrop Grumman will eventually integrate the equipment at its Lake Charles, Louisiana, facility.

Third special ops Osprey
The Bell-Boeing Joint Program Office has been awarded a $61 million modification to an existing contract that covers the conversion of an additional low-rate initial production (LRIP) MV-22B aircraft into a CV-22B special operations configuration. The modifications will be incorporated on an existing aircraft and will be completed by November 2004. The aircraft will join two additional CV-22Bs as part of the flight test programme.

New rescue helo for USAF
The USAF is moving forward with a programme to update its combat rescue and support helicopter fleet. The combat search and rescue (CSAR) mission is currently handled by a fleet of HH-60G Pave Hawks while a smaller fleet of UH-1Ns is used for operational support missions including missile silo support and security, VIP transportation, and search and rescue.

Although the missions assigned to the HH-60G and UH-1N vary greatly, the USAF hopes to replace both helicopters with a single type. The service is shortly expected to release a formal requirements document that will result in the search for a new helicopter beginning in 2006. The new aircraft will not, however, enter service before 2012. The USAF is also expected to complete plans that will provide the HH-60G fleet with a service life extension and allow it to remain in service through 2019.

The replacement aircraft will likely be a development of the Sikorsky S-92 or the AgustaWestland EH101, although the Bell/Boeing CV-22 tiltrotor would likely be considered as well. An Americanised version of the foreign aircraft, known as the US-101, would be built by a consortium that includes Lockheed Martin and Bell Helicopter.

Wearing a temporary UK registration, the first Hawk Mk 120 for the South African Air Force made its maiden flight from Warton on 3 October, with Gordon McClymont at the controls. After three flights in the UK it was disassembled and flown by An-22 to South Africa. The remaining 23 aircraft in the order are being built as kits by BAE Systems at Brough and shipped for assembly by Denel, before assignment to the SAAF's 85th CFS.

Air Power Intelligence

Edwards Open House 2003

Above: Long-serving F-16 VISTA testbed is now used by the USAF Test Pilot School. Its unique flight control system can be varied so that it can simulate other aircraft.

Left: Resplendent in the Air Force Flight Test Center's 'house' colours, this F-16B is one of the aircraft originally built for Pakistan but embargoed due to non-compliance with international nuclear treaties. The AFFTC received six of the batch to augment the test chase/TPS fleet.

A favourite at selected California airshows, the Northrop N-9MB was an important stepping-stone in the development of Northrop's flying-wing designs, which eventually led to today's B-2 Spirit. This aircraft was restored to flight in 1994 by the Planes of Fame Air Museum.

The AFFTC is evaluating an Aero L-39C Albatros trainer (N439RS) that arrived at the facility for a six-month trial on 21 August 2003. Owned by Teton Aviation, the L-39 supports the 412th Test Wing, which will simultaneously evaluate the aircraft's flight capabilities, maintenance reliability and cost performance. The L-39 will initially serve as a partial replacement for the Center's now-retired T-39 fleet, carrying out about 70-80 per cent of the T-39 missions, including USAF Test Pilot School training.

AIR ARM REVIEW

Japan

Transports deployed

The Japan government recently deployed four transport aircraft and some 150 personnel in support of humanitarian operations in Iraq. The aircraft comprised three C-130H airlifters and a single U-4 Gulfstream IV. The aircraft were based in Italy and ferried food and medical supplies to nations bordering Iraq for subsequent distribution to Iraq.

Russia

New base in Kyrgyzstan

On 22 September 2003 Russia and Kyrgyzstan signed an agreement covering the establishment of a Russian air force base at Kant, close to the Kyrgyz capital Bishkek (formerly Frunze). Kant is expected to house a mixed detachment of around 15 Su-25s and Su-27s, seven Il-76s, plus L-39 trainers for training and helicopters. The base is close to the airfield at Manas which is used by US/NATO aircraft policing Afghanistan.

United States

AMC reorganisation

The USAF's Air Mobility Command (AMC) recently undertook a major reorganisation that resulted in the activation of the Eighteenth Air Force (18th AF) at Scott AFB, Illinois, on 1 October 2003 and the redesignation of the command's 15th and 21st Air Forces as 15th and 21st Expeditionary Mobility Task Forces (EMTF). The EMTFs are respectively located at Travis AFB, California, and McGuire AFB, New Jersey. As part of the reorganisation, the 18th AF assumed operational control of the Tanker Airlift Control Center (TACC), both EMTFs, 12 flying wings and three flying groups, and the global en route air mobility system. Tasking and execution of air mobility missions now rests with the 18th AF. The EMTFs retain the heritage of the 15th and 21st Air Forces and command of the four air mobility operations groups and their respective squadrons. The Eighteenth Air Force was originally activated in March 1951 and controlled 16th Troop Carrier Wings that were then assigned to Tactical Air Command (TAC). The organisation was subsequently inactivated in 1958.

Nightingales retired

The last of the 374th Airlift Wing's (AW) four C-9As departed Yokota AB, Japan, on 14 September 2003 and the 30th Airlift Squadron (AS) was inactivated on 25 September 2003. The move, which involved serial 71-0877, was part of the USAF's previously announced plan to retire 20 of its Nightingale transports. As a result of the Nightingale retirement, the KC-135 fleet will take on a portion of the aeromedical airlift duties. The last of five European-based Nightingales, serial 71-0879, left Ramstein AB, Germany, on 27 September 2003 and the 75th AS, a component of the 86th Airlift Wing, was subsequently deactivated on 30 September 2003. Although the 11th AS at Scott AFB, Illinois, had been inactivated a day earlier, the 375th Airlift Wing transferred its last Nightingale to the Aerospace Maintenance and Regeneration Center (AMARC) at Davis Monthan AFB, Arizona, on 1 October 2003. Serial 67-22589 had originally been delivered to Scott in 1968. C-9A serial 71-0878 remains at Scott and was transferred to the Air Force Reserve Command's 932nd AW/73rd AS. The wing will gain two additional Nightingales once the Fiscal Year 2004 Defense bill is approved. The 932nd AW and 73rd AS had been associate units and were not directly assigned any aircraft. Those units had previously partnered with the 375th AW and 11th AS since 1969.

Aggressors return to Nellis

Air Combat Command reactivated the 64th Aggressor Squadron at Nellis AFB, Nevada, on 1 October 2003. The squadron was originally activated at Nellis as the 64th Fighter Weapons Squadron (FWS) on 15 October 1972. On 5 October 1990 the 64th Aggressor Squadron was inactivated and, although reduced in size, its assets formed the Adversary Tactics Division (ATD) of the 414th Combat Training Squadron (CTS). The reactivated squadron assumed responsibility for the adversary F-16C/Ds that had been operated by the ATD. The 414th is tasked with conducting the Red Flag exercises and will continue to carry out these duties.

Hornet squadron news

The 'Warhawks' of VFA-97, currently deployed aboard the USS *Nimitz* (CVN 68) as part of CVW-11, will transition from the F/A-18A to the F/A-18C upon its return. It will also leave CVW-11 and will be the first US Navy Hornet squadron to support the US Marine Corps Unit Deployment Program (UDP) in Japan. As part of the UDP, individual squadrons will support a rotational deployment to MCAS Iwakuni or MCAS Futenma for a normal period of six months.

Navy Prowlers support Marines

The 'Wizards' of VAQ-133, which had deployed to MCAS Iwakuni, Japan, in support of the US Marine Corps Unit Deployment Program (UDP) during July 2003, were replaced by the 'Fighting Phoenix' of VAQ-128 during October 2003. The 'Wizards' were the

Air Power Intelligence

first of four US Navy expeditionary Prowler units to support the UDP.

Viking retirement approaches
The US Navy has confirmed its plans to retire the Lockheed Martin S-3B Viking from the fleet by early 2009. Although it had previously indicated that two squadrons would be inactivated in 2003, this will not take place until early 2004. Although it is no longer used for anti-submarine warfare missions, the Viking is tasked with sea control missions and serves as an airborne tanker. The tanker mission will be taken over by buddy store-equipped Super Hornets over the next few years. The Viking currently serves with 10 fleet squadrons and one fleet readiness squadron.

Squadron	Nickname	Air Wing	Inact.
VS-29	'Dragonfires'	CVW-11	4/04
VS-38	'Red Griffins'	CVW-2	4/04
VS-30	'Diamondcutters'	CVW-17	11/05
VS-24	'Scouts'	CVW-8	9/06
VS-32	'Maulers'	CVW-1	3/07
VS-31	'Topcats'	CVW-7	8/07
VS-41	'Shamrocks'	FRS	9/07
VS-35	'Blue Wolves'	CVW-14	1/08
VS-22	'Checkmates'	CVW-3	6/08
VS-33	'Screwbirds'	CVW-9	2/09
VS-21	'Redtails'	CVW-5	TBD

Patrol Wings reorganised
The US Navy established Patrol and Reconnaissance Wing Two at Marine Corps Air Facility Kaneohe Bay, Hawaii, on 15 October 2003. The wing assumed responsibility for Patrol Squadrons (VP) Four, Nine and Forty Seven, and Special Projects Patrol Squadron (VPU) One at Kaneohe Bay. In addition, Commander Patrol and Reconnaissance Wing One relocated from Naval Support Facility Kamiseya, Japan, to Naval Air Facility Misawa, Japan, and was renamed Commander Patrol and Reconnaissance Force Seventh Fleet (CPRF7F)/Commander, Patrol and Reconnaissance Force Fifth Fleet (CPRF5F). The organisation is responsible for P-3/EP-3 units forward deployed on a rotational basis to facilities in the western Pacific, Indian Ocean and in southwest Asia. As a result of these moves, Commander Patrol and Reconnaissance Force Pacific (COMPATRECFORPAC), stationed at MCAF Kaneohe Bay, Hawaii, was redesignated Director Patrol and Reconnaissance Group Pacific.

Skyhawk squadron deactivated
The last Fleet Composite (VC) squadron on the US Navy rolls was disestablished on 1 October 2003 when the 'Redtails' of VC-8 stood down. The squadron had operated TA-4Js and UH-3Hs in support of fleet training from Naval Station Roosevelt Roads, Puerto Rico. The move was a direct result of the closure of the Navy's Atlantic Fleet Weapons Training Facility at Vieques.

Orions leave Roosevelt Roads
The closure of the Atlantic Fleet Weapons Training Facility in Puerto Rico and the resulting reorganisation of its forces throughout the Caribbean has caused the US Navy to cease scheduled deployments of Atlantic Fleet P-3C maritime patrol aircraft to Naval Station Roosevelt Roads. As a result, scheduled flight operations at the station came to an end on 15 July 2003. The 'Mad Foxes' of Patrol Squadron (VP-5) concluded the final rotation to 'Rosey Roads'. Although the patrol squadrons will continue to support anti-narcotics missions throughout the Caribbean, operations have shifted to forward operating locations at Hato International Airport, Curaçao and Manta Air Base, El Salvador.

Tomcat fleet dwindles
The 'Diamondbacks' of VFA-102 changed their home station from NAS Lemoore, California, to Naval Air Facility Atsugi, Japan, on 1 November 2003. The squadron and 12 F/A-18F Super Hornets will be assigned to Carrier Air Wing Five (CVW-5) replacing the 'Black Knights' of VF-154. The latter squadron relocated to NAS Lemoore, California, and was redesignated as VFA-154 in preparation for its transition from the F-14A to the F/A-18F. Although the effective date of the latter change was 1 October 2003, the squadron's 11 F-14As departed NAF Atsugi on 24 September 2003 and after intermediate stops arrived at NAS Oceana on 28 September. Following its transition VFA-154 will be assigned to CVW-9, replacing the 'Fighting Redcocks' of VFA-22 who will then undergo transition form the F/A-18C to the F/A-18E.

The Naval Strike and Air Warfare Center (NSAWC) recently conducted its last Topgun class using the F-14A Tomcat. Six F-14As had been assigned to NSAWC and two of the aircraft, comprising BuNos 158630 and 162688, have already been stricken and scrapped on site. The organisation's remaining Tomcats, which include BuNos 158617, 161615 and 162592, are destined for display at museums in Pennsylvania, Alabama and Kansas. BuNo. 162608 will be flown to Naval Station Ventura County/Point Mugu, California and will eventually join VC-137C 72-7000 on display at the Ronald Reagan Library in Simi Valley, California.

Tiltrotor test squadron activates
Marine Tiltrotor Test and Evaluation Squadron Twenty Two (VMX-22) was activated on 28 August 2003 at MCAS New River, North Carolina. Reporting operationally to the US Navy's Commander Operational Test and Evaluation Force at Naval Station Norfolk, Virginia, the squadron's mission will be to conduct operational testing of the MV-22B Osprey tiltrotor aircraft. The Osprey is currently undergoing development testing at NAS Patuxent River, Maryland. VMX-22 will be manned by US Marine Corps, US Navy and USAF personnel. Operational testing will not begin, however, until November 2004, and although some testing will be carried out at New River much of it will be conducted at other facilities. The Osprey had been undergoing operational testing with HMX-1 when it was grounded in late 2000 following two fatal crashes. It subsequently returned to the air in May 2002 for developmental testing at Patuxent River.

Attack helicopter named
The US Coast Guard's recently assigned the name Stingray to its Agusta MH-68A helicopters. The fleet of eight armed interdiction helicopters, which are operated by Helicopter Interdiction Tactical Squadron (HITRON) Jacksonville, are based at Cecil Field, Florida. Based upon the civil A109E the helicopter had previously been known by the unofficial nicknames 'Mako' and 'Shark'.

Navy Texan IIs achieve IOC
The first flight of a Raytheon Aircraft T-6A Texan II with a student naval flight officer (SNFO) took place at NAS Pensacola, Florida, on 5 August 2003, when the aircraft achieved initial operating capability (IOC). Although the aircraft will eventually replace the T-34C as the Navy's premier primary/intermediate trainer, it is initially being used for NFO training. The first Texan II was delivered to Training Air Wing Six (TAW-6) at Pensacola on 1 November 2002 and academic

On 3 October 2003 the Belgian Air Component commemorated the 250,000th flying hour of its SIAI-Marchetti SF.260M fleet (the milestone was actually passed in June). Thirty-six were delivered from November 1969 (later augmented by nine SF.260Ds) and over 2,600 students flew the type. Thirteen have been lost, but only two losses were attributable to technical failure. ST-20 was painted in this special scheme, which includes the wolf's head of No. 1 Wing and Egyptian falcon of No. 5 Squadron (the current operator at Beauvechain), and the penguin badge of the Elementary Flying School at Gossoncourt/Goetsenhoven (which flew the SF.260M until 1996).

Above: In late 2003 the JASDF was in the process of retiring its last Fuji T-1Bs. The T-1 was Japan's first indigenous jet aircraft, and first flew on 16 January 1958. The first 46 were T-1As, with Bristol Siddeley Orpheus engines. The T-1B, of which 20 were built, is powered by the locally developed IHI J3 turbojet. The last aircraft were used by the 5th Technical Squadron at Komaki for air traffic controller training.

Right: The JASDF's air combat training camps have traditionally resulted in special markings. These two F-15Js, complete with top-side dragon markings, are from Dai 303 Hiko-tai, part of Dai 6 Koku-dan at Komatsu.

Air Power Intelligence

Recce Super Etendard

Since the retirement of the Etendard IVP, the Aéronavale has used the Super Etendard for its reconnaissance missions. The CRM 280 (Chassis de Reconnaissance Marine) reconnaissance pod was introduced on the Super Etendard Modernisé Standard 4 (SEM 4) from September 2000 onwards. It is equipped with two cameras: a Thomson-CSF Optrosys (now Thales) SDS 250 electro-optical sensor and an Omera 40 wet film camera. The Super Etendard was extensively used in the reconnaissance role over Afghanistan in the framework of Operation Héraclès.

The SDS 250 is designed to perform low- and medium-altitude daytime tactical reconnaissance, either vertically or from a stand-off oblique position up to 20 km (12.4 miles) from the target. It has a 250-mm focal length f/5.6 lens and a 6,000-pixel CCD detector. Its depression angle can be selected by the pilot from any one of seven fixed positions (60°, 30° or 22° to port or starboard, and 90°). The Omera 40 panoramic camera is vertically mounted and has a fixed focal length of 75 mm. A rotating prism in front of the lens gives the camera a field of view of 180° perpendicular to the aircraft's flight path. Its magazine can be loaded with colour or black and white film, and is good for around 300 frames. The Omera 40 is used for low-altitude daytime tactical reconnaissance missions.

Jos Schoofs

Above: This Super Etendard Modernisé Standard 4 is from Flottille 17F, shore-based at Landivisiau when not deployed aboard Charles de Gaulle. In addition to the CRM 280 reconnaissance pod it carries a Barracuda ECM pod; a Magic 2 air-to-air missile is usually carried under the opposite wing.

Right: The CRM 280 system is partly accommodated in the fuselage, replacing the 30-mm cannon.

instruction commenced on 30 June 2003 when four Navy, one USAF and one USMC student began training with Training Squadron Ten (VT-10). The wing currently has 16 Texan IIs assigned but will operate 58 aircraft when the transition from the Turbo Mentor is complete in 2006. Three T-6A flight simulators are currently in place at Pensacola and two further examples will arrive in November 2004.

Fighters stay at Keflavik

The USAF has indefinitely deferred plans, announced in May 2003, to withdraw its four remaining fighter aircraft from Iceland. The small force of F-15 and F-16 fighters, which provides the nation's only air defence, has been attached to the 85th Group/85th Operations Squadron at NAS Keflavik on a rotational basis since the 57th Fighter Squadron was withdrawn in 1995. Although Iceland is a member of NATO it has no military forces and the US has been responsible for its defence since 1951.

Hickam wing renamed

The 15th Air Base Wing at Hickam AFB, Hawaii, has been redesignated the 15th Airlift Wing. The wing reports directly to Pacific Air Forces Headquarters and controls the 65th Airlift Squadron, which operates C-37A and C-40B transports.

USAF's oldest Hercules retired

The last C-130A on the USAF inventory flew its final flight 3 October 2003, ending a test career that spanned 48 years. Nicknamed 'Lone Wolf', NC-130A serial 55-0022 has operated from Duke Field on Eglin AFB, Florida, as part of the 46th Test Wing. Referred to as the NC-130A Airborne Seeker Evaluation and Test System, the aircraft had served as an airborne platform for developmental test and evaluation of air-to-ground and air-to-air seekers and sensors. Although capable of tracking and filming aircraft/weapon separations and weapon intercepts, maintenance personnel could no longer sustain its current systems, many of which were not compatible with the USAF's C-130 fleet. Much of the aircraft's essential equipment will be transferred to a newer NC-130H aircraft that will be assigned to the wing's 40th Flight Test Squadron (FLTS). The new aircraft is currently undergoing programmed depot maintenance (PDM) and it will subsequently undergo modifications that will install a turret and modified floor structure removed from the NC-130A. Additional modifications will enable the new test platform to conduct a wider array of weapon seeker testing. The NC-130A, which conducted its first test flight in 1955, will be grounded but will be used to test installation security systems on the Eglin range.

USAF 2004 force changes

The USAF has released its planned structure changes for Fiscal Year 2004. The realignments are intended to allow the service to better support its highest-priority missions and will result in a total decrease of 5,099 personnel (included are 2,260 military and 2,839 civilians), and 1,055 reservists. These changes affect the Total Force, which comprises active duty, Air National Guard, and Air Force Reserve Command units. As part of these moves the service will retire its fleet of 20 C-9s, along with 54 of the ANG's and AFRC's KC-135E tankers. The latter will partially be replaced by 44 KC-135Rs. The USAF intends to retire all 68 of its remaining KC-135Es by 2006. Details by command are as follows:

Air Combat Command
- Beale AFB, California – the 12th Reconnaissance Squadron will gain three RQ-4A Global Hawk unmanned air vehicles (UAV).
- Davis-Monthan AFB, Arizona – the 355th Wing will gain two HC-130Ps and four additional HH-60G as its recently activated rescue squadrons are brought up toward their assigned strength.
- Pope AFB, North Carolina – the 75th Fighter Squadron will lose six OA-10As.

Air Education and Training Command
- Laughlin AFB, Texas – the 47th Flying Training Wing will exchange 44 T-37Bs for 37 T-6A trainers.
- Little Rock AFB, Arkansas – the 314th Airlift Wing will receive its initial CC-130J, which will eventually replace the C-130E.
- Randolph AFB, Texas – the 12th Flying Training Wing will retire 10 T-37Bs.
- Tyndall AFB, Florida – the 325th Fighter Wing's 43rd Fighter Squadron will acquire of 14 F/A-22A Raptors.

Air Force Space Command
- Vandenberg AFB, California – the 30th Space Wing's 76th Helicopter Flight will be inactivated.

Air Force Special Operations Command
- Hurlburt Field, Florida – the 16th Special Operations Wing's 4th Special Operations Squadron will gain one additional AC-130U gunship.

Air Mobility Command
- Andrews AFB, Maryland – the 459th Airlift Wing will convert from eight C-141Cs to eight KC-135Rs.
- Bangor IAP, Maine – the 101st Air Refueling Wing will lose two KC-135Es.
- Channel Islands ANGS, California – the 146th Airlift Wing will lose a single C-130E.
- Charlotte-Douglas IAP, North Carolina – the 145th Airlift Wing will lose a single C-130H.
- Dover AFB, Delaware – the 436th Airlift Wing will transfer four C-5As to the Tennessee Air National Guard's 164th Airlift Wing.
- Eastern West Virginia Regional Airport Shepherd Field – the 167th Airlift Wing will lose a single C-130H.
- Fairchild AFB, Washington – the

In early December 2003 F-117A 85-0835 (named The Dragon) was given an experimental grey scheme by Det 1, 53rd TEG, the F-117 test group resident at Holloman AFB, New Mexico. The aircraft is undergoing trials to assess its suitability for daytime operations in certain scenarios. Based on the F/A-22's scheme, the grey colour harks back to the earliest days of the Senior Trend programme, when the Full-Scale Development aircraft were painted in an overall light grey scheme.

Air Power Intelligence

New colour scheme for an old veteran – the Pilatus PC-6 Turbo-Porters of the Etablissement Réserve Générale de Matériel ALAT at Montauban have received a smart all-white scheme in place of olive drab. Five are in use with the ERGM, mainly for ferrying spare parts around ALAT bases.

From 6 to 22 October 2003 the French Air Force conducted a large air exercise called Opera at Chateaudun air base. Seventeen nations participated, including nine NATO and four Partnership for Peace countries. The most interesting participant was the Royal Moroccan Air Force, which sent one Mirage F1CH and three F1EHs. They were flown by pilots of the 'Assad' and 'Atlas' squadrons based at Sidi Slimane air base.

92nd Air Refueling Wing will lose 12 KC-135Rs.
■ Forbes Field, Kansas – the 190th Air Refueling Wing will lose two KC-135Es.
■ General Mitchell IAP/JARS, Wisconsin _ the 440th Airlift Wing will reduce its C-130H2 complement by one example.
■ Grissom JARB, Indiana – the 434th Air Refueling Wing will reduce its inventory by four KC-135Rs.
■ Louisville IAP, Kentucky – the 123rd Airlift Wing will lose a single C-130H.
■ March JARB, California – the 452nd Air Mobility Wing will lose two KC-135Rs, reducing its authorised aircraft to eight. The wing will also retire eight C-141Cs.
■ McChord AFB, Washington – the 62nd Airlift Wing will acquire five C-17As.
■ McConnell AFB, Kansas – the 22nd Air Refueling Wing will lose 12 KC-135Rs.
■ McGhee/Tyson Airport, Tennessee – the 134th Air Refueling Wing will lose a pair of KC-135Es.
■ McGuire AFB, New Jersey – the 305th Air Mobility Wing will retire four C-141Bs but will acquire three C-17As.
■ Memphis IAP, Tennessee – the 164th Airlift Wing will convert from eight C-141Cs to eight C-5A aircraft.
■ Minneapolis-St Paul IAP/JARS, Minnesota – the 934th Air Wing will retire two C-130Es but will receive two C-130H2 aircraft.
■ Nashville IAP, Tennessee – the 118th Airlift Wing will lose four C-130Hs.
■ Peterson AFB, Colorado – the 302nd Airlift Wing will lose a single C-130H Hercules.
■ Phoenix-Sky Harbor IAP, Arizona – the 161st Air Refueling Wing will convert from 10 KC-135Es to eight KC-135Rs.
■ Pittsburgh IAP/JARS, Pennsylvania – the 171st Air Refueling Wing will convert from the KC-135E to the KC-135R, however, its assigned complement will be reduced from 20 aircraft to 16.
■ Salt Lake City IAP, Utah – the 151st Air Refueling Wing will transition from the KC-135E to the KC-135R but will reduce its complement from 10 to eight aircraft.
■ Scott AFB, Illinois – the 375th Airlift Wing will retire 10 C-9A Nightingales and the 11th Airlift Squadron will be inactivated.
■ Selfridge ANGB, Michigan – the 927th Air Refueling Wing will transition from the KC-135E to the KC-135R, maintaining its inventory at eight.
■ Seymour Johnson AFB, North Carolina – the 916th Air Refueling Wing will lose a pair of KC-135Rs.
■ Travis AFB, California – the 60th Air Mobility Wing will transfer four C-5As to the Tennessee Air National Guard's 164th Airlift Wing.
■ Willow Grove JARS, Pennsylvania – the 913th Airlift Wing will lose a single C-130E.
■ Wright-Patterson AFB, Ohio – the 445th Airlift Wing will retire eight C-141Cs.
■ Youngstown-Warren JARS, Ohio – the 910th Airlift Wing will reduce its inventory by one C-130H.

Pacific Air Forces
■ Yokota AB, Japan – the 30th Airlift Squadron will retire four C-9As and will be inactivated.

United States Air Forces Europe
■ Ramstein AB, Germany – the 75th Airlift Squadron will retire five C-9As and will be inactivated.

Carrier news
The USS *Enterprise* (CVN 65) carrier strike group (CVSG) departed Norfolk, Virginia on 29 August 2003 for a compressed Composite Training Unit Exercise (COMPTUEX) in preparation for a deployment that immediately followed on 2 October 2003. The deployment is the first for an Atlantic Fleet carrier since major hostilities ended in Iraq in April 2003. The *Enterprise* CVSG is also the first of the Navy's slimmed-down carrier strike groups, which consists of just the carrier, two guided missile cruisers and a combat support vessel. At the conclusion of the training the CVSG immediately steamed for the Mediterranean Sea. The group could be deployed for as much as eight months rather than the normal six-month cycle, and will likely relieve USS *Nimitz* in the Persian Gulf.

The keel for the aircraft-carrier USS *George H.W. Bush* (CVN 77) was laid at Northrop Grumman's Newport News Shipbuilding facility in Virginia on 6 September 2003.

The Navy has elected to name its next multi-purpose amphibious assault ship the USS *Makin Island* (LHD 8) to honour a raid carried out in the Gilbert Islands during August 1942. The ship is currently under construction at the Northrop-Grumman Ship Systems facility in Pascagoula, Mississippi, and will likely be christened in 2006.

US ends operations at PSAB
During a ceremony held on 26 August 2003, US officials returned control of portions of Prince Sultan Air Base in Saudi Arabia to the host nation. The ceremony also marked the inactivation of the 363rd Air Expeditionary Wing (AEW), which had supported operations at the base since 1997. Prince Sultan supported as many as 200 coalition aircraft in support of Operation Southern Watch, but the fall of Iraq's government reduced the USAF's need for the base and the last Americans departed in early September. The aircraft that had operated from Prince Sultan are now operating from other bases in the region.

Zimbabwe

MiG-23s on show
In late July the Air Force of Zimbabwe revealed its rumoured MiG-23s during a firepower demonstration. The number supplied to the AFZ, and their provenance, remains unconfirmed, but it is most likely that they were supplied by Libya. The exact version is also not known, but they are most likely to be MiG-23MS 'Flogger-Es', a downgraded export interceptor.

The 'Pig' at 30

Following the article *21st Century Pigs* in Volume 6, several recent events in Australia's F-111 community require closer scrutiny. The ongoing wing replacement is due to be completed by the end of 2003 and, although as an expediency all aircraft are currently being refitted with the long-span wings, a decision in favour of the short-span wings has now been taken. The short-span wing allows a better operational profile (at the expense of range) and extended fatigue life, but a great deal of flight testing and analysis will now be required. RAAF flight profiles and weapons fits are unique, and testing is to be performed on a time-available basis by the Aircraft Research and Development Unit (ARDU). Integration of the AGM-142 stand-off missile continues, with the prototype aircraft currently undergoing modification, allowing ground trials to commence early in 2004.

The 'Pig' recently celebrated its 30th anniversary, but storm clouds loom on the horizon: the defence budget suffered from recent operations in Iraq and Afghanistan, and several platforms came under consideration for early retirement. A Defence Capability Review released in October stated that the F-111's retirement would be brought forward from around 2015 to 2010. It was considered that the Hornet was sufficient to cover the gap in strike/attack capability until Australia receives its planned F-35s.

Historians will recall that such threats have emerged at regular intervals since the aircraft was ordered off the drawing board in 1963, but it now seems unlikely that there will be a reprieve. However,

No. 1 Squadron's A8-131, one of the first F-111Cs to be delivered, wears special tail markings commemorating the F-111's 30 years in RAAF service.

many within the RAAF (and most enthusiasts within Australia) hope that there is life in the 'Old Pig' yet.

Nigel Pittaway

DEBRIEF

Austrian Hawks

Sikorsky S-70A-42 Black Hawk

In June 2002 the Österreichische Luftstreitkräfte (OLk – Austrian Air Force) became the 25th member of Sikorsky's Black Hawk community. From the mid-1990s efforts were made to purchase new medium-sized transport helicopters, but a lack of funds delayed a final decision. Then, in February 1999, one of Austria's biggest snow avalanche catastrophes at Galltür, Tyrol, dramatically demonstrated the shortage of modern helicopter transport capacity. In the wake of the Galltür disaster, the Austrian government acted fast to remedy the situation.

Four different types were tendered for the competition: EH-101, NH-90, Eurocopter AS 532 Cougar and Sikorsky S-70A Black Hawk. The Cougar and Black Hawk made it into the finals, which was won by the latter following its high scoring in an internal air force evaluation. The Austrian MoD decided to buy nine S-70As at a total value of Euro 196.2 million. An option for three more helicopters was never taken up.

The order included four ESSS (External Stores Support Systems) with eight auxiliary fuel tanks, which will be mainly used for long-range missions abroad, electrically powered winches with 88-m (289-ft) cables and a 272-kg (600-lb) lifting capacity, skids for all helicopters, self protection systems, weather radar and glass cockpits.

On 10 July 2002 the first Austrian S-70 was handed over to the OLk at Stratford, Connecticut. Shortly before, two Austrian Air Force pilots – squadron commander M. Doppler and Mj. H. Santner – started a three-week simulator training at West Palm Beach, Florida. Until Austria's first S-70A-42 Black Hawk (serial 6T-BA) was complete there was no flight training, as no similarly equipped H-60 was in operation at that time. Later, these two pilots performed all check and acceptance flights, before the helicopters were delivered to

Landing skids attach to the wheels to spread the helicopter's footprint – vital for operations in boggy terrain or on snow. This was the first Black Hawk to be handed over, and is seen during the first major Alpine training exercise in the summer of 2003.

Austria's S-70A-42 Black Hawks are among the most advanced yet delivered. They have a full 'glass' cockpit, weather radar and a full defensive suite.

Austria by Antonov An-124 Ruslan between September and December 2002. Home base of Austria's Black Hawk fleet is Langenlebarn AB, some 30 km (19 miles) west of Vienna.

Type conversion for the remaining eight crews – experienced AB 212 pilots – was performed in Austria. Because of the four-month delivery period, and that only the Austrian Black Hawks were employed, it took two courses to complete the conversion process. Training was performed by Doppler and Santner, together with one Sikorsky pilot and three instructors from Flight Safety International, contracted by Sikorsky. Three weeks of theory were followed by 15 training missions, in which each crew logged around 20 flight hours. From the spring of 2003 the crews moved on to more advanced flying and built up experience on the type. Starting in spring 2004, nine more pilots are scheduled to be trained by Austrian instructors. All S-70 candidates have to have IFR rating, as well as AB 212 experience,

Four ESSS kits are available to the S-70A-42 fleet, as carried by this aircraft during Alpine training. The external tanks were purchased primarily to increase the Black Hawk's range during multi-national peace-keeping operations.

plus several hundred hours on the smaller helicopter types in service with the OLk.

Parallel to the training course, service tests took place. In early April 2003, during airborne training for the 25th Infantry Battalion at Klagenfurt, Carinthia, instructional procedures for embarking and disembarking troops with varying equipment were developed. Furthermore, the internal loading of medical equipment, weaponry and ammunition was evaluated, as was the carriage of external loads like heavy equipment and containers.

In the summer of 2003 the number of flights in high alpine regions increased significantly. The main goal of the process is to transfer the OLk's vast experience of Alpine flying from the smaller helicopter types in service to the much larger, more capable and more powerful S-70A. By comparison with existing Austrian helicopters, the S-70 has low ground clearance and a long wheelbase, and these were the two main factors which proved initially unfamiliar to OLk pilots. In autumn 2003 NVG (night vision goggle) training started, as did the use of winch for rescue operations. Fire-fighting training is to begin shortly, and it is anticipated that all crews will reach full operational status by 2006.

For the moment the helicopters are used only in Austria. After the full training programme and operational type fielding has been completed, participation in international peace-keeping and peace support operations are planned. The S-70A-42's main duties in military use will be tactical air transport, support and medevac missions, while they will also support civilian agencies and relief operations.

Erich Strobl; translated by Dr Heinz Berger

Block 30 gets JDAM

Lockheed Martin F-16C Block 30

In July 2003 the USAF's 80th Fighter Squadron, based at Kunsan Air Base, South Korea, became the first active-duty unit flying the F-16C/D Block 30 to become JDAM-capable. The 80th FS, is part of the 8th Fighter Wing 'Wolfpack', which also parents the 35th Fighter Squadron. The 35th FS operates the F-16C/D Block 40, and has been JDAM-capable since 2002. The upgrade of the 80th FS means that all of the 'Wolfpack' aircraft can now drop the GPS-guided munition.

JDAM conversion of the 80th FS was planned for later, but was brought forward because of

Operation Iraqi Freedom, although the squadron was not deployed. The aircraft were fitted with the necessary weapons interfaces in the pylons, and the first was ready for a trials drop on 2 July. Flown by the squadron commander, Lieutenant Colonel Eric Schnitzer, F-16C 86-0308 released two inert GBU-31 2,000-lb (907-kg) training rounds on targets at the Chik-do Island weapons range, with excellent results. The JDAM downloads GPS target co-ordinates from the aircraft's navigation system before launch, and guides itself to those co-ordinates with a precision of tens of feet, unhindered by weather or light conditions. Loading the weapon takes around 15 minutes longer than unguided bombs as it requires a special cable hook-up so that the bomb can 'talk' to the aircraft. The standard loadout time is around 45 minutes.

As well as the JDAM, the 80th FS Block 30s also have a precision capability – albeit one affected by weather conditions – thanks to the Litening laser targeting pod. The co-located 35th FS Block 40s are equipped with the two-pod LANTIRN system.

David Donald

Top and left: Lt Col Schnitzer flies this F-16C Block 30 on the first JDAM drop for this F-16 sub-variant. The aircraft also carries the Litening targeting pod on the starboard chin pylon, and has an ALQ-184 ECM pod on the centreline. The aircraft is marked for the 8th Operations Group, and the fin-stripe carries the two colours of the constituent squadrons – blue for the 35th FS and yellow for the 80th FS.

AETE's 'T-birds'

Canadair CT-133 Silver Star

Although a lot has changed since the introduction of the first ejection seats in the mid-1940s, the basic requirement remains the same, but changes in aircraft and survival equipment technology, together with other factors such as increasing average body dimensions and the introduction of female military pilots, ensures that seat and equipment design remains an important challenge. Furthermore, all survival equipment needs to be quality-controlled, air-tested, validated and certificated as suitable for operational use.

By far the most important part in the long and demanding validation process is the 'live' air-testing of the various elements during simulated ejections. One of the few remaining test centres for this task is Canada's Aerospace Engineering Test Establishment (AETE) – the Canadian flight test centre based at CFB Cold Lake in Alberta. AETE operates Canada's last four remaining Canadair CT-133 'T-Bird' aircraft, of which one is equipped for inflight 'live' ejections. This platform provides AETE with a capability to test survival equipment in realistic conditions that is rivalled in only a very few nations.

AETE: Canada's test centre

AETE was established in 1971 by amalgamating various smaller test centres, and its mission is to provide cost-effective, timely, quality engineering evaluations of the airworthiness and operational effectiveness of the air force's aerospace systems. By becoming the exclusive flight test agency of the Canadian Armed Forces, AETE conducts a wide variety of flight and ground testing involving every aircraft and helicopter type in the Canadian inventory. Furthermore, AETE is also responsible for the evaluation of all new systems to be installed on Canadian military aircraft and helicopters, including weapons.

Based at CFB Cold Lake (Alberta), home base of No. 4 Wing and its three CF-188 Hornet squadrons, AETE uses the vast surrounding airspace for flight test operations. The Cold Lake Air Weapons Range (CLAWR) contains the well-equipped Primrose Lake Evaluation Range (PLER), which is AETE's primary test range. For evaluation and testing purposes, three drop zones for air-launched weapons are available, all 'guarded' by advanced photo- and video-theodolites for data-collection, recording and transmitting. All telemetric data are transmitted in real-time to the Flight Test Control Room in the AETE installations at Cold Lake.

In order to conduct its ground and inflight test work, AETE is staffed by some 230 military and civilian personnel with a wide variety of scientific skills and knowledge. The unit's pilots – originating from Canada's fighter, transport- and helicopter communities – will have all attended one of the four recognised test pilots schools: the Empire Test Pilots School (ETPS) at Boscombe Down, the Air Force Flight Test Center (AFFTC) at Edwards AFB, the Naval Test Pilot School (NTPS) at NAS Patuxent River or the Ecole du Personnel Navigant d'Essais et de Reception (EPNER) at Istres.

One of AETE's four CT-133s (133452) wears this special colour scheme. Three of the aircraft are used for general chase and training purposes, but are also available as platforms for special trials if required.

In common with other test units, AETE employs a wide variety of specially equipped and instrumented fixed-wing (jet) aircraft and helicopters, ranging from modified CF-188 Hornets to CH-146 Griffons. Both types have modern digital data collection systems, providing accurate data to the test crew. AETE also flies unmodified CT-114 Tutors and CT-133 'T-Birds' for photo-chase and training purposes.

Forever young – AETE's 'T-birds'

On 31 March 2002 the Canadian Forces withdrew its last 28 CT-133 Silver Stars from operational service, after an 'illustrious' career of almost half a century with most of the CAF's fixed-wing jet squadrons. Two composite squadrons – No. 414 at CFB Comox (British Columbia) and No. 434 at CFB Greenwood (Nova Scotia), which had used the 'old' T-Birds for multiple support taskings for the Canadian air force and navy, were disbanded, while the other two CT-133 units, No. 417(CS) at CFB Cold Lake (Alberta) and No. 439(CS) at CFB Bagotville (Quebec), re-equipped with helicopters. All remaining CT-133s, some of which had only been updated a few years earlier, were flown to CFB Mountain View (Trenton/Ontario) for storage pending a final decision regarding their future.

The withdrawal of the 'multi-role' and 'cheap to operate' CT-133s deprived the Canadian Forces of a much-needed training asset, since these aircraft were frequently used as (electronic warfare) aggressors during air and naval exercises, and as target-towing aircraft for naval gunnery training. These tasks are now outsourced to civilian contractors.

Four of the withdrawn CT-133s, all modified with upgraded cockpits, were transferred to the AETE at Cold Lake. For many years, the Establishment had used the 'T-Bird' for various test tasks, sometimes necessitating airframe and internal modifications. Although 'T-Birds' of the

Both of AETE's chase/training types (CT-133 and CT-114 Tutor) are retired from operational service with the Canadian Forces. 133610 is one of three general duties CT-133s in use.

Debrief

co-located No. 417 Squadron became redundant and available for AETE, the four 'new' aircraft were chosen after a fleetwide structural inspection. As well as the four in-use machines, two CT-133s (s/n 133546 and 133656) are kept in flyable storage at CFB Mountain View as attrition replacements. Once a week the engines are started and run by personnel of the Aerospace & Telecommunications Engineering Support Squadron (ATESS).

The influx of 'new' aircraft enabled AETE to withdraw its 'old' CT-133s, including the specially modified 133613 and 133413. Still wearing its natural metal finish and extra-long red/white pitot tube, 133613 was used to monitor various structural loads on the airframe and for pitot-static calibration. Recently, the aircraft was sold to a US-based 'warbird' collector. The other modified CT-133 – 133413 – was used as an ejection-seat testbed. Late in 2002 it was withdrawn from operational use and stored at Cold Lake, pending a decision over its fate. Efforts were made to preserve this aircraft, together with a 'standard' CT-133 photo-chase aircraft (133572), at the local museum. At the time 133413 was withdrawn from service, its replacement (133648, formerly used by No. 417 Squadron) was still undergoing modification and rebuild by Kelowna Flightcraft.

Today the three general-purpose CT-133 aircraft are used for a wide variety of test support and training missions. Currently, six AETE pilots are qualified on the 'T-Bird', one of whom is also backseat-qualified. Similar to other test establishments, all AETE pilots are qualified on two aircraft types, with the CT-133 being the most obvious candidate as a second fixed-wing qualification. Together with AETE CT-114 Tutors, the CT-133s are used for photo-chase missions during the various air tests. During these missions, which are flown over the Primrose Lake Evaluation Range (PLER), a qualified chase pilot and a specially trained image system technician fly in the CT-133. In general, a 100-ft (30-m) lateral separation is maintained, enabling the high-speed film cameras in the tip tank and the hand-held camera of the back-seat technician to record the test event on film.

As well as photo-chase work, the 1950s-era 'T-Birds' are used to screen operational pilots applying for test-pilot training. Requiring four new test pilots to be trained every year, AETE candidates are submitted to a low-key test-simulation and screening process. During the course of two weeks these pilots will fly the CT-133 (or CT-114 Tutor) in a simulated test environment, enabling qualified AETE instructors to monitor the test capabilities of the candidates, prior to their eventual selection and training in the United States (NTPS/AFFTC) or Europe (ETPS/EPNER). Finally, the CT-133s are used to familiarise pilots from a transport background with fast-jet flying.

All first-level maintenance and component replacement is done by AETE at Cold Lake, using the vast stocks of 'basic' spare-parts. For regular first and second level inspections, the aircraft are flown to No. 8 Wing at CFB Trenton.

Ejection seat testing

AETE's almost unique developments in escape systems testing has attracted international attention and admiration. An increasing shortage of suitable, efficient and low-cost ejection seat testing hardware triggered AETE's development of some world-class 'home-made' platforms for this role.

For ground and low-speed testing of ejection seats, AETE uses a locally modified Dodge RAM 3500 Series V10 pick-up truck, nicknamed 'Black Thunder'. Fitted with a flatbed, extra external fire extinguishers, up to five high-speed cameras and a control panel in the cabin, the Dodge – officially known as 'Ejection Seat Ground Test Vehicle' – fires test-seats at up to 60 kt (111 km/h; 69 mph) using Cold Lake's outer runway. Costing around $100,000 per firing, these ground firings are cost-effective alternatives to more expensive airborne trials. During the recent Tutor and 'T-Bird' ejection seat modification programmes, six 'Black Thunder' ejections were incorporated into the programme.

Airborne testing of ejection seats and the impact of the ejection process on the pilot is performed by AETE's most remarkable asset: a specially modified CT-133 (133648). Seats for test are bolted into an aluminium 'bathtub' in the rear cockpit, reclined at 12°. Prior to take-off the seat is mechanically armed but electrically isolated until the actual firing, which is commanded by a series of switches. To protect the pilot a blast shield is mounted between the cockpits. Additional modifications are the removal of all flight controls from the rear cockpit, addition of firing circuits in the front cockpit and fortification of the base of the vertical fin to avoid damage from the ejection seat exhaust. Onboard cameras can be mounted in the empty tip tanks.

The seats are fired through a specially developed half-open canopy on command of the pilot in the front. Ejection speeds can vary from 125 kt (231 km/h; 144 mph) to 450 kt (833 km/h; 518 mph), with g forces ranging from -1 to +4, depending on test requirements and objectives. To avoid any interference with recording the ejection and seat separation process, the CT-133 initiates a climb or dive immediately after the ejection, before returning to Cold Lake.

Photo- and video-tracking is performed by image systems technicians flying in the chase aircraft (CT-114 or CT-133), or on the ground using fixed video-theodolites, positioned around the test area. Sometimes AETE's CH-146 Griffon helicopters are used for photo/video collection, especially for filming dummy/seat separation and parachute opening. Since no radar reflector is mounted on the ejection seat

The days of manned ejections are long over. AETE uses dummies, available with three varying degrees of instrumentation. The most sophisticated (and expensive) Hybrid 3 dummies are only used in the final stages of seat testing, when their chances of survival and availability for reuse are high.

This Dodge truck – affectionately dubbed Black Thunder *– provides AETE with a cost-effective low-speed, ground-level seat test vehicle. Runway emergencies during take-off and landing account for a significant proportion of live ejections, and the truck can simulate these conditions well.*

'T-bird' 133452 returns to Cold Lake. In addition to 3 Wing's front-line CF-188 squadrons and AETE, 'Cool Pool' is also home to the advanced/weapons training phases of NFTC and hosts the Maple Flag exercise.

AETE's seat-test CT-133 has a standard front cockpit (left) and an aluminium 'bathtub' in the rear cockpit (right), which can accommodate test seats of varying sizes. The pilot has a blast shield behind to protect him during firings.

and/or dummy pilot, all tracking and videoing is performed manually, demanding high skill levels by all personnel involved.

On average, 10 seats per year have been fired using the CT-133 platform. Since 1995, second-generation seats for the CT-114 Tutor and CT-133 T-Bird and the US Navy Aircraft Common Ejection Seats (NACES) have been tested by AETE. In October 1999 all Canadian CT-133s, including AETE's test aircraft, were grounded after a fatal seat-pilot separation. A nine-month, $800,000 test and modification programme resulted in the installation of a drogue chute on the back of the ejection seat preventing seat-pilot interference after separation. A larger parachute was also provided.

Human impact

Arguably more important than testing the pure technical capabilities and aerodynamic characteristics of the ejection seats, is the in-depth analysis of the impact of the ejection process on the human body. To collect all necessary 'pilot-related' information during seat-firing simulations, AETE uses three types of dummies. For initial trials low-tech 'dumb' (and less expensive) dummies – aka 'CG Guards' – are used. At a later date in the seat development or modification programme, the more sophisticated Hybrid II dummies – equipped with some data-collection devices – are fired to collect more advanced data.

Finally, the data-collection efforts are fine-tuned – especially for validation and certification – by using two 'state-of-the-art' $250,000 highly instrumented dummies. These Hybrid IIIs can measure roll, pitch and yaw rates, and monitor the g forces experienced during the ejection-process. Furthermore, telemetric devices are installed in the head of the dummy, which send real-time data to Cold Lake by using a transmitter located in the body of the dummy.

The introduction of female pilots within the Canadian Forces (and worldwide) required AETE to invest in a 'female' dummy. The physical differences between the two genders presents a technological challenge in the development of a universal escape system (ejection seat, safety harness and parachute canopy).

Stefan Degraef and Edwin Borremans

Until retirement in late 2002 133413 was the ejection seat testbed. Its replacement, 133648, is in essentially the same configuration.

'Tulsa Vipers' re-engine with -229

F-16C/D Block 42

Determined to be the best trained and to have the best combat assets available, the Air National Guard has spent much of the last decade acquiring newer aircraft, as well as upgrading and enhancing existing airframes with new capabilities and equipment. For the 125th Fighter Squadron, 138th Fighter Wing at Tulsa, Oklahoma ANG, this resulted in F-16s being delivered from 1993 to replace A-7s. Even though the F-16s were passed down from active-duty squadrons, the jets were only a few years old and some of the most advanced 'Vipers' available at the time.

Today, the 125th FS has 16 F-16Cs and one F-16D in its inventory. All are Block 42 F-16s, powered by Pratt & Whitney engines. The Block 40/42 version of the 'Viper' was often referred to as the Night Falcon for its ability to carry the LANTIRN system of targeting and navigation pods, as well as its unique HUD and other systems designed for the night attack role. Initially, the ANG did not have LANTIRN pods available for use on any of its F-16s, although several years later the Guard was able to fund the purchase of some targeting pods to be shared between its Block 40/42 units.

The quest to make the Block 42 F-16 even better was not a priority for the active-duty USAF, since the only active units operating the sub-variant are training and test squadrons.

From the outside there is very little to distinguish a -229 jet from a -220, although there are subtle differences in the 'turkey feathers' which surround the jetpipe. This pair of 'Tulsa Vipers' comprises a re-engined aircraft on the left, and one with the original powerplant to the right. Fitting the new engine requires only minor internal work, but the effect on the aircraft's performance – especially at combat weight – is dramatic.

With its Pratt & Whitney F100-PW-220 engine, the heavy Block 42 airframe at maximum payload proved to be slightly underpowered – which is why most of them were relegated to training and test. The General Electric F110-powered Block 40 F-16s have almost 15 percent more thrust than the PW-powered Block 42, so the USAF placed this variant in front-line combat units. During deployments with the -220, the ANG was limited in many

Until 1993 the 125th FS flew the A-7D/K Corsair II, receiving F-16C/Ds as a replacement. It adopted the 'Tulsa Vipers' name when it acquired its new mount.

respects on weapon loads, as well as general aircraft performance. The ANG sought a way to get its jets more power to take full advantage of high gross weight combat configurations with no operational limitations.

Pratt & Whitney, meanwhile, had developed an advanced version of the engine for the Block 52 F-16 – the F100-PW-229. Naturally, the ANG investigated the procurement and installation of these new engines in its Block 42s. From an installation standpoint, the modification to the aircraft was going to be minimal. Aside from a few minor structural modifications, as well as the replacement of a heat exchanger and some environmental control system components, the process would almost be as simple as a routine engine change. Captain Todd Farnsley, one of the maintenance officers assigned to the 138th FW, said, "the process of re-engining a jet from start to finish takes about 5 days. Other 220-powered jets receive the various modifications required for the re-engine programme as time goes on, so that when more engines are acquired and delivered, only a day or two is needed for maintainers to install a new -229."

Inevitably, the most difficult hurdle to overcome in this process is funding. Over $327 million is needed to re-engine three wings of aircraft, so the programme is being performed in increments. The process of getting congressional additions in the annual defence budget started with the FY 2001 budget cycle. The ANG secured almost $50 million in the first year, and the programme was initiated upon receipt of funding in early 2001. Less than 18 months later, the 'Tulsa Vipers', along with two other Block 42 units in the ANG, were deployed to Turkey to fly combat missions over northern Iraq with Dash 229-powered F-16s.

The 138th FW assumed the centre slot in a 90-day deployment to Turkey that lasted from August to December of 2002. The concept of splitting the normal three-month AEF rotation between Guard and Reserve units is a regular practice commonly referred to as 'rainbowing'. The ANG unit from Toledo, Ohio, made the initial deployment of aircraft and associated personnel to Incirlik, and the 132nd FW from Des Moines, Iowa, redeployed the jets back to the United States after fulfilling the commitment in early December.

Personnel from Tulsa arrived at Incirlik AB in September 2002 to assume the ONW mission for a month-long stint as part of AEF 6. During the following 30 days, the 138th FW flew combat missions over northern Iraq using 12 Block 42s (four from each of the three units) that were all powered by new -229 engines. Pilots and maintainers who had been on similar deployments in previous years commended the new engines as being the single most important upgrade their jets had received over the past decade. With an additional 5,000 lb (22.25 kN) of thrust available, the 'Viper' had excess power even at the maximum payload allowable for a Block 42. The extra thrust was significant, since the three units were called upon to fly with heavy payloads on almost every mission during the deployment. The 'Tulsa Vipers' had the distinction of dropping the first JDAM on a target in the northern 'No-fly' zone over Iraq during the 2002 deployment. It was also noteworthy that every sortie flown returned 'Code One' for the engine, and no engine issues ever interfered with combat operations.

Upgrades for the 138th FW and the other two members of the ANG Block 42 coalition are far from over. Currently, each of the three units has five F-16s fitted with the new engines, as well as a few spares for support. Since the three fighter wings operate a combined total of 51 aircraft, a further 36 'Vipers' need to go

A visible difference between the two engines is the colour of the afterburner plume. The Dash 229 produces a pale blue plume, whereas that of the Dash 220 is thicker and yellow.

through the re-engining programme. By the end of FY 2002, the programme had received around one-third of the necessary funding. Plans call for an additional 15 aircraft to receive the engine upgrade by 2005, and the full complement of all 51 should be complete by 2007. As more ANG F-16s receive the new -229 engines, the -220s are not going to waste. F-15s of the Hawaii ANG are receiving the excess -220s to replace the much older -200s that were installed originally. Though the -220s do not increase thrust for the F-15s, they do improve reliability and maintainability.

Another upgrade planned for the 'Tulsa Vipers' in the near future is the Common Configuration Implementation Program (CCIP). CCIP is an avionics upgrade that will bring the entire USAF/ANG Block 40/42 and 50/52 fleets to a common upgraded avionics configuration. All Block 50/52 F-16s are going through this depot level modification at the Ogden Air Logistics Center in Utah. The Block 40/42 'Vipers' are next to get the CCIP mod package and will gain many new features, such as colour cockpit displays, Joint Helmet Mounted Cueing System, Advanced Identification Friend or Foe and Link 16, as well as full integration of all the latest weapons and sensors. With a full squadron of PW-229 powered Block 42s that have additionally received the CCIP upgrade, the 'Tulsa Vipers' will then boast some of the most advanced F-16s available.

Andy Wolfe

Four 'Tulsa Vipers' carry loads of inert free-fall Mk 82 bombs during a training session. Experience from the ONW deployment showed that the Dash 229-engined Block 42s could fly combat patrols with a heavy weapon load at much higher altitudes than was possible with the Dash 220 engine, with a consequent reduction in vulnerability to anti-aircraft defences.

TECHNICAL BRIEFING
ERIEYE
Sweden's Eyes on the World

The Erieye phased-array radar is arguably the most advanced operational airborne early warning and control system in the world today. It certainly offers the best combination of price and performance. Coupled with the Embraer EMB-145 jet platform, the Erieye has grown to be a major force in the rapidly growing market of airborne C^{4}ISTAR systems.

The Erieye airborne early warning (AEW) system has now been bought by four countries around the world. Although at first glance that might not look like a large number, it is a significant slice of today's air forces that actually own and operate an AEW system. The Erieye is a high-value asset – and the nations that possess it are set to grow in number as several new customers nail down their AEW requirements in the near term. The Erieye has taken up this position as a major market force thanks to some sound design, good technology, a far-reaching operational concept – and a little luck. With the Erieye, radar maker Ericsson found itself in the right place at the right time. The Erieye was born into a scenario that any defence staff will recognise. Its original customer, the Swedish air force (Svenska Flygvapnet), needed an effective airborne radar system that was utterly reliable in wartime but which had to be supremely affordable in peacetime. A new set of technologies was developed to make the Erieye a reality while, at the same time, Sweden had to invent a new theory of airborne early warning operations so that it could have its vital radar platforms – and be able to afford them too.

From the earliest days of the Cold War, Sweden was a believer in air power and built up a substantial independent aviation industry, led by Saab, to equip its air force. After World War II Sweden consolidated a policy of armed neutrality, because it had been caught ill-prepared for war and scrambled to find meaningful numbers of modern combat aircraft to press into service. This disturbing experience was one factor behind the powerhouse of Swedish self-sufficiency that grew from the late 1940s and beyond. Sweden also learned valuable lessons from the Finnish Winter War, where a

Left: The original implementation of Ericsson's Erieye phased-array radar was in the Swedish air force's S 100B Argus airborne early warning platform. Six of these aircraft were delivered in the mid-1990s.

Right: With its adoption for Brazil's SIVAM programme, the Embraer EMB-145AEW&C took the Erieye system to a completely new level of jet performance. This aircraft was involved in 2003 testing of the self-defence system that will be delivered as part of the Greek configuration.

Below left: Brazil adopted the EMB-145AEW&C as the R 99A, part of a whole new special missions family developed by Embraer from the the Regional Jet design.

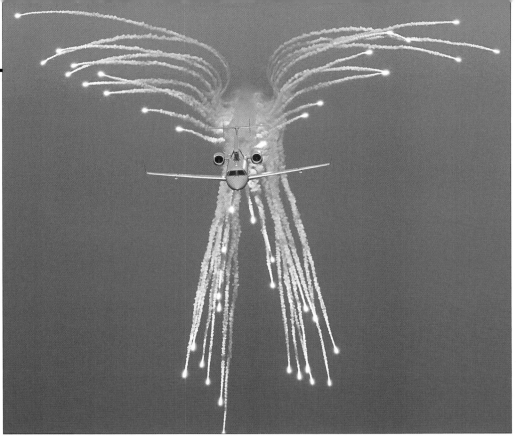

'volunteer' Flygvapnet wing was sent to help defend Finland against the invading Russian forces. Post-war, Sweden found itself facing a similar threat from the Soviet Union and the new Eastern Bloc in Europe.

When it came to planning for the 'next war' Sweden had only a few decisions to make. An invading force could come from the north – but the rough terrain there made large-scale ground manoeuvres almost impossible, especially during winter. Any Russian thrust southwards through Sweden would have to cover many hundreds of miles and could not be easily sustained. The most likely route for an invader would be an air and sea assault across the narrow waters of the Baltic, directly into Sweden's southern heartland, to hook rapidly around into the rest of northern Europe. The Swedish armed forces trained to fight this invasion. However, no matter how well-equipped or motivated the defenders were, in the event of a Soviet attack they would always be facing numerical superiority.

To give its forces the edge, Sweden turned to innovative tactics and new technology. The front-line of defence would be its air force. Flygvapnet developed a unique doctrine of dispersed basing, abandoning its conventional airfields – which were little more than large fixed targets – for wartime road-strip bases, hidden in the forests and hills. From these bases small groups of fighter and attack aircraft would make rapid series of attacks before moving to new concealed locations. To tie this dispersed force together Sweden developed a wartime command and control (C^2) network using underground command centres and a series of hidden radars that were difficult to target and even harder to destroy.

Another key national warfighting capability was an investment in military datalinks. Sweden's work in this field stretched back to the Draken era (the first two-way datalink was operational on J 35B/Ds by 1963) and, during the 1970s and 1980s, Swedish datalink technology – developed and fielded in extreme secrecy – became the most advanced and widely implemented anywhere in the world. Confidence in this secure C^2 net led to another innovative concept, distributed airborne early warning.

Preparing the defences

By the late 1960s Sweden was examining an airborne early warning (AEW) solution. There was never any doubt from which direction the threat would be coming and that, to a great extent, made the defenders' life a more straightforward affair. Rapid detection of a Soviet invasion force crossing the Baltic by air and sea would be crucial. A chain of about 15 fixed long-range PS 65/66 air surveillance radar sites gave good overlapping coverage across the country, but only against high-flying aircraft. Hampered by the inescapable mathematics of the Earth's curvature, the radars that looked out to sea were even more limited. To improve their line-of-sight the PS 15 maritime surveillance radars were mounted on 100-m (328-ft) towers that extended effective range out to around 50 or 60 km (31 or 37 miles). These sites still did not provide 100 per cent coverage, they were fixed and inflexible, and always vulnerable to attack. By the 1980s a series of improved maritime radars had been developed that were more mobile. In turn, the air defence radar system deployed networked antennas that were mounted on 'pop-up' masts which could be hidden in hardened silos if the site was attacked. The late-model PS 870 radars had a maximum detection range of around 150 km (93 miles), but this fell to about 50 km (31 miles) against low-flying targets – and, increasingly, the greatest threat was from low-flying aircraft and cruise missiles.

As a footnote, a manual optical tracking system called OPIS was available to back up the radar chain. OPIS personnel could observe enemy aircraft and phone their reports into the national command system. One of Sweden's many Cold War secrets, the OPIS system remained available into the early 1990s, but it was purely a fallback option and, of course, it could only function once Sweden had actually been invaded. The system worked but the results it achieved were out of all proportion to its costs. As aircraft performance increased, contacts were invariably flying too fast or too high to be tracked by telephone. One former air defence Viggen pilot ruefully recalled, "It was better than nothing, but very many times I found I was chasing myself."

What Sweden needed most of all was a fast-moving gap-filling system that could cover any holes punched in the main radar net. Ground-based radar was an imperfect tool – radars only monitor one specific sector and the Swedish planners estimated that two-thirds of the time they would not be transmitting, to protect against anti-radiation missiles. The emerging

From the moment of their inception the Erieye radar and S 100B aircraft were designed to slot into Sweden's datalinked national command network, while operating from dispersed bases hidden across the country.

AEW requirement sought the detection of low-flying targets at ranges of about 60 km (37 miles) from the forward line of national defences. This meant that the (airborne) platform would need significant stand-off radar range to ensure its own security, in the order of 200 km (124 miles).

The first serious radar studies began around 1982/83, and already an electronic phased-array radar was seen as the best solution – even though that technology did not actually exist yet. The only extant AEW options were the Boeing E-3 Sentry and the Grumman (now Northrop Grumman) E-2 Hawkeye. The gold-plated Sentry was not even considered, but some serious thought was given to the more affordable E-2. Its APS-138 radar was seen as an unhappy compromise for Sweden. It had excellent over-sea performance but this fell away steeply when working over land. With many of Sweden's future war scenarios rooted in the dense archipelago of many thousands of islands off the south coast, the Hawkeye would not do the job properly. However, that was not

The Swedish air force christened the S 100B as the Argus, but of course the crews never use that name. In service the aircraft are usually referred to as 'FSR:en' (FSRs) – an abbreviation for Flygspaningsradar.

seen as an insurmountable problem. What finished the Hawkeye in Sweden was that it was just too big and heavy to operate from Flygvapnet's network of wartime roadbases. Attention was also given to the UK's Nimrod AEW programme. Observing the British experience, the Swedes could again see the huge difficulties in taking an over-water radar system and making it work in a littoral environment.

Defining the Swedish system

Swedish attention turned to building a brand-new national programme and to making some firm decisions on how its new AEW system would function. From the outset developing one with a full 360° coverage was discounted. For reasons of cost and technology the new system would have a sector-looking radar.

Ericsson's Carl-Gilbert Lönroth, one of the key figures in the Erieye story, remembers, "There was very little value in covering 360°, there was never any argument about it – and today it's not difficult to convince the (new) customer about this either. Most national defence scenarios are similar to the Swedish one. You know where the threat is. Sector-looking radar can't look 'around corners' it's true, so to remedy that you just get two aircraft – which you'll always have. We started with a requirement for 15 aircraft with a short-range radar, but as a better radar came along we needed fewer aircraft. Now one aircraft covers a 500-km (310-mile) coastline."

Above: During the 1980s and 1990s some essential development work for the Erieye was undertaken by this Swedish air force Metro (Tp 88) testbed.

Below: The first Saab 340AEW&C took to the air, with its radar antenna installed, in July 1994. Note the new rear section with added ventral fins and APU housing.

The next question was, if such a new radar could be built, what aircraft would carry it? Saab was the obvious provider. By the early 1980s it had embarked on the joint design, with Fairchild, of a twin-turboprop regional airliner that would become the SF340 (and later the S340 following Fairchild's departure from the programme). The first aircraft flew in March 1984 and by November 1985 deliveries had started to airline customers. Almost automatically it was selected as the radar platform for

Brazil's force of five SIVAM-dedicated R 99A AEW&C platforms represents a major investment for the Força Aérea Brasileira – and a huge boost for the Erieye programme as a whole. Brazil was Ericsson's first export customer

Flygvapnet. By 1989 Saab had the 'hot-and-high' S340B version available, with uprated General Electric CT7-9B engines, and it was this variant that was finally adopted. For the transformation into an AEW&C platform several changes were made to the airframe. These included: a 9.7-m (31.8-ft) dorsal antenna for the radar, two stabilising strakes below the rear fuselage, vortex generators on the fin and rudder, revised engine mounts to offset the thrust line to starboard, an enhanced air cooling system for the cabin-mounted electronics and a new auxiliary power unit in an extended tailcone.

The Saab 340 was an affordable, Swedish-sourced solution that was also compact enough to join the rest of the air force hiding in the trees, if necessary. Ericsson has always championed availability over coverage and so the S340B, designed for high-tempo operations in daily airline service, was a good choice. Sweden could get away with employing such a small aircraft because its concept of operations called for no onboard operators. Instead, the radar platform would be flown by two crew up front with all the radar information transmitted by secure high-speed datalink into the StriC command and control network. The AEW aircraft would not operate as a stand-alone battle management system, like the AWACS. Instead, it would be another link in the chain mail of Sweden's C^2 armour. The downside of using the Saab 340 as a radar platform was its modest performance and cruising altitude.

As early as 1985 development work was underway on the Erieye radar. In 1986 came the first aerodynamic trials of the basic antenna configuration. These involved a modified Fairchild Metro III trials ship (Flygvapnet designation Tp 88). One of the Ericsson development team remembered, "It was crucial that we started work with a 'demonstrator', rather than a real radar. We wanted to avoid the British experience of getting in too deep, too quickly. There's no way we could have done any of this without our fighter radar experience. It would have been impossible if we'd just been a ground-based radar company. Our mix of radar and microwave technology, plus military and civil experience, is pretty unique."

By 1991 the Erieye radar was undergoing ground testing. By the end of that year and on into 1992 it was flying aboard the Tp 88 testbed. Ericsson's phased-array technology turned out to be better than anyone had hoped for. The radar testbed was expected to have merely a coastal surveillance range but it soon surprised the entire team with its level of performance. The radar installed in the Metro was single-sided only, but it proved invaluable in solving some of the early technical hitches that affected the phased-array design. For example, it became essential to reduce the radar's sidelobes – not merely because of the jamming resistance that this imparts but because (under certain conditions) returns from the ground were found to be jamming the radar.

Go-ahead for Erieye

In December 1992 the Swedish Defence Materiel Administration (FMV) awarded Ericsson the production contract to build six Erieye radars for installation on Flygvapnet Saab 340s. The Saab 340AEW&C programme was born. In Swedish service the Erieye radar was designated PS 890, so the entire airborne system would be the FSR 890 (Flygburen Spaningsradar 890, airborne reconnaissance radar 890). Under the original schedule, the first aircraft was due to be delivered by the end of 1995. The order for the initial Saab 340AEW&C

Phased-array radars

A new breed of antennas are at the cutting edge of today's radar technology. Instead of the familiar dish or flatplate antenna, they incorporate arrays of individual transmitter/receiver (T/R) modules, that can each be independently controlled. Each T/R module operates as a separate 'mini radar'. The modules can be grouped together to operate as one large radar or several smaller radars – all looking in different directions and at different targets. These groups are controlled in phase, to either transmit or receive. Therefore, they can be actively 'looking' for targets like a normal radar, or passively 'listening' to detect the emissions from other, hostile, electronic emitters. Because the T/R modules are arranged in rows these radars are often referred to as planar arrays, but because of the way they operate (using several simultaneous phases instead of just one) they are most commonly referred to as phased-arrays.

The first generation of phased-array radars were largely passive phased-arrays. Examples include the B-2's APQ-181 (developed by Hughes), the Rafale's RBE2 (developed by Thomson-CSF) and the MiG-31's Zaslon (developed by Phazotron). Passive arrays are essentially single arrays, with one transmitter driving all the elements of the array. The phase of the transmissions from each element is then delayed through a beam-forming computer to switch the radio frequency (RF) energy along different delay paths, producing the required phase changes in each module.

In an active phased-array, the mass of smaller, individual T/R modules (typically in their hundreds) does away with the need to manipulate a single radar beam. As in a passive array the electronic scanning in the horizontal and vertical planes is controlled by the phase of the individual radiating elements. However, in the active array, each of these has its own transmitter, receiver and antenna. Each module transmits radar pulses individually, controlled in phase so that the complete array will produce a beam of transmitted energy or a receive beam of the required shape, all directed in the desired direction.

Electronic scanning allows the user to look in any direction at any time, to acquire near-simultaneous target updates from several different directions. The Erieye's S-band radar offers extremely sharp and narrow main beams, with low sidelobes, compared to a the UHF wavelength of other phased-arrays.

Technical Briefing

Left: So important was Greece's Erieye order to Ericsson that the authorities back home agreed to lease one third of Sweden's S 100B fleet to bridge the gap before the Greek air force's EMB-145AEW&Cs could be delivered. Two Argus aircraft operated from Elefsis until July 2003 where they provided valuable lead-in training for the new AEW squadron, 380 Mira (badges inset).

Right: Swedish industry has made much of the synergies between Brazil's Erieyes and the Gripen, in the chase to win the FAB's FX-BR next-generation fighter order.

between them to even out flying hours across the fleet while retaining maximum 'peacetime' utility for the aircraft. Each Argus can be converted to a 30-seat transport in about 24 hours. Today Sweden's fleet of S 100Bs is operated by F17 Specialflygenhet Malmen. The unit is a part of F17 Wing but is based at Malmen, near Linköping – not at F17's main base in Ronneby. The S 100Bs were formally transferred to F17's control on 1 January 2003 following the decision to close Uppsala as an operational base (and disband F16) by the end of 2003.

Argus in operation

The Swedish air force conducted its first full-scale integrated exercise with the Gripen and Argus early in 1999. The FSR 890 (S 100B) is an integral part of Flygvapnet's FV2000 (Air Force 2000) plan for a fully-integrated network-centric warfighting capability. Under the Swedish concept of operations, the S 100B is controlled by the national network of underground StriC (Stridsledningscentral) control and reporting centres. Data is transmitted from the air using the secure high-speed datalink element of the TARAS digital tactical radio system. The StriC operators fuse the information from the FSR 890 with that from the rest of the national radar and sensor network to build a complete picture of the battlespace. From the StriC, FSR 890 data can be uplinked to other aircraft, such as the JAS 39 Gripen, or across to the Navy's own command centres for transmission to ships at sea. The six S 100Bs represented maximum value at minimum cost. By eliminating onboard operators from the equation, Sweden also did away with the need to recruit, train and maintain a corps of personnel to operate the aircraft. Sweden already had a highly integrated C^2 system – and the air force was entering a period of heavy cutbacks when every resource had to be maximised.

In Flygvapnet service the Erieye has demonstrated an instrumented range of 450 km (280 miles) – and Ericsson points out that this figure is a software limit set by the Swedish customer. Some company demonstrations have indicated an actual detection range of 500 km (310 miles). Cruising at 8,000 m (26,246 ft) the S 100B can detect a fighter-sized target at around 350 km (217 miles), a cruise missile or similar low radar cross section target (with an RCS well below 1 m^2) at around 150 km (93 miles) and larger ships at between 320 to 350 m (199 to 217 miles). There is an unspoken acknowledgement that, in some areas, the Erieye's ground functions were deliberately limited to dissuade army and navy access to the system. Patrolling at around 160 kt (296 km/h; 184 mph) the S 100B has an on-station endurance of six hours.

Sweden has examined the possibility of adding onboard operator stations to its S 100Bs, to support possible deployed operations. The

was signed in February 1993 (notionally a prototype) and, 10 months later in December 1993, this was increased to include the complete batch of six S340Bs. By 1994 Saab was gearing up for the delivery of its first Saab 2000 high-speed turboprop – following on from the maiden flight of March 1992. The Saab 2000 was everything the Saab 340 was not. Not only was it a larger aircraft with higher operating weights but it also boasted record-breaking high-speed performance plus longer range/endurance and a higher cruising altitude. However, because its size did not fit easily into the air force's basing plans, it would also be more expensive, bring a delay to the overall schedule and none of the existing engineering studies for radar integration were applicable. Some thought was given to adopting the Saab 2000, especially for export customers, but it did not make it into the Swedish programme. The Saab 340AEW&C did benefit from one element of the Saab 2000 design, however – the APU tail section of the larger aircraft was integrated into the S340 to give it a fully autonomous engine start and power generation capability when deployed in the field.

The first Saab 340AEW&C (which would become the first for the air force, serial 100002) flew on 17 January 1994, wearing the Saab test registration SE-C42. It flew again with its antenna housing installed on 1 July 1994. The FSR 890 programme slipped from its original 1995 delivery date, and it was not until 5 November 1997 that the FMV (acting in its role as the official Swedish 'customer') handed over the first four FSR 890s to Flygvapnet. In fact, two 'vanilla' aircraft, without radars, had already been put into service by this point and had been used to provide crew training experience. The FSR 890 was initially taken on charge at F16 Wing, based at Uppsala.

In 1990 a single Saab 340 entered Flygvapnet service as a VIP transport, under the Tp 100A designation (it has since been modified to serve as an Open Skies verification aircraft). When the first Saab 340AEW&C took to the air the service variant for the air force was being referred to as the 'Tp 100B'. In 1995 the new radar platforms were allocated the designation S 100B (S = Spaning, reconnaissance). The name Argus was drawn from Greek mythology and given to the S 100B. Argus, the son of Arestor, was a beast with 100 eyes that never closed. He was a guardian, charged by Juno to watch over Io (whom Jupiter had transformed into a heifer). Unfortunately for Argus he was killed by Mercury who then eloped with Io – but Argus lived on, after a fashion, when Juno placed his eyes on the tail of the peacock.

Flygvapnet maintains a policy of keeping two of its Argus airframes in a passenger configuration, swapping the radars around

The Erieye described

The Erieye is an active phased-array radar, operating in the S-band (3.1 to 3.3 Ghz). Its solid state hardware uses 192 transmitter/receiver (T/R) units, that are arranged in a row at the centre of a carbon fibre antenna unit that is mounted above the carrier aircraft. The antenna plates for the radar run along both sides of the housing. Two cooling ducts run above and below the radar modules, and between the antenna plates, fed by a ram air inlet at the front of the antenna unit. The radar is a multi-mode pulse-Doppler system that has a high bandwidth and a flexible waveform. The beamwidth (in azimuth and elevation) is 0.7° and 9°. It has a selected 3-D capability and uses an adaptive sidelobe cancelling technique to improve the performance of what is already a low sidelobe antenna design.

The Erieye radar has an instrumented range of around 450 km (279 miles). It can detect a (high-altitude) fighter-sized target at around 350 km (217 miles), a surface ship at around 300 km (186 miles) and a low-flying cruise-missile-type target at about 150 km (93 miles). The ESM system has a detection range of 450 km (279 miles) against a fighter radar.

The Erieye's surveillance search area is defined by the operator and can be concentrated across a broad front or in a constant, specific area. The search pattern can be aircraft-stabilised, to search along the track flown by the aircraft, or ground-stabilised; i.e. always fixed on a particular area of interest no matter where the aircraft goes.

Radar modes include: (air target surveillance) air target track-while-search, support air surveillance, helicopter surveillance, high-performance air tracking, extended early warning, primary air surveillance, secondary air surveillance; (sea target surveillance) sea search

Erieye's active phased-array radar uses a technique known as adaptive radar control. This allows for the intelligent uses of radar energy that can be concentrated on specific targets or areas of interest. Unlike a rotating radar antenna, the Erieye is not limited to scanning a fixed volume of airspace over a fixed period of time. As soon as a target of interest is detected one of the radar's multiple beams (generated by the multiple T/R modules) can be allocated to lock on to that target, with tracking initiated immediately after first detection. By concentrating on a specific area the Erieye delivers a high update rate that can be prioritised as more information on the likely threat emerges. The first radar 'hit', or target detection, in a surveillance scan is followed immediately by a higher-energy, shaped radar beam that establishes a track confirmation far faster than a conventional radar – minimising the time in which a target can be lost switching from detection to tracking. Any target that begins to manoeuvre will immediately attract a higher measurement and update rate. The rapid updating allows for effective tracking of a target that is manoeuvring hard, perhaps in an attempt to evade radar detection or to gain an advantageous position for weapons release.

The Erieye can track several targets, or groups of targets, in its surveillance area using individual radar beams – while all the time maintaining an ongoing search scan. At the same time the radar operations can be interleaved to offer simultaneous air priority, air surveillance and sea surveillance modes.

The onboard mission system, as selected by all the customers outside Sweden, uses an open architecture system design with COTS (commercial off the shelf) hardware and operating systems. A MIL-STD 1553B databus connects the Erieye radar, its IFF/SSR and ESM subsystems plus the navigation system, to the main command and control and data management computers. These computers are tied into datalink and other tactical communications equipment and drive the aircraft's onboard workstations.

The aircraft can be equipped with the NATO-standard Mk XII IFF/SSR (Identification Friend or Foe/Secondary Surveillance Radar) that offers Mode 1, 2, 3/A, C and Secure Mode 4 operations. The (optional) ESM system provides coverage in the 2-18 Ghz range. This system is designed to operate in a dense RF (radio frequency) environment with an automatic analysis and identification process, correlated with an onboard threat library. The system will deliver high DF (direction-finding) accuracy for localisation and targeting, with high sensitivity for long-range detection. For Elint tasks target tracks and pulse descriptions can be recorded, and exploited on the onboard consoles. A self-protection suite with an integrated threat warning system and countermeasures dispenser can be fitted.

---

aircraft already has a 'technical operator's station' in the main cabin (used largely for flight test purposes) but there is an acknowledgement that two or three Argus plus a squadron of Gripens could function like a small independent air force, if Flygvapnet chose to do so.

When Ericsson started to develop the Erieye there was no other phased-array AEW radar available – or even a plan for one. Since then the Israeli-developed Phalcon system has come to the market. There is only one user of a single system (Chile's Condor aircraft) although a deal has now been struck to supply the Phalcon to India, using an Il-76 platform (US pressure on Israel blocked an earlier Phalcon deal with China). The Phalcon uses a 1-Ghz L-band transmitter. This has a direct effect on the size of the platform aircraft, because longer wavelength radars need a corresponding larger antenna to produce their given beamwidth.

One assessment of this is that longer wavelength radars benefit from an uncomplicated design but are very easy to jam. Higher frequency radars, such as the 3-Ghz S-band Erieye, have a narrower beam-width. Using an 8-m (26-ft 2-in) antenna, for example, an L-band radar with have three times the beamwidth of an equivalent S-band transmitter. The wider a radar's beam, the easier it is for hostile jamming to isolate it and crack it open. The Erieye produces a 1° beam that is very narrow, focused and hard to jam. By way of comparison, a typical UHF beamwidth could be around 10 times that.

The Swedish version of the Erieye covers an arc of 120° on either side of the aircraft. For Brazil's R 99As this coverage was increased to a 150° arc (still maintaining the 1° beamwidth). While Ericsson has always been dismissive of the criticism that its radar's basic design does not afford a full 360° coverage, it has quietly moved to provide just that. The radar fitted to Greece's EMB-145AEW&Cs delivers (compensated) 360° coverage. Sweden's FSR 890 system can track 300 air targets and 300 maritime targets. For export customers that capability has been significantly expanded. The Greek aircraft, for example, are capable of tracking 1,000 air targets and 1,000 sea targets.

The Erieye system has proved to be adaptable, with great growth potential. It is blessed (some might say cursed) with a technology that is relentlessly moving forward. Even by the late 1990s the radar's transmitter/receiver modules had already been updated three times. Ericsson is conducting far-reaching work on the development of new phased-array radars and techniques, much of which is shared between the advanced radar programmes for the Gripen fighter. Like the Gripen, the Erieye benefits from being a software-driven design and

improvements to the system have been made through new software editions – not changes to the radar hardware. This is not to say that radar development is a simple or cheap job. In the words of one Ericsson engineer, "even though capability modifications and improvements are software-driven, it takes a tremendous amount of time and money – so while the programming task is not impossible, taking the man-hours to do it would just not be viable unless you have a customer paying for it." The key to the Erieye's success was to make it a success outside Sweden. While Ericsson had a good track record in selling ground-based military radars, sales of airborne systems were a whole new world to it. Perhaps such a revolutionary new system needed a revolutionary programme to help take it out to the wider market. That is certainly what happened when the Erieye flew south to Brazil.

Going to Brazil

Brazil's SIPAM/SIVAM programme is a hugely ambitious project that is building a giant protective surveillance network across an area of 5.2 million km^2 (2,007,877 sq miles). One third of the world's tropical rainforests lies within this vast Amazonian region of Brazil, the Amazônia Legal. It, in turn, makes up 61 per cent of Brazil's total territory (although it is home to just 12 per cent of its population). During the 1980s the Brazilian authorities had to admit that they simply didn't know what was going on out there. It was clear that a precious natural resource was under threat from illegal logging and mining. More fundamentally, Brazil's national security was at risk from the explosive growth in narcotics production and trafficking, smuggling and the potential for armed groups linked to the drugs trade or cross-border insurgency to lurk under the jungle canopy. The SIPAM concept (Sistema de Proteção da Amazônia, system for the protection of the Amazon) was launched in 1992 and is now administered by an agency of the Brazilian government. SIPAM controls the SIVAM programme (Sistema de Vigilância da Amazônia, or the Amazonian vigilance system) which ties in other ministries and agencies with expertise in the fields of natural resources, science, defence, industry and technology, meteorology and economics. SIVAM is an expanding network of satellites, aerial sensors, ground-based radars, earth resources monitoring,

The first EMB-145AEW&C, in toned-down FAB marks, arrives at the FIDAE 2000 air show, in Santiago, for its public debut. It was displayed alongside the EMB-145RS multi-sensor platform.

regional co-ordination centres plus an inter-linking telecommunications network. Its aim is to build a multi-layered monitoring system around the Amazon – and most fundamentally a surveillance capability in the air and on the ground, backed up by a military-level C^2 network.

After a competitive evaluation, Raytheon was selected to act as the SIVAM co-ordinator and integrator. In 1997, Raytheon signed the contracts that launched SIVAM as a reality and this included a final agreement with Embraer that was central to the airborne component of the concept. SIVAM is a programme with teeth. The Força Aérea Brasileira (FAB) is fielding its new family of A/AT-29s Super Tucanos (ALX) specifically to patrol the Amazon and engage hostile elements, if required. Flying above this combat component is an eight-strong surveillance and monitoring force based on specially modified versions of the EMB-145 jet platform (Embraer's ERJ-145 Regional Jet). The SIVAM system went 'live' in July 2002 when an initial operating capability was officially declared.

The 'other' Erieyes

It now seems certain that only six Saab 340-based Erieyes (Sweden's S 100Bs) will ever be built. All of Ericsson's marketing efforts are conducted in parallel with Embraer's EMB-145 and this is the level of aircraft performance that future customers will demand. However, the Erieye radar system is compatible with a very wide range of carrier aircraft. When the SIVAM programme became a reality in 1992, the first non-Swedish Erieye platform was to be the Embraer EMB-120 Brasilia. In 1994 a contract was agreed to acquire five Erieye-equipped Brasilias, each priced at $25 million. The first two of these aircraft were originally scheduled for delivery in 1996. The AEW Brasilias were to be equipped with three onboard operator consoles – two for the radar operators and one for a dedicated Elint system (to be supplied by Watkins-Johnson). All five of these aircraft were due to be in service by 2000, but the EMB-120 was dropped (probably very wisely) when the EMB-145 was launched. In March 1997, when the final Erieye/SIVAM contract was signed, the deal was for five EMB-145AEW&C aircraft.

In the Netherlands, Fokker drew up plans for an AEW version of the Fokker 50, under the KingBird Mk 2 programme. This aircraft was to be fitted with an Erieye radar, and followed on from the previous KingBird Mk 1 design that mated an F27 Friendship with a derivative of the F-14 Tomcat's AWG-9 radar. Neither of the KingBirds ever flew.

Together with Lockheed (now Lockheed Martin) Ericsson has considered a C-130 AEW platform using the Erieye radar. This was one of several C-130 AEW concepts actively investigated by Lockheed during the 1980s and into the 1990s. Studies have been made using both the C-130E/H and the latest-generation C-130J. In January 1997, Ericsson Microwave Systems announced that it had signed an MoU with (what was then) Lockheed Martin Ocean, Radar and Sensor Systems for the joint marketing of advanced AEW&C solutions. Under this agreement an Erieye/C-130 combination is understood to have been proposed for both Australia's Air 5077 (Project Wedgetail) competition and the similar AEW requirement in South Korea (E-X). In the case of Australia an alternative US-supplied radar was selected for the bid, and all other C-130 plans have fallen in abeyance.

Of all the potential Erieye platforms that have been proposed over the years perhaps the strangest was another mid-1990s concept drawn up by a US firm, Special Mission Aircraft (SMA). This company proposed to modify Convair 580s (or ex-military C-131s) to serve as the EC-131K AEW platform. SMA also offered the option of modifying 'new-build' Kelowna Flightcraft Convair 5800s. Both the Erieye and the Lockheed Martin AN/APS-145 radars were offered as possible mission fits for the EC-131K, which would have had eight operator consoles in the main cabin. The EC-131K was to be further modified with air-to-air refuelling equipment, winglets to improve flight performance and four underwing hardpoints to carry weapons, if required. Press reports from 1996 cited potential customers including Canada, Thailand and the UAE, but the programme disappeared without a trace.

Two Erieyes that never were, but to which serious consideration was given, were the EMB-120EW SIVAM proposal (below left) and a C-130 concept which underwent several design iterations, including this artist's impression (below).

In case there was any doubt that the Swedish air force really does deploy its S 100Bs 'in the field', these aircraft are seen at the Färila war base, near Ljusdal in central Sweden, during the Flygvapenövning 03 exercise held in May 2003. The S 100Bs worked with Gripen fighters in the ongoing expansion of Sweden's net-centric warfare capabilities. The Argus flight crews maintain the same level of roadbase operational skills that all of Sweden's military pilots are required to have. One aircraft can be seen landing (below) in the transport configuration, without its Erieye radar fitted.

In the early days of SIVAM/Erieye planning the Embraer EMB-120 Brasilia had been earmarked as the radar platform. Significantly smaller than the Saab 340, the Brasilia was selected to support local industry. It was never a happy choice for the SIVAM Erieye mission and there were doubts about its performance. It was suggested that the aircraft might have to be fitted with a RATO (rocket-assisted take-off) system to operate from some of the smaller airfields in the Amazon region – something that did not appeal to the FAB. However, by the late 1980s Brazilian engineers were adding the final touches to the EMB-145 design. Some of those working on the SIVAM programme inside Embraer could see that the jet would be a much more appropriate platform for the Erieye, and for SIVAM. By early 1994 a preliminary deal to acquire the Erieye Brasilia had been agreed – but in October that same year Embraer began to assemble the EMB-145 prototype. A reassessment of SIVAM plans followed and a move to the larger, more capable jet became obvious.

The Brazilian authorities had already specified an onboard C^2 capability for their AEW platform. This was compromised by the small cabin of the EMB-120, but could grow to undreamt-of levels in the EMB-145. In March 1997 Ericsson was formally selected as the AEW radar supplier for SIVAM and was contracted to supply five Erieye systems to be integrated with an airborne C^2 system plus an IFF system. This was the first export order for the Erieye, and Ericsson's single largest defence electronics order to date. It also launched Embraer's EMB-145AEW&C programme which became the linchpin of all future Erieye sales.

In FAB service the EMB-145 carries the designation R 99. Brazil's AEW&C aircraft is the R 99A. The first was handed over in 2001 and deliveries are ongoing. Five R 99As are tasked to operate alongside three R 99B (EMB-145RS) remote sensing variants – the latter equipped with a range of multi-spectral sensors and synthetic aperture ground surveillance radars. The R 99A is based on the ERJ-145LR airframe, the Long Range variant introduced by Embraer in 1998. It is powered by a pair of uprated 7,430-lb (33.1-kN) Rolls-Royce AE3007A1 turbofans that deliver improved 'hot-and-high' and climb performance. The R 99As have also been fitted with a set of distinctive winglets, first seen on the ERJ-145XR (Extra Long Range) variant that was unveiled in 2000. The winglets were not part of the original configuration but were added in 2001 as a result of flight test experience during the development phase. Other aerodynamic modifications were made, such as the strakes on the tail fin. The R 99As cruise at a 6° nose-up attitude (instead of the 4° adopted by the Swedish S 100Bs).

Both the R 99A and the R 99B are based at Anápolis AB in the state of Goias, central Brazil. They are flown by the 2°/6° GAV (Grupo de Aviaçâo, aviation group). When in the air their call sign is 'Guardião' (guardian). The FAB appears to be very pleased with their performance so far. Towards the end of 2003 the Brazilian press reported that an R 99A had played a crucial role in the rescue of 70 Argentinean captives that were being held by Peruvian guerrillas. There are no official details of the mission but it is understood that R 99As were used to locate suspicious air traffic that pin-pointed the group's location.

The integration of the Erieye with Embraer's jet had a radical effect on the Erieye AEW&C programme. The Swedish radar has been mated with a larger, faster, higher flying platform – that is affordable to purchase and economical to operate. Ericsson claims that the Erieye system has the fastest reaction time of any AEW&C platform. The EMB-145AEW&C can be airborne and operational seven minutes after the order to scramble is received. It takes about two minutes for taxi and take-off. Five minutes

Technical Briefing

The first EMB-145AEW&C for the Greek air force was the first 'NATO-standard' aircraft to be produced by the Erieye programme, and heralded a coming of age for the Swedish and Brazilian team. Swedish industry takes its NATO-compatible credentials very seriously because of how keenly customers look for that stamp of approval.

later the aircraft has climbed to FL100 (10,000 ft) and is able to start radar operations. It takes another 10 minutes to climb from FL100 to FL200. According to the Ericsson model, three aircraft are needed to maintain a 24-hour AEW patrol station for 30 days. Using the sizeable body of dispatch reliability statistics built up from years of regional jet operations, the manufacturer says just two aircraft can support a patrol station 75 nm (138 km; 86 miles) from base 99 per cent of the time (over 30 days); 150 nm (277 km; 172.5 miles) from base 98 per cent of the time; and 300 nm (555 km; 345 miles) from base 97 per cent of the time. The transit time to reach a patrol station 75 nm distant is 17 minutes and the aircraft can spend seven hours on patrol there (followed by an 11-minute return leg). To reach an operational station at 150 nm from base takes 28 minutes of flight time, with six hours endurance on station. Four aircraft can maintain two patrol stations, with two in the air and one on ground alert at all times. Ericsson says that the Erieye costs between one eighth and one tenth of an E-3 to operate and has quoted a cost of $500 per flight hour for the Erieye, compared with $2,700 for an E-2C and $8,300 for the E-3.

Brazil's experience with the R 99A has served as a shop window for the world and led directly to the sales that followed. Selling the Erieye in its pure Swedish form is difficult. Few customers have the luxury of Sweden's integrated defence network within which to remotely operate an airborne radar 'node'. Some customers, particularly those that showed an early interest in the system, saw it as 'too defensive' and there were those at Ericsson who feared that the Swedish concept of operations would not win favour with those to whom an onboard C^2 capability was axiomatic to any AEW&C platform. This assessment has been born out by all subsequent Erieye sales. Ericsson would probably have gone on to develop a larger and more capable Erieye platform in any case. However, the arrival of the well-funded SIVAM programme brought a step-change in the entire Erieye concept.

Sales around the world

By the mid-1990s customer interest in the Erieye was growing and Ericsson was conducting sales campaigns in Europe, Asia and Latin America. One of the hottest prospects was Greece and, with this programme, came a new member to the team. For the first time in an AEW&C sale Ericsson was dealing with a NATO customer, and so it needed a recognised systems integrator that could reliably fit the NATO-specific systems that Greece required. In March 1998 an MoU was signed with Thomson-CSF Radars and Countermeasures (now Thales) to develop, manufacture and market the 'interoperable' elements of the system – such as secure communications and IFF. Thales is also involved in supplying system enhancements such as the ESM (electronic surveillance measures) fit. Interestingly, the MoU with Thales cited "future upgrades to the Erieye system plus the possible development of an entirely new generation of AEW&C systems." There was a very clear reason for selecting the French-led company – it was not American. One senior figure in the programme noted, "once you work with the US industry you become beholden to them." By teaming with Thales, the programme gained a successful partner with proven technical competence – but more importantly it maintained the strategic independence of the overall system.

In December 1998 Greece selected the Erieye to be its future AEW&C system against stiff competition from the Northrop Grumman Hawkeye 2000 and a C-130J-based proposal

The Erieye family today

S 100B Argus: AEW platform for Flygvapnet, based on modified Saab 340B (**Saab 340AEW&C**). Six aircraft acquired, two later leased to Greece to prepare for EMB-145AEW&C. No onboard operators, all radar data downlinked to the Swedish national command and control system. First aircraft flew in 1995 and delivered in 1997.

R 99A: Referred to by Embraer as the **EMB-145SA** (surveillance aircraft). For Brazil's national SIVAM programme to provide surveillance over the Amazon region. Five Erieye-equipped R 99As acquired in tandem with three R 99B multi-sensor aircraft. Original intention was to mount radar on EMB-120 Brasilia but EMB-145 airframe selected instead in 1997. Aircraft equipped with independent command and control capability, with three onboard operator stations. Datalink capability offers air-to-ground functionality plus air-to-air capability for fighter control. R 99A also integrates SSR (secondary surveillance radar) and ESM (electronic support measures). First aircraft flew in 1999.

Hellenic Erieye programme: Four **EMB-145AEW&C** aircraft ordered in 1998 (also identified as **EMB-145H**, Hellas), each equipped with five onboard operator stations – four radar operators, one ESM operator. Full NATO interoperability with Thales TSB 2500 Mk XII SSR/IFF (secondary surveillance radar/identification friend-or-foe, Thales DR 3000 ESM (electronic support measures) and secure NATO communications fit provided by Thales. 20-in colour displays. The MIDS-compatible datalink will interface with the NATO Air Defence Ground Environment (NADGE) system and Greece's own Hermes national communications network.

This is the updated multi-function mission console fit that the Swedish air force has developed for its S 100Bs. Although it is not standard practice for Sweden's aircraft to fly with operators onboard, this capability is central to every other Erieye export customer. Greece and Brazil are both fielding what is essentially a common mission systems fit, but with a different number of operators aboard. In another change to their operational approaches, the crew inside the Brazilian aircraft sit facing forward, but inside the Greek aircraft they face to the side.

Above: The EMB-145AEW&C (R 99A) partners the EMB-145RS (R 99B) in FAB service. Embraer is marketing both aircraft to a range of international customers.

from Lockheed Martin. The deal was valued at $575 million and reportedly included a 149.7 per cent offset agreement. Four Erieye aircraft were ordered, for delivery in 2001 – although the final contract included an option for a further two. The Greek order went through several phases of negotiation, being 'confirmed' again the following year before being finalised on 1 July 1999. At the end of these lengthy discussions Greece had arranged to lease two Swedish S 100B aircraft in 2001 (adjusted from 2000) to bridge the gap before its own EMB-145 platforms would become available in 2002. Since then the Greek timeline has slid further to the right. The first full-standard EMB-145AEW&C aircraft is not expected to arrive until 2004, to enter service fully perhaps in 2005. Some sources have referred to these jets as the 'EMB-145H' (H, Hellas = Greece) but Embraer says there is no official designation other than EMB-145AEW&C.

In March 2001 Embraer announced that Mexico had become an Erieye customer with an order for one AEW&C aircraft to be delivered in 2004. This would partner two EMB-145MP maritime patrol aircraft (a new variant for which Mexico was the launch customer).

Greece gets underway

The first of the Swedish S 100Bs (100004) on interim lease to the Greek air force arrived in-country in June 2001. It began familiarisation training straight away, in conjunction with the EGIS (Erieye Ground Interface Segment) base stations supplied by Ericsson. The second Swedish Argus (100003) arrived at Elefsis on 23 September, in time for the formal hand-over ceremony of both aircraft, the next day. This occasion marked the first time that a European-built AEW&C platform had ever entered NATO service. To mark the delivery the Swedish team undertook one of their typical tactical demos – launching an Argus on an 'operational' mission and down-linking real-time radar data to a large video screen in front of the assembled guests. Two Mirage 2000 interceptors were vectored on to a pair of 'intruding' F-16s out over the Aegean sea. With two Erieyes in service Greece now had a 24-hour airborne surveillance capability. The S 100Bs (described to in some places as 'S340Hs') were taken on charge by 380 Mira ASEPE (AEW&C squadron) 'Uranos' (meaning sky), a newly-formed unit within the 112th Combat Wing at Elefsis air base near Athens –

This Saab 2000 Erieye concept was quietly unveiled in July 2003, pointing to the obvious conclusion that some secretive plans are being made by Saab and Ericsson to revive this combination of radar and airframe.

this unit will also be the EMB-145AEW&C operator, once they enter service.

In September 2003 the first of the Greek air force's purpose-built EMB-145AEW&Cs was flown to Sweden for Erieye testing and performance checks – expected to last between nine and 10 months. The new aircraft was based at Halmstad, near Ericsson's main Mölndal plant, in Gothenburg. According to the official schedule, deliveries of the three remaining Greek air force aircraft are due in intervals from January 2004 onwards. Some Greek sources have reported that the second example will actually be handed over before the end of 2003 and that all four will be delivered by June 2004. The installation of the aircraft's NATO-secret components is to be undertaken in Greece and work should begin on the second delivered EMB-145AEW&C in early 2004. This means that aircraft No. 2 will be the first to arrive in Greece, as its sister-ship will still be undertaking radar trials in Sweden. It is hoped that both aircraft might be available for use by mid-2004. The Greek authorities are known to be keen to have a capability in place for the Athens Olympics. However, reports from Greece towards the end of 2003 suggested that some US permissions for the transfer of NATO cryptographic equipment had not yet been agreed, leading to more delays.

A radar looks to the future

Ericsson and Embraer maintain a policy of no comment on their current Erieye marketing campaigns. In the past Ericsson has acknowledged sales efforts in Chile, Argentina, Venezuela and the successful attempt in Mexico. There are also hopes of follow-on sales in Brazil. Other countries with ongoing applicable requirements include South Korea, Malaysia, Indonesia, Thailand, Taiwan, Singapore, Bahrain, Kuwait, Oman, the UAE, Italy and Spain. Ericsson has talked about the potential for 20 Erieye sales over a five-year period. At the 2003 Paris air show Saab exhibited a model of a Saab 2000 fitted with an Erieye radar. There was little further information on this surprising new development. It can be assumed that, with several ex-airline Saab 2000s now available at the end of their operating leases, the Saab 2000/Erieye combination is being seriously considered for at least one potential client.

The sale of Swedish military equipment has long been restricted by severe government export regulations – and public opinion. The Swedish regulatory framework has changed significantly in recent years, as has the mood of the nation. The country's defence exporters are no longer having to operate with their hands tied behind their backs. Also, because the Erieye is not a combat system it has more freedom of movement in the market than the Gripen, for example. At the same time the Erieye and the Gripen were born and bred together, and have been offered to several customers (such as Brazil) as complimentary systems that share the same high-tech pedigree. Both aircraft also offer a degree of independent sourcing and technology transfer that their rivals cannot match. While it is true that the Erieye is a niche product with none of the glamour of shiny fighters, any air force with an interest in 'pointy-nosed' jets needs a system like the Erieye if it hopes to fight and win the 'information war' of today, and tomorrow.

Robert Hewson

PHOTO FEATURE

When Finland retired its las[t]
Drakens in 2000, th[e]
Österreichische Luftstreitkräft[e]
(Austrian air force) became th[e]
final military user of this classi[c]
1960s fighter. Scheduled to retir[e]
by the end of 2005, the Drake[n]
remains in service with th[e]
Überwachungsgeschwade[r,]
Fliegerregiment 2

Last of
the Dragons

Photographed by Luigino Caliaro

Austria's Drakens

Austria evaluated the Saab J 35 Draken as early as 1967, but there was insufficient money available to buy any fighters at the time. Interest in a supersonic fighter resurfaced in the early 1980s, and discussions with Sweden began in July 1982. After nearly three years of public debate concerning the need for a fighter, and studies of other options (including the Lightning offered by British Aerospace) the Draken was selected in March 1985, the contract being inked on 21 May. The $550 million deal comprised a complete package including spares, simulator, ground equipment and training, to be conducted by the Swedish air force's F10 wing at Ängelholm. Austrian and Saab engineers selected 24 J 35Ds to be updated to J 35OE standard, the work being accomplished at Saab's main Linköping plant. Upgrades included structural work to ensure a 1,000-hour fatigue life, modifications to the radar and avionics, and the fitment of a J 35F-style clear-view canopy. Deliveries to Austria were made between 25 June 1987 and 18 May 1989. In the late 1990s a further five Drakens (J 35Fs) were acquired from Svenska Flygvapnet to act as spares sources.

The ÖLk allocated its Drakens to two squadrons of the Überwachungsgeschwader, headquartered at Zeltweg where 1. Staffel was located. 2. Staffel was formed at Graz-Thalerhof, also in southern Austria. At the time Austria was forbidden by treaty from owning guided missiles, and Draken weaponry was initially restricted to the 30-mm ADEN cannon. In 1991 Yugoslav aircraft made repeated incursions into Austrian airspace during the brief fighting over Slovenia: as a result the treaty was waived to allow the delivery of AIM-9P-3 Sidewinders in 1993 to provide the Drakens with a meaningful air defence capability. The Draken fleet was scheduled to be replaced in 1996, but the procurement process dragged on for years until the Typhoon was selected. In the meantime, the J 35OEs received RWR and chaff/flare dispensers, using equipment taken from retired Danish Drakens which was installed by Valmet of Finland. The planned out-of-service date is late 2005.

One of the Draken squadrons is due to disband in 2004, with the other following suit in 2005, barring any unforeseen problems with the aircraft. Of the 23 remaining Drakens, around 10 are undergoing maintenance at any one time, and flying hours are now very expensive due to the nearly 40 maintenance man-hours required per flying hour. The contract for support from Sweden's FMV expires in 2003.

Above: The AIM-9P-3 Sidewinder is the Draken's principal armament, represented here by inert training rounds. The two underfuselage 500-litre (110-Imp gal) tanks are near-permanent features, and a further pair can be carried on the wing pylons for ferry flights. Either side of the engine nozzle, scabbed on behind the afterburner intakes, are the chaff/flare dispensers installed by Valmet.

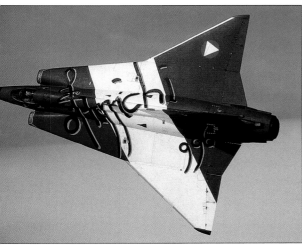

Above and below: Austria's J 35OEs have worn the same three-tone air defence grey scheme throughout their careers, although overwing Dayglo tactical codes were adopted at an early stage. In 2001 this aircraft (08) received a special scheme inspired by the national flag, which is still worn in 2003. Typhoons are due to be delivered in 2007, although Austria is looking for an earlier, interim delivery of around six to ten Tranche 1 aircraft.

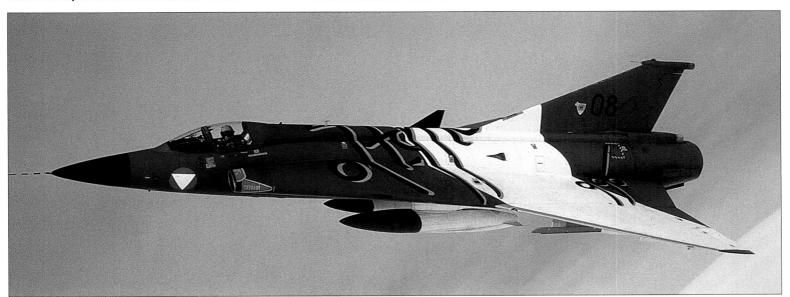

FOCUS AIRCRAFT

Super Hornet production Lot numbers follow on from those of the 'legacy' Hornet, but to all intents the F/A-18E/F is no more than a derivative design, sharing few common parts apart from the baseline avionics suite, which is in the process of being largely replaced by newer items. The new aircraft has gained the nickname of 'Rhino' on account of its larger size and grey colour. Given enough imagination it can also be suggested that the Super Hornet has a horn, in the shape of the AIFF fairing forward of the windscreen. VFA-14 'Tophatters' was the second squadron to get the single-seat F/A-18E, and one of its aircraft is seen here in the 'five-wet' tanker configuration. Note the spinning turbine which provides power for the A/A42R-1 refuelling pod on the centreline.

Boeing
F/A-18E/F
Super Hornet

Born out of the procurement mire of the late 1980s/early 1990s which saw the cancellation of several important naval aircraft programmes, the Super Hornet has not introduced any radical new technology, nor any quantum leap in performance. Instead, it has been developed on time and to a tight budget, and is a highly adaptable and versatile aircraft for the US Navy to operate in the post-Cold War combat environment. Furthermore, it represents a considerable improvement over its predecessor in terms of range/load characteristics, reliability and affordability. It is slated to receive the latest in sensor, weapons and communications technology, and has the room for growth which allows it to be at the cutting edge of the development of netcentric warfare, and a key platform in the race to cut 'sensor to shooter' times to a minimum.

A familiar shape, but somehow different – the Super Hornet retains many of the fine qualities of the original, but addresses the F/A-18C/D's shortfalls while providing greatly expanded combat capabilities. 'Home' to the Super Hornet is NAS Lemoore in California, where new squadrons and aircrew convert to the type. This pair is from VFA-122, the Super Hornet Fleet Readiness Squadron.

Providing an excellent comparison between the Super Hornet and its predecessor (now known as the 'baby' or 'legacy' Hornet), the prototype F/A-18E flies alongside an F/A-18C. Immediately apparent are the overall larger size of the Super, its leading-edge 'dogteeth', and the massively enlarged leading-edge root extensions. The tailplane exhibits a sharp-edged planform, one of the many treatments applied to the type which dramatically reduce the Super's RCS compared to that of the original. The larger size (and increased fuel capacity) allows the F/A-18E/F to fly considerably further than the C/D, redressing to some extent the principal drawback of the original design.

Now the centrepiece of naval aviation, the Boeing F/A-18E/F strike fighter is flying fighter, attack, reconnaissance, refuelling and suppression of enemy air defences (SEAD) missions for the US Navy's carrier fleet. Derived from an already versatile and flexible platform – the F/A-18 Hornet – the Super Hornet is the culmination of years of work to perfect the Hornet design. Indeed, the Super Hornet in many respects represents what the original Hornet should have been, as it specifically addresses and resolves deficiencies in range, payload and bring-back capability that plagued the earlier machine, while at the same time providing the Navy with a renewed growth clock to incorporate new technologies.

The Super Hornet is now operational with several fleet squadrons and has already seen combat in operations over Iraq and flown patrol missions over Afghanistan. Although still not representative of what the Super Hornet will ulti-

mately bring to the table, these aircraft performed outstandingly and confirmed the predictions of their proponents. According to Captain B.D. Gaddis, NAVAIR F/A-18 Program Manager, "All of our debriefs from the commanders and CAGs are saying that the Super Hornet is exceeding all expectations in performance and capability." Captain Gaddis pointed out, "These reports are even more significant in the context of the Super Hornet because they are coming not only off the aircraft's maiden deployment, but also from its first combat deployment, and an extended one at that." As planned systems such as the Advanced Tactical Forward-Looking Infra-red (ATFLIR) pod, the Multi-functional Information Distribution System (MIDS), the Joint Helmet-Mounted Cueing System (JHMCS), and SHAred Reconnaissance Pod (SHARP) enter service over the next five years, the Super Hornet will prove even more lethal, making it the definitive strike fighter – one that no opponent will want to meet in combat.

The Super Hornet will also serve as the Navy's new electronic warfare/electronic attack platform, offering a combination of speed and functionality far surpassing that of the EA-6B. Scheduled to enter service in 2009, the EA-18G will fly the same mission profiles as its -E/F stablemates and also carry a robust self-defence capability. Moreover, the incorporation of new technology allows the two-person EA-18G crew to perform the mission currently handled by four EA-6B crew members.

The advent of the Super Hornet brings with it a change of face for naval aviation. The aircraft-carriers of 2015 and beyond will be stocked with at least three strike fighter squadrons of F/A-18E/Fs flying a range of missions including strike, fighter, SEAD, reconnaissance and sea control, as well as providing organic tanking services to carrier aircraft. A squadron of five or six EA-18G aircraft will provide jamming and offensive electronic warfare support, while a squadron of the new F-35 Joint Strike Fighter (JSF) will give the carrier stealthy, deep-strike capabilities. Long-range airborne early warning duties will be flown by the latest version of the E-2C Hawkeye, and anti-submarine duties will be filled by the MH-60R Seahawk.

F/A-18E/F Super Hornet

Captain Gaddis told *International Air Power Review* that he considers two features to be key to the Super Hornet design. "The Advanced Mission Computer and Displays are the heart of the aircraft that allow it to connect to and fuse the data. The ability to move data from sensor to display and process that data into a usable format has vastly increased. The second aspect is the APG-79 AESA radar. In terms of raw capabilities, the radar is phenomenal. But the real capability comes when you mate the radar with the various sensors – ATFLIR, SHARP, the EW suite – and from those build a high-resolution map and get that map to the commanders on the ground. These systems literally revolutionise how we fight today." Captain Gaddis added, "[the] advanced sensors are critical to developing the powerful network that the Navy is trying to achieve."

The Navy is currently in the fourth year of its first multi-year purchase contract, and approximately 87 of the 222 ordered Super Hornets have been delivered. A second multi-year contract is approaching, which will result in another 210, 90 of which will be built as EA-18Gs.

Birth of the Super Hornet

The late 1980s and early 1990s were very turbulent times for military aircraft procurement, characterised by ever-growing budgetary constraints and a major refocusing of US war-fighting strategy and force structure. These forces worked together to produce the programme that today is the F/A-18E/F Super Hornet. As the 1980s began, the Navy found itself with several ageing airframes in need of replacement. Many aircraft, such as the F-4, RF-8 Crusader and A-7 Corsair II, had been flying since the late 1960s and were approaching, if not surpassing, the end of their useful lives. Other airframes, such as the A-6E – although performing well at the time – were not technologically advanced enough to survive in the projected air defence climate of the 21st century. Moreover, their airframes were limited in growth capacity. What was needed were follow-on aircraft for all three major missions: attack, fighter and strike fighter.

Cold War plans called for high-end follow-ons to both the F-14 Tomcat fleet defence fighter and the A-6E Intruder attack aircraft that would incorporate high levels of stealthiness and ensure continued US naval air dominance well into the 21st century. A navalised F-22, called the Naval Advanced Tactical Fighter (NATF), was slated to succeed the F-14, and the flying-wing A-12 Avenger II was to replace the venerable Intruder. Yet, both programmes were costly and technologically a long way from fruition.

In July 1987, the Department of Defense ordered the Navy and the Air Force (which at the time was pursuing the proposed F-22 ATF) to investigate possible derivatives of the F/A-18 and F-16 as a stopgap measure until the F-22 and A-12s began to enter service in the early 2000s. McDonnell Douglas responded with the Hornet 2000 programme in 1987, which was essentially a further evolution of the original F/A-18, and which further corrected some of the deficiencies of the baseline F/A-18A such as its poor range and payload. The Hornet 2000 featured a larger wing and stabilators, additional internal fuel and more powerful engines. The programme was actually marketed abroad, with little interest, and was temporarily shelved. At about the same time, and in anticipation of the NATF and A-12, the Navy scrapped plans to introduce the improved A-6F Intruder II and limited production of the advanced F-14D to just 54 aircraft.

Shortly after McDonnell Douglas began its Hornet programme, two related events occurred that perhaps are the most significant factors in the ultimate decision to pursue the F/A-18E/F – the collapse of the Soviet Union and the end of the Cold War. With these events came both a major restructuring of the US military and massive budget cuts that ultimately spelled the end for both the NATF and the A-12. Of course, the latter programme had significant cost issues of its own that contributed significantly to its demise. With America's chief enemy now gone, strategy shifted from fighting open seas, 'blue-water' engagements with the Soviet fleet to operations in 'white water', or the littoral regions – targets within a few hundred miles of the coastal areas. This, in turn, negated the need for a naval long-range strike aircraft. Likewise, with no threat posed by vast fleets of missile-carrying Soviet bombers or submarines, the need diminished for a long-range high-end interceptor. Moreover, in addition to strategic considera-

A bigger, smarter, stealthier Hornet – this view of the prototype F/A-18E highlights the enlarged wing of the Super, and the trapezoidal intakes which help reduce the type's RCS.

With two extra pylons, enhanced weapons capability and the routine use of asymmetric loads, the number of weapon loadouts for the Super Hornet is vast. Two EMD aircraft were assigned to initial weapons separation trials to investigate some of the Super Hornet's huge repertoire. E5 and F2 were sent to the Navy's main weapon test base at China Lake for this work, the pair being the first Super Hornets with a working avionics suite that included APG-73 radar. Both were liberally covered with photo-calibration marks, and here E5 carries inward-facing cameras on wingtip-mounted test rigs, and an aft-facing camera under the forward fuselage.

Focus Aircraft

EMD aircraft

E1 – BuNo. 165164
F/f – 29 November 1995

Charged with investigating flying qualities and flight envelope expansion. Tests on E1 confirmed the aircraft's compliance with flutter safety margins. E1 concluded its test flights on 23 October 1998, and tallied 408 sorties and 757.9 flight hours during the EMD.

E2 – BuNo. 165165
F/f – 26 December 1995

E2's EMD tasks included engine and performance testing; the aircraft flew 486 sorties and 847.9 flight hours in the period from 19 February 1996 through 30 April 1999. Here the aircraft is seen during trials at Patuxent River.

E3 – BuNo. 165167
F/f – 2 January 1997

Last of the EMD aircraft to take to the air, E3 handled the load testing portion of the EMD, during which it flew 354 sorties and 552.5 flight hours. Like E1 and E2, the aircraft did not have radar installed and was fitted with an air data instrumentation boom in the nose.

E4 – BuNo. 165168
F/f – 2 July 1996

E4 was assigned to high angle-of-attack trials. During EMD it amassed 402 sorties and 570.4 flight hours. The aircraft received a red and white scheme for maximum conspicuity during its aerodynamic trials, and for most sorties carried a spin recovery chute mounted on a gantry above the jetpipes.

E5 – BuNo. 165170
F/f – 27 August 1996

E5 was the first Super Hornet to possess mission capability, and was fitted with APG-73 radar. The aircraft made numerous 'firsts' during weapons evaluations, and ended the EMD phase with 480 flights and 594.1 flight hours. It initiated the stores separation trials on 19 February 1997 when a fuel tank was jettisoned.

F1 – BuNo. 165166
F/f – 1 April 1996

The first two-seat aircraft was the third EMD aircraft to fly. As well as aerodynamic evaluation of the two-seat aircraft, it was assigned to carrier suitability and weapons testing. F1 tallied 508 sorties and 601.8 flight hours, the most flight time of all EMD aircraft.

F2 – BuNo. 165169
F/f – 11 October 1996

F2 was also assigned to some carrier qualification duties, joining F1 in an intensive trials campaign on *Truman* in March 1999. As it was the second Super Hornet with a full avionics suite, F2 spent much of its time flying alongside E5 performing weapons testing at China Lake, including launches of AIM-9 and AIM-120 AAMs.

tions, the defence budget was curtailed significantly as the US began a force draw-down, resulting in less money.

Thus, as the 1990s began, the Navy found itself with a tough choice. Although it needed to replace both its fighter and attack aircraft, it simply did not have the funds to pursue a new aircraft for each mission. The Navy therefore decided to pursue one major system upgrade and one new aircraft. Navy strategy at that time still focused on power projection as outlined in the paper ... *From the Sea*, so the Navy sought a new aircraft to replace the Intruder; this came through a new programme, dubbed the A-X (later redesignated A/F-X). In turn, the Navy undertook a major upgrade to replace the fighters, and McDonnell Douglas's Hornet 2000 provided the catalyst for the aircraft that became the F/A-18E/F.

As the Navy moved from power projection to littoral missions, the NATF and the A-12 were no longer deemed viable equipment. Moreover, the Bottom-Up Review of 1993 concluded that the US simply could not afford all of the high-end programmes that had emerged from the Cold War days of the 1980s. This meant that costly and inefficient programmes had to be trimmed or cancelled altogether. The Navy's A/F-X (a planned replacement for the A-6E and F-14) and the Air Force's Multi-Role Fighter (its planned replacement for the A-10 and F-16) programmes met with a similar fate. Navy officials, as well as their counterparts in the Air Force, then had to select which aircraft system they would pursue. In the end, the Navy chose the F/A-18E/F upgrade based on the original Hornet 2000, to be complemented by the Joint Service Fighter (later known as the Joint Strike Fighter and now the F-35). The Air Force decided to support the F-22, with a complementary JSF force.

Super Hornet comes to fruition

Navy support for the Super Hornet came in late 1991, followed by a formal declaration of an 'intent to procure' on 12 May 1992. Total cost of the programme was estimated to be $US63,090 million (in FY96 dollars), with $US5,783 million slated for development costs. Initial Operational Capability (IOC) was scheduled for 2000, with the first carrier-based squadron to deploy in 2003. Congressional approval was received in June 1992, culminating in a $US4.88 billion (in FY92 dollars) contract for engineering and development (EMD). Of this contract, $US3.7 billion went to McDonnell Douglas; the remaining amount went to General Electric for development of the F414 engine. Congress placed a cap on the overall Super Hornet cost of 125 per cent of an F/A-18C. The Navy signed the final F/A-18E/F contract on 7 December 1992, authorising McDonnell Douglas and Northrop Grumman to produce three ground test airframes (ST-50 for static tests; DT-50 for drop tests; and FT-50 for fatigue tests), five single-place F/A-18Es, and two dual-place F/A-18Fs.

The E/F programme underwent a successful Preliminary Design Review from 28 June through 2 July 1993, resulting in only 53 relatively minor action items, all of which were subsequently resolved. A more thorough Critical Design Review was held from 13 to 17 June 1994, with the Super Hornet passing all of the Navy's schedule, cost, technical, reliability and maintainability requirements. Northrop Grumman commenced production of the centre/aft fuselage for E1 (BuNo. 165164) on 24 May 1994 at its Hawthorne, California, facility and McDonnell Douglas began work on the forward fuselage in St Louis, Missouri, on 23 September that same year. The first Super Hornet, E1, debuted in a highly-touted roll-out ceremony at McDonnell Douglas's St Louis facility on 18 September 1995, and made its first flight on 29 November. Speaking at the ceremony, then-Secretary of the Navy John H. Dalton said that the aircraft represented a "remarkable achievement" that had come in "on schedule, on budget and under weight". Company test pilot Fred Madenwald piloted the maiden voyage. E1 left for Patuxent River, Maryland, in February 1996 to begin its flight tests.

VX-9 'Vampires'

Air Test and Evaluation Squadron (AIRTEVRON) Nine was formed at China Lake on 29 April 1994 by combining the operations of VX-4 (established 15 September 1952) and VX-5 (established 18 June 1957) – hence the new unit number. VX-4 was previously at Point Mugu, where it undertook air-to-air evaluation work, while VX-5 at China Lake was the Navy's main air-to-ground and tactical evaluation unit. Although the China Lake squadron now oversees both elements, it maintains a detachment at Point Mugu. VX-9 was the first Navy squadron to have Super Hornets assigned, and was responsible for conducting the OPEVAL. Markings consist of the bat and lightning bolts from VX-5's old badge, with four stars from VX-4.

Above: BuNo. 165533 was the first Super Hornet from Low-Rate Initial Production. It first flew in unpainted state (with 'F/A-18E6' on the tail) on 6 November 1998, six weeks ahead of schedule. It was subsequently painted in VX-9's markings, and was handed over to the US Navy on 18 December.

A procurement contract for low-rate initial production (LRIP) was signed in March 1997. This agreement called for a total of 62 aircraft to be delivered in three lots (12 aircraft in LRIP 1 during FY97; 20 in LRIP 2 during FY98; and 30 in LRIP 3 during FY99). The last of the LRIP-1 aircraft (eight Es and four Fs) was delivered on 9 November 1999 – ahead of schedule. LRIP-2 aircraft (8 Es and 12 Fs) were delivered between January and October 2000, and LRIP-3 deliveries (14 Es and 16 Fs) ended in July 2001. These aircraft were flown during the initial Operational Evaluation (OPEVAL) and were also used to form the new fleet readiness squadron (FRS), VFA-122, and the first operational Super Hornet squadron, VFA-115. Full-rate production began in September 2000, with 36 Super Hornets (15 Es and 21 Fs), followed by 39 aircraft in 2001. Of this latter production, 14 aircraft were Es and 25 were F models. The five-year contract signed on 15 June 2000 authorised an additional 222 aircraft through FY04.

Although a total of 548 Super Hornets was planned, this procurement was trimmed in 2003, with production scheduled to drop from 48 aircraft per year in 2002 to 45 in 2003 and 42 in 2004. In part, the uncertainty arises from the Marine Corps' initial decision not to adopt the Super Hornet as a replacement for its F/A-18C/Ds. It further stems from the decision to build the EA-18G as a replacement for the EA-6B electronic combat aircraft. As discussed later in this article, the Navy plans to purchase as many as 90 reconfigured two-place Super Hornets to perform fleet jamming and electronic attack missions. The Marines may follow suit as well, although they are also evaluating a proposed EW variant of the F-35.

Compromises on mission and stealth

Before discussing the aircraft itself, a few points must be made concerning the overall F/A-18E/F programme. During the late 1980s, a major re-thinking occurred as to how to approach aircraft design. Budget constraints were making it very difficult to develop a high-end, all-around aircraft that maximised all available technologies, such as stealth. Cost had to be a driving factor. In an interview in late 2000, Vice Admiral Joe Dyer, then-Commander Naval Air Systems Command and former F/A-18 Program Manager from 1994 through 1997, explained, "[there] was a recognition in the late 1980s that we simply could not do things the same way in procurement." Given this realisation, Dyer maintained, the Navy needed to replace a

Although assigned to VX-9 at China Lake, the first production aircraft was initially dispatched to NAWC-AD at Patuxent River (in the background) for flight trials. After a few months at the Maryland base it was transferred to China Lake to begin VX-9's OPEVAL work.

In its two-seat F/A-18F form, the Super Hornet is replacing the F-14 in the fleet. Providing a comparison between the two is an F/A-18F from VX-9 and an F-14D from the squadron's Point Mugu detachment. The Tomcat is the latest in a line of VX-4/VX-9 aircraft to be painted gloss black, which operate with the callsign VANDY 1.

Focus Aircraft

VX-9 currently flies a number of Super Hornets at China Lake on weapons and tactics evaluation duties.

Three VX-9 aircraft fly with the CAG and CO jets from VFA-122. The Super Hornet training squadron worked very closely with VX-9 during its early months to establish a training syllabus for the new type, and to train up a cadre of pilot and WSO instructors. VFA-122 was in existence for nearly a year before it received its own aircraft.

certain number of aircraft within a certain price. "They looked at how much money was available at the time and at how many aircraft were needed to do the job the Navy needed done. This allowed us to fix a number, then work to develop a design that fell within it."

Coupled with these considerations, Dyer continued, there had been an overall move away from single-mission aircraft to more of a force composition structure wherein aircraft could rely on other assets to accomplish the overall mission. This was demonstrated during Operation Allied Force over Kosovo in 1999, and in later operations over the Persian Gulf region in Operation Iraqi Freedom. Today, strike fighters co-ordinate missions with F-15Es, EA-6Bs, tankers, and reconnaissance aircraft, and even stealthy F-117s and B-2s. Vice Admiral Dyer stressed that "by focusing on a design in isolation, without reference to the systems at large, we took affordability out of the picture."

Vice Admiral Dyer added that the acceptance of these concepts led to a more open attitude towards trade-offs, and said, "We applied this to the question of stealth and survivability. We saw the crucial question as 'how much stealth is enough' for this design? Our answer was simple – enough to deliver stand-off weapons and survive to fight another day." Certainly, the admiral's comments are well taken. Stealth can be achieved in a number of ways, at varying costs, and this is exemplified by the different approaches taken with low-observability by the so-called 'stealth designs', the F-117, F/A-22 and B-2. The F/A-18E/F achieved its stealthiness through a combination of radar-absorbing materials, electronics and innovative use of stand-off weapons. Admiral Dyer concluded, "[the] key is finding the right balance, within the confines of affordability."

These comments dove-tailed with those made by Rear Admiral Dennis V. McGinn, Director of Air Warfare in the

Office of the Chief of Naval Operations, who commented in 1998 concerning the 'perishability' of stealth. "If you try to place your mission effectiveness in one specific area, you can get yourself in a situation ... if the enemy comes up with a countermeasure..." McGinn noted that the Super Hornet's design offers "the flexibility ... to shift your strategy ... more toward electronics, or more toward missile performance, or more toward onboard or offboard sensor components. That's the key to staying two steps ahead of potential adversaries."

This compromise approach also surfaced in the design of the overall Super Hornet mission. Conceived as a strike fighter, the proposed design had to accommodate both roles, much as the venerable F-4 did during the 1960s in the skies over Vietnam. Yet, as a strike fighter, it is obvious that no design can perform all missions across the operational spectrum with equal proficiency. Many of the attributes that make for a quality high-end interceptor make for a poor attack aircraft, and vice versa. "What the Super Hornet does," Dyer stated, "is to stake out the middle of the spectrum, expanding out as far towards the ends as possible." This creates the flexibility that air commanders want to prosecute their air campaigns.

F/A-18E/F versus the F/A-18C/D

The Super Hornet represents a design compromise of five variables, each intended to cure perceived shortcomings of the original Hornet – range, payload, bring-back, survivability and growth room. Of these five, perhaps the two most important variables were range and bring-back. Hornets had long been criticised for having 'short legs'. The Navy Operational Requirements (OR) called for the original F/A-18A to have an unrefuelled 400-nm (741-km) range for fighter and a 450-nm (833-km) range for attack missions. The -A/B performance, however, came in at 366 and 415 nm (678 and 769 km), respectively, figures that only got worse over time as new components added more and more weight to the airframe. In all, the Super Hornet promised an increase of approximately 40 per cent in range over the current F/A-18C/D. The Navy's 1992 OR called for ranges of 410 nm (759 km) in the fighter role and 430 nm (796 km)

in the attack role, figures which the -E/F has beaten at 420 and 490 nm (778 and 907 km).

To create room for the extra fuel, the -E/F fuselage was stretched by 2 ft 10 in (0.86 m) and its wings enlarged by 25 per cent. This represented an increase of 4 ft 3.5 in (1.31 m) and 100 sq ft (9.29 m^2) over the C/D, and provided an additional 3,000 lb (1361 kg) of fuel (approximately 33 per cent more than the F/A-18C). According to projected figures (and subsequently confirmed), in an air defence role the -E/F has 40 per cent greater combat radius than the -C/D (1.8 hours versus 1.0 hours) while operating 200 nm (370 km) from the carrier. Similarly, the -E/F in an escort role posts an increase in range of nearly 38 per cent. Range for self-escorted strike missions improved from 277 to 475 nm (513 to 880 km), an increase of over 70 per cent. These were confirmed through subsequent test flights.

The larger wings brought additional benefits to the aircraft, including two new weapons stations and an overall payload rating of 17,000 lb (7711 kg) (an increase of 20 per cent). Rated at 1,250 lb (567 kg) each, the additional hardpoints (Nos 2 and 10) permit more flexible mission loads, which translates into fewer missions, fewer aircraft per mission and more target options. All of these factors contribute to a more survivable and more flexible platform.

Top: VFA-122's first aircraft were from the LRIP batches. The squadron spent the first 18 months of its existence hammering out the syllabus and training instructors so that it could open for business in June 2000.

Above: As well as providing a shore base for fleet squadrons, NAS Lemoore is the centre of the Super Hornet training effort. The airfield has been a major training location for decades.

Located in the Californian Central Valley, Lemoore is surrounded by open, flat land – ideal for initial conversion training. For more challenging flying the mountains and desert are not far away.

VFA-122 'Flying Eagles'

Attack Squadron 122 was created by the renumbering of VA(AW)-35 on 29 June 1959. The unit was first assigned to its present base at NAS Lemoore in January 1963, when it was acting as the West Coast FRS for the A-1 Skyraider. In 1966 it began training A-7 Corsair II pilots, a role which ended with the disestablishment of the squadron on 31 May 1991. The nickname 'Flying Eagles' was in use from 1971. The squadron was reactivated as VFA-122 on 1 October 1998 to act as the Super Hornet FRS, and was ceremonially stood up on 15 January 1999. It did not receive its first aircraft until November, allowing the task of training sufficient instructors to staff the squadron to begin. VFA-122's motto is 'We train the experts', and it uses the EXPERT callsign. The badge consists of a bald eagle clutching three arrows, and the tail marking consists of the bird's head. The 'NJ' tailcode is that of the Pacific Fleet training wing.

VFA-122 has 37 Super Hornets assigned, of which 11 are single-seaters. Of the F/A-18Fs, eight are configured as twin-stickers to provide initial type conversion training before students progress to mission-based training on the E or single-stick F. The squadron has two 'boss-birds' (Modexes 101 and 201) which wear high-visibility markings. Due to the Super Hornet's improved aerodynamics and more forgiving handling, basic conversion from older Hornets or Tomcats is considered straight-forward. However, learning to use the more advanced weapons system takes some time due to the capabilities of the avionics.

Opposite page, bottom: This VFA-122 F/A-18F carries a test round of the AGM-154 JSOW. Guided by GPS, the AGM-154A version provides the Super Hornet with a stand-off attack capability against area targets. The A-model JSOW dispenses 145 BLU-97 sub-munitions as it glides over the target area. JSOW has been widely used against Iraqi air defence sites.

A Super Hornet can now fly with a load of two AIM-9s, an AMRAAM, a FLIR pod, two high-speed anti-radiation missiles (HARMs), two precision-guided munitions (PGMs), and two 480-US gal (1817-litre) fuel tanks. F/A-18Cs would be limited to a single PGM and HARM, or two of either, and two 330-US gal (1249-litre) tanks, and have reduced overall range.

The strengthened wings also gave the Super Hornet added lift, thereby allowing a slower approach speed. At 128 kt (237 km/h), Super Hornets handle better behind the carrier. The larger wings further permit use of the larger 480-US gal (1817-litre) external tanks, which current models could only carry in a land-based role. The larger wings and strengthened fuselage also helped alleviate another significant criticism of the Hornet: bring-back capability. As new systems were added to the Hornet, its heavier weight decreased operational warloads and, more importantly, decreased the amount of unexpended ordnance that the Hornet could safely bring back to the carrier. One company official estimated as much as 0.33 lb (150 g) per day was added to the Hornet over its life. Although Hornets could overcome this problem on take-off by launching with less fuel, then immediately tanking, there were few options when aircraft returned from missions with full loads and had to jettison unexpended ordnance into the ocean. Such practice was common during operations over Bosnia, Kosovo and the No-Fly Zones in Iraq. In fact, during Allied Force operations over Kosovo, an F/A-18C launching with four AGM-154 Joint Stand-Off Weapons could bring only two back onboard. This wasted significant amounts of money. An F/A-18C has a bring-back capability of approximately 5,500 lb (2495 kg); the E can bring back 9,900 lb (4490 kg) and the F about 9,000 lb (4082 kg). This increased ability factored large in recent operations over Afghanistan and Iraq.

Enhanced survivability

Increasing overall survivability was another chief concern of the Super Hornet's design team, and it has been placed by some analysts at more than five times that of the original -C/D. Interestingly, the increase was obtained not by stealth, but by a combination of factors that includes incorporation of radar cross-section reduction techniques, improved electronic countermeasures, reduced system vulnerability and improved stand-off weapons delivery tactics. While not depending on F-117 stealth design techniques, a significant radar cross-section reduction was obtained by a combination of radar-absorbing materials (RAM) and the redesign of panels and engine inlets. Some 154 lb (70 kg) of RAM is used throughout the aircraft, most notably on the leading-edge surfaces, the pivot point of the tail-hook and the aileron actuator fairings and hinges.

Access panels and landing gear doors now feature jagged or saw-toothed edges to redeflect radar waves. The formerly D-shaped engine outlets of the -C/D were reconfigured to an angled box (similar to those of the F/A-22). These inlets not only help deflect radar waves, they permit greater airflow to the engines. Other measures included tightening tolerances and better aligning planforms. Boeing also looked at ways of hiding the tailhook, but ultimately declined to vary from proven prior practice. Although figures are classified, radar signature has been reduced by an order of magnitude.

Improved countermeasures systems were also added to increase the -E/F's effectiveness against surface-to-air and air-to-air missiles. Centred around the Integrated DEfence CounterMeasures (IDECM) system designated ALQ-214, this suite comprises the enhanced ALR-67(V)3 radar warning receiver, the ALQ-214 countermeasures system and the fibre-optic towed ALE-55 deceptive jammer. Super Hornets have initially deployed with Raytheon's ALE-50 towed decoy pending the final development of the ALE-55. The number of flare/chaff dispensers was doubled from 60 to 120 units, using the BAE Systems Integrated Defence Solutions (formerly Tracor) ALE-47, which is also capable of dispensing POET and GEN-X active expendable decoys.

Reduced vulnerability

Loss figures will be lower because vulnerable areas are decreased by more than 14 per cent compared to the -C/D. Dry-bay fire suppression gear was added to reduce the incidence of fire and explosion. Consisting of 14 optical sensors and seven extinguishing heads that automatically sense heat then release inert gas to counter fires, the dry-bay system is located in the bottom of the fuselage near the hydraulic lines and flight control circuits. Boeing designers consider this the single greatest factor in lowering overall -E/F vulnerability. "It makes the -E/F harder to bring down when it is hit," one spokesperson said. A quadruply-redundant fly-by-wire flight control system was also included to ensure the retention of flight control in the event of a missile strike. These improvements are maximised when coupled with new stand-off weapons such as the JSOW (Joint Stand-Off Weapon), JDAM (Joint Direct Attack Munition) and JASSM (Joint Air to Surface Stand-off Missile), which allow Super Hornet crews to stay far outside the effective range of various enemy air defences. Also, as mentioned earlier, the two additional hardpoints mean that Super Hornets can carry more ordnance and so, perhaps, fly fewer missions.

Overall, improved survivability systems reduced -E/F combat losses by 87 per cent in a simulation run by Boeing, compared against F/A-18Cs. For that simulation, 10 -E/Fs flew anti-air defence system strike missions. In an interview reported in *Aviation Week & Space Technology*, Boeing officials noted, "[aircraft] participating in the strike simulation had 45 per cent fewer encounters with hostile aircraft during missions due to signature reductions and improved jamming, and improved countermeasures reduced the lethality from ground-launched missiles by 80 per cent."

Finally, one of the most desired features of the Super Hornet is its expanded room for growth. Provisions for upgrades have been designed into the Super Hornet such that it will enter service with about 40 per cent growth capacity in electrical power, cooling and equipment volume. The -C/D has less than 0.2 cu ft (0.0056 m³) left for systems growth; the -E/F entered service with 17 cu ft (0.4813 m³) of growth space. This new lease on life essentially restarts the growth clock that had all but stopped in the original model. The lack of growth space in the -C/D was one of the chief reasons for looking at a new variant.

Several other areas were modified as a result of these primary changes. To accommodate the heavier aircraft, the more powerful General Electric F414-GE-400 engines were installed, providing 35 per cent more thrust than the F404.

The F/A-18E/F has full night capability and the training syllabus includes a fair proportion of night work. Both cockpits of the F have a large multi-purpose display (above, showing a moving map) and two flanking displays. The small screen at lower left shows fuel and engine information. Low-voltage 'slime' lights and the use of night vision goggles allow pilots to fly in close formation (top) even in darkness.

Focus Aircraft

A key part of the VFA-122 syllabus is carrier qualifications. From its Lemoore base the squadron regularly sends small training detachments of aircraft and pilots to carriers operating off the southern California coast. Carriers usually sail specifically for these training periods, which also involve other aircraft types from the other FRSs on the West Coast. These F/A-18Fs were conducting CarQuals in October 2000 aboard USS Carl Vinson (CVN 70). At the time VFA-122 was training its first students – an advanced guard of pilots fresh from flight school (CAT 1) and destined for service with VFA-115.

Each capable of delivering up to 21,890 lb (97.36 kN) of thrust, the F414s were derived from the F404 used in the original Hornet and modified with technologies from the F412 engine developed for the cancelled A-12. The new engines have a 16 per cent higher airflow than the F404, a 30:1 pressure ratio and a 9:1 thrust ratio, all within the same length and aft diameter as the F404. The increased performance is due largely to use of an integrally-bladed disc in the compressor, called a blisk. The -E/F is the first system to use dual-channel full authority digital engine control.

Additional structural changes included enlarging the rudders by 54 per cent, increasing the horizontal stabilators by 36 per cent, enlarging the vertical tails by 15 per cent and expanding the LEX by 34 per cent. The enlarged wings, discussed above, were expanded to 500 sq ft (46.45 m²). Internally, commonality with the -C/D is just 10 per cent. Most changes concerned redesigning features in order to save weight and cost. Carbon-fibre composite use has doubled compared to the -C/D, to 22 per cent of the structural weight, most of which is applied to the centre and aft fuselage, wings, and leading- and trailing-edge flaps. IM7, an improved stiffness/strength carbon-fibre, was used extensively in the wings and tail skins. Use of aluminium alloys decreased from 50 per cent in the -C/D to 29 per cent in the -E/F. The landing gear was strengthened by using Aermet 100 steel, which has a higher damage tolerance.

Overall, the -E/F has 8,100 structural parts compared to the -C/D's 14,100. In fact, easier assembly and parts reduction was a guiding design principle of the product definition teams. One example can be found in the nose barrel bulkhead, where a formerly 90-piece assembly in the -C/D was reduced to a one-piece machined part in the Super Hornet by using high-speed machining tools. Fuselage splicing was also modified in order to reduce tolerance problems; assembly is now accomplished using a new Boeing-designed splice tool, which uses a laser tracker to guide the two sections together for a nominal fit.

Cockpit improvements

The cockpit gained new features as well. Cockpit displays were modified by substituting a new Kaiser 8 x 8-in (20.3 x 20.3-cm) flat-panel active matrix LCD display for the C/D's 5 x 5-in (12.7 x 12.7-cm) central display and by replacing the C/D's two 5 x 5-in (12.7 x 12.7-cm) MRDs with two multipurpose CRT touch-screens. The up-front control panel display was also replaced with a monochrome touch-sensitive screen. A new feature included an engine/fuel display featuring a programmable monochrome active-matrix LCD that graphically displays nozzle positions and fuel tank capacity, and fuel tank 'bingo' in pounds. At the time the first Hornet was delivered, the F/A-18C Lot 19 served as the avionics baseline for the E/F. Thus, avionics retained about 90 per cent commonality with the C/D models. Controls remained at 85 per cent commonality and the flight control systems at 67 per cent. As for avionics, the decision to begin with commonality represented Boeing's effort to contain costs and reduce programme risks.

As production of the seven contracted EMD Super Hornets went on, programme focus shifted to demonstrating that the aircraft could perform as advertised. The primary objectives of the engineering, manufacturing and development phase were to "translate the most promising design into a stable, producible, cost-effective design; validate the manufacturing processes; and demonstrate system capabilities through testing". Essentially, the goal was to "detect what was not predicted and to fix what goes wrong". After initial test flights in St Louis, the Super Hornet headed to the Naval Air Warfare Center at NAS Patuxent River, Maryland, for the development flight test programme. Seven Hornets – five single-seat E models and two dual-place F models – were used for these tests, as were three ground test articles. An F/A-18D was fitted with new avionics and served as an avionics testbed. Clearly indicative of the programme's attention to detail, the Super Hornet entered EMD flight testing 1,000 lb (454 kg) under projected weight.

Pax River adventures

The development flight test phase of the EMD began on 4 March 1996 under the guidance of the Navy/Industry Integrated Test Team (ITT). The -E/F's EMD phase differed from prior test programmes in that it combined Navy and contractor flight testing under one programme, rather than having two independent tests proceed consecutively. The Navy's Operational Test and Evaluation Squadron, VX-9, also provided support for testing activities. Having engineers and test pilots from both groups working side by side, interacting directly with their impressions and findings, reduced the EMD test flight phase by as much as one year. Overall, the EMD Super Hornets accumulated 3,000 flight hours and expanded the flight envelope to 49,500 ft (15088 m), speeds greater than Mach 1.5, and +7/-1.7 g.

The ITT's work at this stage sought to show that "the aircraft can fly, that it can fight, and that its systems can work together". Each aircraft in the test programme was given specific duties: aircraft E1 made its maiden flight on 29 November 1995 and was charged with investigating flying qualities and flight envelope expansion. E1 made its initial flight with stores on 21 February 1997; its load consisted of three external tanks, two Mk 84 bombs, two AIM-9 Sidewinders, and two AGM-88 HARMs. Total weight reached 62,400 lb (28304 kg). E2 first flew on 26 December 1995 and was used for engine and performance testing. E3 handled the load testing portion of the EMD, while high angle-of-attack (AoA) evaluations were assigned to E4. E5 was the first Super Hornet to possess full mission capability and was used for weapons work.

The two dual-place Super Hornets were also busy. F1 was assigned carrier suitability and, later, weapons testing. First flying on 1 April 1996, F1 racked up 508 sorties and 601.8 flight hours, the most flight time of all EMD aircraft. F2 (BuNo. 165169; first flight 11 October 1996), was also assigned to some carrier qualification duties, but spent much of its time with E5 performing weapons testing at China Lake. F2 was the second Super Hornet to have a full avionics suite.

Static tests were performed on ST50 commencing in August 1995. It was later sent to Lakehurst, New Jersey, for emergency barricade testing. The airframe completed three tests before being damaged on 23 September when a restraining cable failed and the aircraft overturned into a wooded area. ST50 was subsequently repaired at Boeing and transferred to China Lake for live-firing testing using large armour-piercing incendiary projectiles. DT50 began its shock loading assessment in February 1996, and FT50 began its fatigue testing in June 1997. FT50 completed its first lifetime (6,000 hours) one month ahead of schedule. The second ran from January 1997 through November 1999.

The first segment of the Super Hornet's sea trials began on 6 August 1996 at NAS Patuxent River, Maryland, when Commander Tom Gurney made the Super Hornet's first catapult launch from a land-based, steam-powered MR-7 catapult in F1. Fifteen days later, F1 made its first arrested landing. F/A-18F1 headed out to USS *John C. Stennis* (CVN 74) in January 1997 for the Super Hornet's initial sea trials. In cold and snowy weather, F1 made the two-hour flight from the Naval Air Warfare Center at Patuxent River and landed in what were considered 'marginal' weather conditions for an aircraft at this stage of its flight tests. The

VFA-115 'Eagles' landed the honour of becoming the fleet's first Super Hornet squadron, and the first unit to pass through the VFA-122 'schoolhouse'. VFA-115's aircraft were from the third and final LRIP batch, the last aircraft built before full production got under way. Several features of full production aircraft were missing, and their avionics suite was roughly comparable to that of the F/A-18C – lacking MIDS, SHARP and JHMCS capability. The squadron did receive four LRIP ATFLIR pods, but they were not used when the squadron went into combat due to reliability issues associated with the early-production equipment.

Above: VFA-115's 'boss-bird' traps aboard Lincoln *in August 2002, during the squadron's first operational deployment. The carrier was sailing off Hawaii at the time. The commander's aircraft wears less colourful markings than the CAG jet.*

Left: VFA-115 F/A-18Es share deck space on Lincoln *with a VF-31 F-14D. At the time Air Wing 14 was undergoing a COMPTUEX prior to its operational cruise (the first with the Super Hornet).*

Focus Aircraft

After its work-up phase, VFA-115 deployed to Abraham Lincoln on 24 July 2002 for the Super Hornet's first operational cruise. This took the carrier and Air Wing Fourteen to the Arabian Sea for operations over Afghanistan. At the time US forces were still conducting large-scale 'mopping-up' exercises against Taliban/al Qaeda forces, and VFA-115's Super Hornets provided top cover for these operations, although they did not drop any weapons. During their patrol in the Arabian Sea the 'Eagles' suggested changing their name back temporarily to the original 'Arabs'.

first carrier landing was made by Navy Lieutenant Frank Morley on 18 January at approximately 10:00 a.m. EST off the coast of Cape Hatteras, North Carolina. Commander Tom Gurney then switched places with Morley and piloted F1 on its first carrier catapult launch, at approximately 2:30 p.m. that same day.

A total of 61 daytime launches and recoveries was made during the six-day deployment. Flights explored flying qualities from behind the ship, dual- and single-engine handling, and trim and crosswind effects while launching from the bow and waist catapults. A total of 54 touch-and-go landings confirmed the -E/F's landing approach speed to be 10 kt (18.5 km/h) slower than the -C/D's. The pilots described the Super Hornet as having "great hands-off fly-away characteristics off the catapult" and commented that it "[flew] well on approach, as expected, despite challenging wind conditions".

Weapons separation tests

Super Hornets E5 and F2 were assigned to weapons separation testing duties. The first such test occurred on 19 February 1997 as F2 successfully released an empty 480-US gal (1817-litre) fuel tank from 5,000 ft (1524 m). Two days later, E1 flew an Aero Servo Elasticity stores configuration comprised of three 480-US gal (1817-litre) tanks, two AIM-9s, two Mk 84 iron bombs and two HARMs, marking the first time a Super Hornet carried a simulated warload. At some 62,400 lb (23800 kg), it also represented the largest gross weight of the programme to date. Northrop Grumman test pilot Jim Sandberg, who flew E2 during the test, stated that "[the] airplane flew effortlessly throughout the flight" and "performed as if it were flying clean".

F2 launched the first air-to-air missile – an AIM-9 Sidewinder – in April, followed by the launch of an AIM-120 on 5 May. Additional tests followed, and by the end of May the Super Hornet had successfully released a sampling of the typical Hornet warloads – AIM-7, ALE-47 flares, Stand-off Land Attack Missile (SLAM), Harpoon, a ripple of 10 Mk 82s, Mk 83s, a dual load of CBU-100s – and ejected 480-US gal (1817-litre) tanks from both wing and the centreline stations. Tests were also performed with the ALE-50 towed decoy. Live firings of HARMs followed in December 1998, then a Harpoon launch against a moving ship in January 1999. Twenty-five missiles and over 500,000 lb (226796 kg) of ordnance had been expended by the squadron before the end of the EMD weapons separation phase. Twenty-nine weapons configurations were cleared and made available for the OPEVAL.

Final CQs

The Super Hornet's final carrier qualifications before OPEVAL took place in February and March 1999 aboard USS *Harry S. Truman* (CVN 75) off the coast of Florida. This time, both -Fs were taken to sea. Navy pilot Lieutenant Commander Lance Floyd made the F/A-18F's first night-time carrier landing. The two -Fs successfully launched off the bow with 15 kt (28 km/h) of crosswind and off the waist catapult with 10 kt (18.5 km/h) of crosswind. Launches and traps were also made with various asymmetrical weapons configurations and using the automatic carrier landing approach system from distances of 4 and

VFA-115 'Eagles'

VA-115 was formed by the renumbering of VA-12A on 15 July 1948. It flew TBM Avengers, AD Skyraiders and A-6 Intruders before it transitioned to the F/A-18C in 1996, redesignating as VFA-115 in the process. From the 1950s the squadron used the nickname 'Arabs', although this was changed to the potentially less controversial 'Eagles' in 1979. The unit's time with the 'baby' Hornet was brief, as it was chosen to become the first front-line Super Hornet unit. Conversion began in late 2000 and was completed in the first half of the next year. The squadron badge, approved for use on 17 September 1956, consists of a stylised wing and globe design. The 'NK' tailcode signifies assignment to CVW-14 (formerly *Abraham Lincoln*/CVN 72, but from mid-2003 assigned to *John C. Stennis*/CVN 74).

8 miles (6.5 and 13 km). Other flights included minimum end speed tests with military power and full afterburner. At its maximum gross weight (66,000 lb/29937 kg), the -F was able to launch with full afterburner at 142 kt (263 km/h), and reportedly sank only 10 ft (3 m) below the bow before recovering. The aircraft was noted to handle superbly and was considered very responsive to last-minute corrections.

EMD problems

As is not unusual in a test programme, the ITT soon identified over 400 deficiencies, the most significant of which concerned the aircraft's flying qualities, service life, engine performance and weapons separation. Fortunately, only two delays occurred. The first came during the summer of 1996 and involved the delivery of the final three EMD aircraft, due to a three-month machinists' strike at one of the major contractors' facilities. The second delay followed an inflight engine failure, which ended all Super Hornet flight tests (except those of F1 to determine carrier suitability) for two months.

Perhaps the most publicised problem of the EMD was the wing drop phenomenon. First appearing in March 1996, this phenomenon is formally described as "an unacceptable, uncommanded abrupt lateral roll that randomly occurs at the altitude and speed at which air-to-air combat manoeuvres are expected to occur". In its simplest terms, wing drop is caused by airflow separating on one wing before the other and typically occurred when the Super Hornet was manoeuvred at relatively high angles of attack and high g forces. First noticed during the 1950s during test flights on the F-86, it is a common event associated with high-performance, swept-wing aircraft.

Work towards resolving the problem continued until mid-1997, and isolated the wing drop in the centre of the Super Hornet's flight envelope – between Mach 0.70 and 0.95 at altitudes of 10,000 to 40,000 ft (3048 to 12192 m) and AoA between 6° and 12°. ITT members working with Navy and Boeing engineers, as well as NASA, eventually settled on an interim solution based on modifications to the leading-edge flaps and flight control software. The final solution was achieved through porous wing-fold fairings, which, according to Boeing documents, comprise "many small holes that influence the airflow over the wing, eliminating wing drop throughout the manoeuvring envelope".

Early wind tunnel tests conducted during July and August 1993 showed that some stores would collide with the side of the fuselage or with other stores when released. This problem resulted from adverse airflow created by the aircraft's airframe. To cure the problem, the pylons were redesigned and canted outward at 3°. Testing during the EMD confirmed the redesign had corrected most of the problem. Another related deficiency concerned unwanted

The two-seat F/A-18F enjoys the same combat capability as the single-seater, although its heavier weight reduces its 'bring-back' capacity slightly. Early production aircraft have coupled cockpits, but F/A-18Fs are now being produced with advanced displays in the rear cockpit and the ability to decouple the two work-stations, allowing independent targeting by front- and back-seaters. In most cases this would entail the pilot concentrating on the air threat while the WSO focuses on air-to-ground attacks. The F/A-18F expands further on work done previously by Boeing in this field on the F-15E and the F/A-18D for the USMC.

The capability of the Super Hornet in the tanker role cannot be overestimated, and in the future it will become the air wing's only organic refuelling capacity when the S-3 is retired. The Super Hornet offers a greater offload ability, although its endurance time is reduced, and pound for pound it is more expensive to operate. However, it can stay with the strike force, and during operations over Iraq routinely 'crossed the beach', taking strike aircraft to about 150 miles from their targets. Here a VFA-41 F/A-18F buddy refuels another tanker-configured aircraft.

Focus Aircraft

During a visit to the UK for the Farnborough air show in 2002, VFA-41's 'boss-bird' flies past the cliffs of Beachy Head, with Eastbourne Pier in the background. A concerted sales effort has seen the Super Hornet attend most of the world's major trade shows, where its displays have focused on the type's excellent low-speed manoeuvrability and high-Alpha controllability.

With the arrival of VFA-14 and VFA-41, CVW-11 became the first and only air wing to operate all four operational US Navy Hornet versions (A, C, E and F), and the first to operate the Navy's preferred mix of two Super and two 'legacy' squadrons. Once aboard the operational carrier (Nimitz), Air Wing Eleven headed for the Persian Gulf for operations over Iraq.

noise and vibration created with certain stores configurations. To prevent structural damage in the short-term, speed limitations were placed on the Super Hornet while carrying certain weapons stores. Although a few of these limitations remain in place today (these were noted during the subsequent OPEVAL), the Navy and Boeing have come upon a more economical solution – redesigning the weapons to minimise the noise.

OPEVAL – the last step

Following completion of the successful EMD phase, the Super Hornet moved into the Operational Evaluation phase, the last test prior to fleet introduction. From a conceptual standpoint, the OPEVAL has a different focus than EMD flights, which seek to "test and confirm the aircraft's performance, as well as test new parameters". The OPEVAL, conversely, "looks to test the aircraft's ability to perform operationally and tactically in a realistic wartime environment". Former VX-9 OPEVAL Test Director, Commander Jeff Penfield, USN, described the process as follows: "[we] evaluate the aircraft [in OPEVAL] to determine how it will fit into real world operations. Operational pilots look at things differently than flight test pilots. If I took a Super Hornet into battle tomorrow, how would the aircraft perform?" The best rating for any unit or system entering an OPEVAL is a finding of 'operationally effective' and 'operationally suitable'. Operationally effective means that the aircraft is able to perform its prescribed mission in a fleet environment, and in the face of unexpected threats; operationally suitable means that the aircraft, when operated and maintained by typical fleet personnel in the expected numbers and of the expected experience level, is supportable when deployed.

VFA-41 'Black Aces'

The 'Black Aces' (callsign FAST EAGLE) transitioned from F-4Ns to F-14As in April 1976, and subsequently took the Tomcat into action against Libya (downing two Su-22s), and in Operations Desert Storm, Allied Force and Enduring Freedom. The final Tomcat cruise, which included combat action over Afghanistan, was made with CVW-8 aboard USS *Enterprise* (CVN 65). After its return in November 2001, the squadron was stood down for conversion to the F/A-18F, the first fleet unit to be equipped with the two-seat version. The squadron is now assigned to CVW-11 (tailcode 'NH') aboard USS *Nimitz* (CVN 68) alongside VFA-14.

The Super Hornet OPEVAL officially began on 27 May 1999 and ran through 19 November that year. Guidelines for the OPEVAL stemmed from the 1991 Navy Operation Requirement Document (ORD) for the F/A-18E/F Upgrade, which mandated numerous areas of improvements over the existing F/A-18C/D: (1) increased mission radius; (2) increased payload flexibility; (3) increased carrier recovery or bring-back; (4) increased survivability; and (5) decreased vulnerability. Improvements in combat performance over the Lot XII F/A-18 C/D (turn rate, climb rate and acceleration) and growth capability for general avionics (electrical, environmental control, flight control, and hydro-mechanical systems) were considered vital.

The OPEVAL was conducted by Test and Evaluation Squadron Nine (VX-9) based at NAWC China Lake, California, using three F/A-18Es and four F/A-18Fs. These aircraft were the first delivered under the LRIP Lot 1 contract, and all incorporated the modifications resulting from the EMD. The OPEVAL tests were flown by a team of 14 pilots and nine WSOs with diverse backgrounds, including the F/A-18A/B/C/D, F-14, A-6E, A-7E and S-3B communities; all had a significant amount of flight time and were regarded as outstanding crews. Approximately 70 Navy maintenance personnel were assigned to evaluate the aircraft's maintainability during the OPEVAL.

F/A-18E/F evaluated 'as is'

The F/A-18E/F OPEVAL was performed without reference to any of the new capabilities penned for the Super Hornet. Thus, systems such as the Active Electronically Scanned Array (AESA) radar, the AIM-9X Sidewinder off-boresight air-to-air missile, or the Joint Helmet-Mounted Cueing System (JHMCS) were not factored into the overall

VFA-14 'TOPHATTERS'

Despite only becoming VF-14 on 15 December 1949, the 'Tophatters' claim to be the oldest US Navy squadron with a lineage tracing back to 1919. With Tomcats, VF-14 (callsign CAMELOT) flew in Operations Desert Storm, Deny Flight and Enduring Freedom. For the last cruise aboard *Enterprise* the squadron partnered VF-41 in CVW-8, and was heavily committed to 'Bombcat' operations over Afghanistan. At the end of the cruise in November 1991 the squadron began conversion to the F/A-18E, one of two Tomcat units planned for transition to the single-seat Super Hornet. After re-equipment and training, VFA-14 was assigned alongside VFA-41 to *Nimitz*'s CVW-11 ('NH'), the first wing to have its full complement of Super Hornets allocated.

all rating. Moreover, no consideration was given to follow-on systems – such as the Advanced Tactical FLIR (ATFLIR) or SHAred Reconnaissance Pod (SHARP) – designed to replace ageing legacy systems carried over from the current -C/D models. Commander Dave Dunaway, one of the lead VX-9 test pilots and a liaison to the EMD Integrated Test Team, described the OPEVAL as "taking an immature aircraft, one in its infancy, and pitting it against established threat systems". Given this, it is clear that the aircraft as tested did not represent the full range of the Super Hornet's tactical capability.

Multi-phase test program

The OPEVAL consisted of a five-phase test programme designed to evaluate the Super Hornet under realistic operating conditions to determine the effectiveness and suitability of the aircraft, its systems, and its weapons for combat. With the exception of reconnaissance (which was evaluated in a subsequent follow-on OPEVAL during 2002), all principle missions of the -E/F – interdiction, war-at-sea, fighter escort, combat air patrol (CAP), alert interceptor, suppression of enemy air defences (SEAD), close air support (CAS), tanker, and forward air control-airborne (FAC-A) – were evaluated. The OPEVAL aircrews initially familiarised themselves with the aircraft at China Lake before completing carrier qualifications refreshers aboard USS *Abraham Lincoln*.

■ **Air-to-ground phase:** Evaluations of various air-to-air weapons and air-to-air sensors began on 27 May at China Lake. VX-9 aircrews also evaluated the E/F's air combat manoeuvring (ACM), defence suppression capabilities and the aircraft's overall survivability. Twenty-nine of the weapons planned for the E/F were cleared for OPEVAL, which represented those planned for the aircraft's initial deployment. As a comparison, the original Hornet had just two configurations at OPEVAL. The Super Hornets deliv-

These photos show VFA-14 F/A-18Es in the 'five-wet' configuration. Typically an embarked E squadron would have four aircraft configured as tankers. Standard procedure would be to launch the tankers twice during a strike, once to escort the outbound strike aircraft and then again to provide cover for the recovery.

The first operational deployment for the Super Hornet came in the latter part of the Enduring Freedom campaign over Afghanistan, although the aircraft did not drop any weapons. On 29 October 2002 VFA-115 began flying Southern Watch duties over Iraq. Here one of the squadron's aircraft refuels from a USAF KC-10A during an OSW mission. Armament comprises AIM-9 Sidewinders, AIM-120 AMRAAM and GBU-12 LGBs.

Lieutenant John Turner taxis on the deck of Lincoln after returning from a mission in which the Super Hornet had dropped its first bomb drops in anger. The date was 6 November 2002 and the target was a command centre at Tallil. Two Super Hornets released four GBU-31 JDAMs in the attack, and all guided precisely.

the aircraft's flight characteristics around the boat and assessed its overall integration with a carrier air wing during routine operations. During the two weeks from 12-28 July, a VX-9 detachment took a group of Super Hornets to USS *John C. Stennis* (CVN 74). After spending the first week qualifying the aircraft, VX-9 personnel operated the Super Hornets as a small squadron with other Air Wing Nine (CVW-9) aircraft, allowing the evaluators to critique the Super Hornet's performance and its ability to integrate with other carrier assets. Missions flown included simulated deck-launched intercepts, tanking, mining and war-at-sea strikes.

The anti-carrier and mining operations were conducted off the southern California operations area, while the long-range offensive strikes were flown from the *John C. Stennis* against the Fallon and China Lake ranges. Aircrews stated that the Super Hornet integrated well and fulfilled all tasked missions. Moreover, its greater range and flight time allowed planners to increase cycle time, which means that the Super Hornet will not be as limited as its predecessor in cyclic operations.

ered Mk 82 (500 lb/227 kg), Mk 83 (1,000 lb/454 kg), and Mk 84 (2,000 lb/907 kg) iron bombs during this phase, as well as cluster bomb units (CBUs). A variety of range profiles was also flown to verify the flight performance database predicted by Boeing. The Super Hornet's ability to serve as a tanker was also explored during day and night tanking missions, with the Super Hornet performing as expected.

■ **Air combat phase:** This was conducted during a two-week detachment at NAS Key West, Florida, from 14-25 June. VX-9 evaluated the Super Hornet in a variety of fighter escort and CAP profiles and ACM regimes, assessing both tactics and survivability. At Key West, the evaluators focused mainly on air-to-air critical operational issues (COIs) – fighter escort and air combat manoeuvrability. "The tactics COIs specifically tasked the evaluators to determine if the F/A-18E/F could execute current tactics." Adversary services were provided by F-16Cs from the 185th Fighter Squadron, Air National Guard of Sioux City, Iowa, which flew a series of realistic threat tactics emulating the latest-generation MiG-29. Scenarios pitted up to four Super Hornets against an equal or larger number of threat adversaries. In others, mixed sections of Hornets and Super Hornets were flown to compare the performance of the two aircraft under similar conditions.

■ **Carrier operations performance:** Given that the Super Hornet will live and fight from the decks of an aircraft-carrier, a crucial aspect of the OPEVAL evaluated

■ **Combined joint operations:** The final OPEVAL detachment went to Nellis AFB, Nevada, from 16-27 August. There, Super Hornets participated in a multinational Combined/Joint Red Flag Exercise with over 60 aircraft from the Air Force, Marine Corps, Navy, and several foreign countries. Flying strike, SEAD, fighter escort, interdiction, and FAC-A missions representative of current NATO operations, most flights used instrument pods that allowed later analysis. A limited number of flights used live ordnance. During the exercises, the Super Hornet's tanking ability was again assessed.

■ **Survivability, air-to-air missiles and smart weapons:** The final stage of the OPEVAL was conducted at China Lake from September through November 1999, and focused on survivability. Operationally representative flights were flown against actual and simulated threat surface-to-air missile (SAM) systems, followed by air-to-ground gunnery and air-to-ground sensor flights. Air-to-surface weapons tested included the Mk 80 series iron bombs, Rockeye, SLAM, Harpoon and Maverick. VX-9 crews found the -E/F's additional fuel significant because it allowed the use of routing alternatives, lower altitudes and more frequent use of afterburner to maintain energy during combat manoeuvring. During this phase, an actual side-by-side comparison was made between the -E/F and the Lot XII or later F/A-18C. Prior segment comparisons had been either quantitative (with a specific number in mind) or qualitative (in which the issue was whether the -E/F could execute in the particular mission area). Test data confirmed that the -E/F was more survivable than the F/A-18C/D.

The OPEVAL officially ended on 19 November. Some 850 sorties and 1,233 flight hours had been amassed and approximately 400,000 lb (181440 kg) of ordnance expended over the course of the six-month test programme. Rear Admiral John B. Natham, then-Director of Air Warfare in the Office of the Chief of Naval Operations, announced the OPEVAL results on 15 February 2000 during a ceremony at the Pentagon in Washington, DC. The Super Hornet was "operationally effective and operationally suitable". VX-9 therefore recommended the aircraft's introduc-

Right: As well as refuelling from USAF tankers and Air Wing 14's S-3B Vikings (VS-35 'Blue Wolves'), VFA-115's Super Hornets also 'buddy'-tanked during Operation Southern Watch missions. Following the 6 November combat debut, F/A-18Es were in action again the following day, and then three days later. The attacks were made in response to Iraqi sites threatening OSW aircraft.

Even though CVW-14 had Viking tankers, VFA-115 put up four F/A-18Es in tanker configuration during OIF. Their use permitted over 430 sorties to be flown that would otherwise not have been possible.

tion into fleet service. Commenting on the results, Rear Admiral James Godwin III said, "[there] were no surprises with what we saw out of the OPEVAL report, and what we predicted we would see during the developmental testing that we had ongoing prior to OPEVAL is precisely what we saw during the OPEVAL." Godwin continued, "We did not experience anything that was unexpected."

Areas of significant enhancement

The OPEVAL concluded the -E/F possessed several areas of "significant enhancement" over the F/A-18C/D, ranging from increased tactical and payload flexibility to carrier performance and survivability. The Super Hornet's ability to fly a wide variety of missions creates more options for war-planners in executing both combat and support missions. The evaluators found that when flying as a tanker, the Super Hornet can match the altitude and speed of the strike package, which leads to optimisation of flight profiles and resultant fuel efficiency.

Payload flexibility addresses the Super Hornet's ability to carry a wider variety and greater number of weapons than the current Hornet. The -E/F's two additional weapons stations increase mission effectiveness in two ways: more offensive weapons means that an F/A-18E/F can achieve the desired probability of destruction with fewer sorties flown, and the two additional weapons stations can be used to carry more self-defence weapons. Both options further enhance the -E/F's survivability and serve to reduce the number of sorties needed to reach a desired objective.

Two additional areas were noted concerning the Super Hornet's performance around a carrier – the -E/F's slower approach speed and its increased bring-back capability, both of which make the aircraft safer during landings. The enhanced bring-back amounted to an additional 3,400 lb (1542 kg) of ordnance for the -E and an additional 2,400 lb (1087 kg) for the -F, as compared to the Lot XIX F/A-18C. These increases are ever-more significant in air operations such as those over Iraq and Afghanistan, where US pilots fly dozens of ground support missions and routinely return to the carrier with unspent ordnance.

The -E/F's manoeuvring and handling qualities were rated high and the aircraft was noted to resist departure even under aggressive high-AoA manoeuvring. Evaluators also relished the aircraft's 'positive nose pointing' – how

Above: VFA-115's CAG-bird peels away from the camera platform during Operation Iraqi Freedom. The aircraft carries two JDAMs and the AAS-38B Nite Hawk pod. This view highlights the 3° outward canting of the weapons pylons, introduced to cure a separation problem.

Above: Carrying two JDAMs, VFA-115's 'boss-bird' heads for Iraq.

Left: A VFA-115 jet launches from Lincoln, carrying an asymmetric load of one GBU-16 LGB, one AIM-120 AMRAAM plus three 500-lb Mk 82 GP bombs. Such loads were commonplace during the latter part of Iraqi Freedom, when aircraft were often launching without any pre-planned targets, and carried a mix of ordnance to cover several eventualities. Free-fall weapons were used mainly against troop concentrations, where precision was not required.

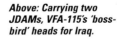

Focus Aircraft

Above: VFA-115's 'war duck' goes into action during OIF. For most missions the Super Hornets carried AIM-9s for self-defence, plus an AMRAAM on the fuselage.

Above right: A VFA-41 F/A-18F tanker catches the wire on Nimitz shortly before the squadron entered the fray over Iraq.

Right: For VFA-115 operations continued round the clock in OIF. Most aircraft carried nose markings relating to the Twin Towers attacks.

Below: The CAG jet is about to launch on a combat sortie with a mixed weapon load. Twin-rail adapters are used with a variety of the smaller diameter weapons, including Mk 82 bombs (as here) and AIM-120 AMRAAMs, of which 12 can be carried in the maximum loadout.

quickly a pilot can put the aircraft's nose on a 'bogie' – which translates into who can take the first shot during an engagement. This quality has been described as outstanding by OPEVAL pilots. Weapons delivery accuracy was reported to be excellent, exceeding the ORD accuracy requirements and achieving results equal to or better than those of the F/A-18C.

Noted areas of concern

The OPEVAL identified nine areas of concern, although by that time most were being addressed or had been remedied. One concern noted the limited number of specific stores configurations that had been cleared and argued for the expedited removal of the restrictions placed on the carriage and release of those configurations. Although only 29 weapons configurations were available at the time of OPEVAL, this compared favourably to the two configurations available during the F/A-18A/B OPEVAL. Phillip E. Coyle III, Director, Operational Test & Evaluation for the Navy, in his statement to the Senate Armed Services Committee, AirLand Forces Subcommittee on 22 March 2000, noted, "Air-to-air missiles could not be employed if they were carried on a store station adjacent to air-to-ground ordnance. Numerous munitions could be carried and/or employed only from selected stores stations, although the plan is to bear these munitions from other stations as well."

Because of this, Coyle observed, "many of the load advantages planned for the F/A-18 E/F were not demonstrated during OPEVAL". Even so, many of the configurations which were presented were significantly beyond the capabilities of current F/A-18Cs.

Other areas concerned were minor and can be summarised as follows: (1) more severe noise and vibration environment under the wing compared to F/A-18 C/D; (2) a diminished specific excess power (Ps) rating; (3) buffet during 1g transition flight; and (4) residual lateral activity, albeit minimum. From a performance standpoint, the -E/F was considered comparable or superior to the -C/D in turns, climbs and deceleration at subsonic speeds. Large decelerations (airspeed bleed-off), however, were experienced in the transonic/supersonic regime during manoeuvres. From a tactical standpoint, this is insignificant, because most manoeuvring during an engagement quickly migrates to the 'current of the flight envelope' (approxi-

Left: On 30 March 2003, four Super Hornets (two Es and two Fs) leapfrogged ahead of Nimitz as it sailed towards the war zone, to operate from Lincoln. This not only bolstered the latter's air wing but also gave the Nimitz crews a taste of combat action before the carrier and the rest of the air wing arrived in the theatre. Here a VFA-14 F/A-18E launches from Nimitz at the start of its journey to Lincoln.

mately Mach 0.86 at 15,000 ft/4572 m). Moreover, since the Super Hornet is virtually immune to departure, it is extremely capable during a close-in fight.

KPPs confirmed during OPEVAL

Three key performance parameters were identified in the Congressional mandate for the F/A-18E/F programme which required it to exceed the F/A-18A-D's range and payload performance in specified associated flight profiles. Profiles evaluated included the E/F's combat radius in the fighter escort mission and interdiction mission roles, using two and three 480-US gal (1817-litre) external fuel tanks. The -E/F exceeded all threshold range requirements, posting a 425-nm (787-km) radius in the fighter escort mission and 400 nm (741 km) (two external tanks) and 450 nm (833 km) (three external tanks) in the interdiction mission.

Although the Super Hornet was projected to surpass these threshold requirements by Boeing and Navy officials, based on a flight performance database constructed from theoretical calculations, wind tunnel and engine-run tests, and developmental flight test data, these numbers needed to be confirmed. VX-9 aircrews undertook a calibration of the performance database by using a 'flight segment' approach. Under this system, fuel consumption data was collected in small, dedicated portions of many flights using a variety of aircraft configurations, gross weights and flight loads. As testament to the methodology employed by Boeing and the Navy, the deviations between actual and projected fuel consumption were insignificant, and the OPEVAL crews were able to confirm the database numbers as accurate. Based on this confirmation, OPEVAL crews were able to compute the range performance of the F/A-18E/F for the three profiles defined by the ORD and the nine profiles defined by the Chief of Naval Operations. The nine CNO-defined operational missions were computed using a 4,000-lb (1815-kg) fuel reserve; the ORD-defined operational missions were computed using a 2,000-lb (907-kg) reserve.

Several recommendations stemmed from the OPEVAL, the most significant of which involved the correction of underwing noise and vibration. Addressing such problems is critical, Coyle noted in his statement to the Senate, because if left uncorrected it can impede the attainment of full weapons life. OPEVAL further recommended the immediate introduction of new systems such as the all-aspect high off-boresight AIM-9X missile and the JHMCS, the electronically scanned array radar (now designated APG-79), the ATFLIR, a positive identification capability such as the

Lightly armed, a VFA-14 F/A-18E refuels over Iraq during a policing mission. The single-seat Hornets used the old AAS-38B Nite Hawk designation pod during OIF, rather than the latest ASQ-228 ATFLIR.

A VFA-14 F/A-18E folds its wings shortly after its arrival aboard Lincoln. During the war the US Navy was hampered by a lack of tanker capacity, and the Super Hornet was routinely used in the role.

Right: This VFA-41 F/A-18F is seen on 1 April, the day it arrived on Lincoln after deploying forward from Nimitz. During their sojourn on Lincoln the four extra Super Hornets were attached to VFA-115. In the foreground is a GBU-16 Paveway II bomb, one of the weapons employed by the Super Hornets during OIF.

Below: F/A-18Cs refuel from a KC-135R during OIF as the VFA-41 CAG-bird looks on. The Super Hornet is configured as a tanker with two wing tanks, which appears to be a standard loadout for the two-seat F/A-18F. The F/A-18E single-seater routinely operates with four wing tanks.

Below: Nimitz deck crew conduct a FOD sweep during OIF operations on 17 April. Most of the Hornets in this view are Supers, with a cluster of VFA-14's F/A-18Es round the fantail. The four Super Hornets deployed to Lincoln returned to Nimitz on 6 April after the carrier entered the Persian Gulf. Air Wing Eleven flew combat missions up to and after the stated end of the combat phase on 1 May.

combined interrogator transmitter, decoupled cockpits for enhanced flexibility, multi-function information distribution system, and further enhancements to the planned integrated defensive electronic countermeasures system. Each of these systems plays a key role in the overall Super Hornet road map. Until such systems are fielded, however, the Super Hornet will not fully realise either its potential or the operational capabilities with which it was envisaged.

The radar – APG-73 and APG-79 AESA

A key feature of the E/F is its radar. Baseline Super Hornets feature the APG-73 radar installed in late model F/A-18C/Ds (Lot XVI and beyond) and retrofitted into early model F/A-18C/Ds. The APG-73 features the same air-to-air and air-to-ground modes as its APG-65 predecessor but offers a 10-fold increase in processing capabilities and greater memory; it is also easier to maintain and is more reliable. The APG-73, which was introduced in May 1994, started as a joint US-Canadian effort to improve the APG-65's electronic counter-countermeasures (ECCM) system. It uses the same antenna and transmitter as the APG-65, but incorporates all new electronics.

Later versions of the radar, used primarily in the F/A-18D, incorporated a high-resolution synthetic aperture radar (SAR) for mapping during reconnaissance missions, autonomous targeting for JSOW and JDAM weapons, and the ability to track up to 24 targets, prioritising the top eight threats. The APG-73 uses colour to distinguish enemy aircraft from friendlies and unknown aircraft, and also displays target information. One feature shows the number of degrees the target must manoeuvre in order to evade a fired missile. An azimuth track capability was added in 1998 which allows differentiation of aircraft stacked on top of one another.

AESA for Block II

Beginning in 2007, the Super Hornet will incorporate the next-generation agile-beam APG-79 Active Electronically Scanned Array (AESA) radar built by Raytheon as part of the aircraft's planned Block II upgrade programme. According to the Navy, the APG-79 represents "a revolutionary leap" in capability and provides the Super Hornet with a state-of-the-art radar. Indeed, Boeing AESA Program Manager Don Thole said, "AESA is the key component of the Super Hornet that gives us the capabilities we sought." Thole continued, "This represents the first time in over 25 years where a brand new radar technology has been introduced." APG-79 builds on and improves the AESA technology featured in certain Alaska-based F-15C squadrons. APG-79's AESA features a new open architecture that incorporates the ability to expand software systems. It also uses a new fibre-channel interface that permits faster distribution of information.

According to Navy officials, the radar uses a search-while-track system that significantly improves the track quality of multiple targets with "little or no degradation of the search capability of the radar". For operators, one of the principle benefits is target updating, which means that crews receive information more quickly on their target's actions. With traditional radars, operators have to wait while targeting information is updated between radar sweeps. Boeing's Don Thole says, "You essentially have single target tracking imagery with a multi-target tracking capability." Moreover, it searches while it tracks the target on a need-to-know basis. Once a target is located, the radar automatically and periodically looks back to keep track of the target, at a much higher level of fidelity and frequency than mechanical models.

NAVAIR APG-79 Program Manager Commander Dave Dunaway said, in a 2002 interview, "The radar will change the way we do business. We are currently limited by the radar's ability to detect air-to-air targets at range. In other words, we have to wait and let the radar catch up before we can shoot the missile. With the AESA, we'll be able to shoot the missile even before the target comes within the missile's range. The APG-79 has been designed to enable the aircrew to detect and process the target well before it enters the maximum range of the Super Hornet's air-to-air missiles, allowing missile launch at maximum range."

Air-to-ground resolution and range are vastly increased, with estimates of three times the detection range for large targets and two times the detection range of current systems for smaller targets. The APG-79 also presents higher-resolution SAR maps. "The aircrew can see minute runway details on the map and can identify aircraft." The APG-79 can also scan air-to-air and air-to-ground nearly simultaneously, which gives a huge advantage in multiple target attack capability and is part of the reason why the radar is touted as revolutionary. With the APG-79, a pilot can still maintain a track on a ground target while simultaneously detecting, locking and attacking an air threat. Other benefits derive from combining AESA-generated information with that from other sensors to give a clearer picture of the target.

From an airframe perspective, two modifications were required to incorporate the APG-79. First, pursuant to ECP 6038R1, the forward fuselage has been extended to accommodate the slightly larger frame. These changes began with Lot 26 aircraft now under production and are intended to allow an easy retrofit once the AESA comes on line. Moreover, the new forward section features 40 per cent fewer parts and 51 per cent fewer fasteners, and can be built in 31 per cent less time than the prior -E/F fuselage section. The second modification involved a more powerful cooling system upgraded by Northrop Grumman to help maintain a constant temperature for the APG-79. The system provides 16 kW of liquid cooling capacity, with room for additional capability. Boeing says that AESA uses all but 1 kW of this additional cooling power, but notes that designers believe that they have considered all aspects and possibilities.

Boeing's Don Thole stated that cooling is essential to the radar and is a key to its reliability. "Keeping the radar cool, and the ability to hold the temperature constant, are key to the unit's reliability." Thole also noted that in terms of reliability, the APG-79 is "phenomenal". Test results are currently showing a five-fold decrease in mean time between failure rates. Part of the reliability comes from the fact that the radar utilises fewer parts; another factor is the lack of a traditional radar dish. Estimates are that the AESA will not need array maintenance for 10 to 20 years.

Three aircraft are ultimately designated for AESA test flights. The first flight of an AESA-equipped Lot 21 F/A-18E took place on 30 July 2003 at China Lake, during which work was performed on real beam mapping. A total of 18 flights has already been made, the tests focusing on various radar modes and SAR and real beam maps. A production-equivalent -E and -F are scheduled to join the test programme in 2004.

The AESA system is currently under full-scale development and should begin LRIP incorporation in 2005, followed by introduction into the fleet -E/Fs from 2007. Eight units have been approved for LRIP I, with 12 units scheduled for LRIP II and 22 units for LRIP III. TECHEVAL is now scheduled for February 2005, followed by OPEVAL in February 2006. APG-79 units will be installed in production aircraft beginning with Lot 30 in 2006, and IOC is scheduled for 2007. The first AESA-equipped squadron is scheduled for deployment in September 2007. Eventually, some 400 F/A-18E/F and EA-18G aircraft will be fitted with the APG-79, including approximately 135 retrofits of post-Lot 26 aircraft. F/A-18E/Fs prior to that Lot cannot be modified for AESA due to the absence of fuselage modifications associated with ECP 6038R1.

In the aftermath of the main fighting, US warplanes kept up a constant patrol over Iraq to root out pockets of resistance and to provide on-call air support should ground forces come under attack. Armed with GBU-12 LGBs, this VFA-41 F/A-18F is patrolling over Baghdad International Airport.

Weapons

Another advantage of the Super Hornet is its weapons carriage capability. Not only can the E/F carry vastly more tonnage than the C/D it replaces, it can carry a more robust and flexible assortment of weapons, including the long-range and highly-accurate 'J'-series weapons such as JDAM, AGM-154 JSOW and AGM-158 JASSM. Although the Super Hornet can carry more air-to-air ordnance (namely AIM-9 Sidewinders and AIM-120 AMRAAMs) than its predecessor models, the real advantage comes in the air-to-ground arena. Currently, the Super Hornet can carry various combinations of the JSOW, AGM-84 Harpoon, GBU-10/-12/-16, Mks 82 and 83 iron bombs, GBU-29/-30/-31 JDAM, AGM-65 Maverick and AGM-88 HARM. The 2,000-lb (907-kg) GBU-24 Paveway III LGB is at present awaiting clearance.

According to crews, the added stations, combined with the increased range and bring-back, make the Super

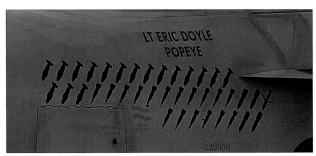

After nearly 10 months at sea on a cruise which was extended to cover operations throughout the combat phase of Iraqi Freedom, VFA-115 returned to Lemoore on 1 May 2003. Its aircraft wore impressive tallies of mission marks (left) recording the type of ordnance dropped, in this instance a mix of JDAMs, LGBs, unguided bombs and a single JSOW. In November 2002 Lt Eric 'Popeye' Doyle had shared with Lt John Turner the distinction of dropping the Super Hornet's first bombs in anger.

Focus Aircraft

In late 2001 the US Navy retained 12 Tomcat squadrons, but by the end of 2003 five had converted or begun conversion to the Super Hornet, and the last is due to transition in 2007. With the exception of VFAs-11 and -14, all will operate the F/A-18F. This aircraft wears full-colour squadron markings and the '102' Modex for the squadron's Executive Officer.

Squadrons converting to the Super Hornet pick up their new aircraft at Lemoore, fresh from the factory. They use the aircraft for part of the conversion process, before transferring to their own squadron training syllabus, taking their new aircraft with them. This work-up includes an air wing COMPTUEX at sea, and a weapons deployment at Fallon before deploying for an operational cruise. Here a VFA-102 aircraft returns to Lemoore at the end of a training sortie.

Hornet a very lethal aircraft. A section of Super Hornets can now handle virtually any threat that materialises – be it long-range air-to-air, surface-to-air or precision strike – while at the same time providing the flexibility to hit targets of opportunity as needed. Moreover, with the new AESA radar, and the better targeting offered by the ATFLIR, the Super Hornet can take full advantage of stand-off weapons.

Weapon stations

As noted in the introductory comments about the Super Hornet, one of the benefits of larger and stronger wings is the addition of two new hardpoints, stations 2 and 10. Rated at 2,150 lb (975 kg) each, these new pylons are capable of carrying air-to-air and air-to-ground weapons. The additional stores greatly increase the Super Hornet's lethality and mission flexibility, permitting a more effective precision strike with a self-escort/self-protect capability. In the air-to-air role, the Super Hornet can carry as many as 12 AIM-120s and two AIM-9s, plus a centreline fuel tank. In the strike role, a Super Hornet can launch with three fuel tanks and still have enough open stations left to carry a warload of two HARMS, a JDAM, two Sidewinders and a laser-guided bomb (LGB).

JHMCS

One of the major systems upgrades from the EMD and baseline E/F is the addition of the revolutionary Joint Helmet-Mounted Cueing System. A shared effort between Boeing and Visions Systems International, JHMCS uses a magnetic head tracker to synchronise the pilot's head movements such that a crew member can train the aircraft's radar, air-to-air weaponry, infra-red sensors and air-to-ground weaponry merely by pointing the helmet at the target and pressing a button on the control stick. The technology also exists for interleaving SHARP imagery, although Boeing has not been asked to adopt this ability. JHMCS represents a joint-service, multi-aircraft system that will be incorporated into the F-15C, F-16 and the F/A-18C/D/E/F. Currently, four squadrons (two F-15C and two Super Hornet) operate the system. A modified F-16 is currently being evaluated at Ogden ALC (Hill AFB), and JHMCS is expected to become operational on F-16s in the next 12 to 18 months.

Mike Reitz, Boeing's Program Manager for the JHMCS, says, "A critical application of JHMCS is when it is mated with the new Raytheon high off-boresight AIM-9X Sidewinder." The AIM-9X allows engagements of targets at greater than 90° off boresight, which, according to Reitz, means that "crews can now attack almost any targets that can be seen". He says that AIM-9X/JHMCS "eliminates the need to manoeuvre the shooter aircraft into the effective seeker arch of the missile, which makes the Super Hornet even more lethal". Reitz adds that JHMCS means "if the pilot can see it, he can kill it".

JHMCS also offers aircrews the ability to view critical aircraft data on a helmet-mounted visor display so that they no longer must look back into the cockpit during critical engagements. Located on the right visor, this monocular mini-HUD allows users to view information on airspeed, altitude and target range without ever unlocking their view of the enemy aircraft. Target acquisition is obtained by directing a small + (plus sign) on the visor onto the target. Early tests found that this system presented some limitations, in that the helmet and seat configuration prevented the pilot from looking up or behind the aircraft. Thus, two additional indicators are found at the top of the visor, left

VFA-102 'DIAMONDBACKS'

VF-102 (callsign DIAMONDBACK) converted from F-4Js to F-14s in 1981, and participated in Desert Storm. On the squadron's last Tomcat cruise it flew combat missions in F-14Bs over Afghanistan from the deck of USS *Theodore Roosevelt* (CVN 71). Returning in spring 2002, the squadron moved to Lemoore to begin conversion to the F/A-18F. In November 2003 VFA-102 departed for its new home at NAS Atsugi in Japan for assignment to CVW-5 ('NF' tailcode) and the carrier USS *Kitty Hawk* (CV 63). All squadron aircraft wear the badge of a diamondback rattlesnake coiled around the globe on the fin, and CAG/CO/XO aircraft wear the squadron's multi-diamond nose stripe. The tailcode is in Japanese-style script to signify the squadron's new assignment.

F/A-18E/F Super Hornet

These images were recorded during VFA-102's CarQuals aboard USS John C. Stennis (CVN 74) in August 2003. The squadron converted to the Super Hornet alone, rather than as half of an E/F pair, and later deployed to Atsugi (CVW-5) as a direct replacement for VF-154. It will be joined in time by an F/A-18E squadron, although none of Air Wing Five's current F/A-18C units is yet slated to convert to the 'Rhino'.

and right, which give the pilot an additional 30° field of view. Once a target is designated, the JHMCS remembers its location, thereby allowing a crew member to look away and then return to the target using a small arrow that gets shorter as the pilot's view moves towards the target.

The system was originally intended only for pilots, but the Navy is now planning to provide the unit to weapons systems officers (WSOs) as well, which will enable independent targeting and tasking for crew members: for example, a pilot would be able to prosecute an air-to-air threat while the WSO executed a ground attack. Boeing's Mike Reitz says that the dual-cockpit capability is also important for talking crew members onto targets. "With both crew members fitted with the JHMCS, the WSO can locate a target, then visually direct the pilot to that target almost instantaneously. Once a crew member designates the target, a special symbol appears on the other crew member's visor that allows him or her to look to what the other is seeing. This significantly enhances crew co-ordination." Currently, crews have to 'talk' one another on to a target, which often wastes valuable time.

JHMCS completed OPEVAL in mid-2002 and is now fitted to F/A-18E/Fs in VFA-14 and -41. It may also be with VFA-102, as that squadron has taken VFA-41's aircraft to Atsugi, Japan. According to Boeing and Navy officials, the JHMCS is slated to be retrofitted into some F/A-18C/D models beginning in 2006. According to Mike Reitz, one of the upgrades being pursued for the JHMCS is a panoramic night-vision goggle capability. Currently, pilots use their standard JHMCS visor during air operations but don NVGs at night or at dusk, especially around the carrier. The new visor would incorporate an NVG capability with a panoramic field of view.

SHARP

Raytheon's SHAred Reconnaissance Pod (SHARP) system joined the fleet with VFA-41's deployment aboard USS Nimitz (CVN 68) as the Navy's new tactical reconnaissance pod. SHARP will eventually replace the TARPS (Tactical Air Reconnaissance Pod System) unit currently used by F-14s

Below: One of VFA-102's F/A-18Fs lands at Atsugi on 13 November 2003 after its ferry flight from Lemoore. The 'Diamondbacks' have replaced VF-154's Tomcats in Air Wing Five, which is permanently based in Japan. VF-154 had returned to the US at the end of September to commence its conversion to the F/A-18F.

Focus Aircraft

VX-31 'Dust Devils'

Weapons R&D work has been undertaken at China Lake for many years, originally organised under the Naval Weapons Center. On 1 January 1992 a streamlining of the US Navy's test organisations resulted in the creation of the Naval Air Warfare Center Weapons Division, which controlled the weapons test activities at its headquarters at China Lake (handled by the newly created Naval Weapons Test Squadron China Lake – NWTSCL 'Dust Devils') and Point Mugu (NWTSPM 'Bloodhounds'). On 1 May 2002 Naval Air Systems Command introduced another change, with five of its squadrons acquiring 'VX' (Air Test and Evaluation Squadron – AIRTEVRON) designations. NWTSCL became VX-31, retaining its nickname and 'Dust Devil' badge.

Above: Part of Naval Air Systems Command (NAVAIR), VX-31 undertakes developmental testing on weapons and aircraft systems, and as well as the F/A-18E (illustrated) and F, it operates the F/A-18A/C/D, AV-8B, AH-1W, HH-1N and T-39D. Once development work has been completed by VX-31, the new equipment is then allocated to VX-9 for operational evaluation.

Right: An F/A-18F from VX-31 taxis in at China Lake after a test mission. Most of the units aircraft are 'orange-wired' – that is, they are configured to accept a variety of test and recording equipment. China Lake offers extensive ranges of over 1.1 million acres (445170 hectares) for the testing of air-to-ground weapons, in near-perfect year-round flying conditions. Air-to-air tests are conducted over the Pacific.

This particular F/A-18E was the fourth production aircraft, and was assigned to the NWTSCL 'Dust Devils' and painted in full-colour markings. Here it is seen testing a rocket pod, with cameras mounted on test installations under triple ejector racks to record weapons separation.

and provides day and night capability, with day stand-off ranges out to 45 nm (83 km). According to Rear Admiral James Godwin III, Program Executive Officer for Tactical Aircraft, the pod represents a "quantum leap in capabilities" for the F/A-18 fleet. The pod, which is approximately the size of a 330-US gal (1259-litre) tank, mounts on the centreline station of F/A-18C/D/E/F aircraft using a standard 30-in (76-cm) attachment, and features a Mil-Std 1760 interface. SHARP was designed with a rotating mid-section to enhance visual coverage and to protect the camera window.

The system was approved for LRIP in 2002 and the first two pods (of 10) were received on 2 April 2003 during a ceremony at Raytheon's Technical Services Company facility in Indianapolis. Captain Charles Wright of CVW-11 aboard USS *Nimitz* commented on the pods' use during work-ups, noting that "the pictures are great". Four pods were deployed with VFA-41 aboard *Nimitz* and are reported to have provided excellent images.

ATFLIR

Hornet operations during the 1991 Gulf War demonstrated that the F/A-18 needed an autonomous laser designation system if it was to take the lead in the Navy's strike mission. At that time, Hornets carried only the ASQ-173 (which allowed tracking but not designation) and had to rely on other aircraft, such as A-6Es, for laser target designation. The answer came in the AAS-38A, but only limited numbers were then available and served only with the Marine Corps F/A-18D squadron, VMFA(AW)-121. Subsequently, all F/A-18Cs were equipped with the improved AAS-38B, which added a self-tracking capability to the unit's designator.

The Super Hornet uses a much-improved third-generation system known as Advanced Tactical FLIR. Made by Raytheon Corporation, the ATFLIR includes integrated FLIR, IR and laser spot trackers, and a laser designator which allows target identification and tracking in virtually all environments. Tests and fleet usage have demonstrated a four-fold improvement in performance over current Navy FLIRs, allowing greater stand-off distances for weapons employment. Designated the ASQ-228, ATFLIR uses 680 x 480-element mid-wave staring focal plane technology and offers a field of view of either 0.7°, 2.8° or 6.0°. It also features a continuous auto-boresight alignment capability that enables first-pass kills. The ATFLIR's targeting range will be augmented by new radar modes for the APG-79, which will give a high-resolution synthetic aperture radar ground imaging capability that is "fantastic", says Boeing's Paul Summers.

According to Navy officials, the real benefit of ATFLIR is its extended range, which allows maximum use of current and planned weapons systems such as the GBU-24 LGB and JDAM. ATFLIR is also more reliable than earlier models, due in part to the fact that it has about two-thirds fewer components, none of which is considered a high-failure item. Mean time between failures is over 300 hours. ATFLIR completed OPEVAL in 2003 and is now operationally deployed with VFAs-14, -41 and -115. Captain Gaddis noted, "One of the signs that the ATFLIR OPEVAL was going well was that the squadron testing the system turned in almost half of its allotted ordnance unexpended." Gaddis added, "The system was performing so well, and was so reliable, that the test and evaluation crews were giving spare parts to VFA-41 to use operationally with their

VX-23/NSATS operates no fewer than 26 Hornets, of which several are Es and Fs. Super Hornet tests have been a prominent part of the squadron's activities in recent years, and they continue today. Many of the Super Hornet's key systems have been tested by the unit, including SHARP, JHMCS, MIDS and the inflight refuelling pod, while the squadron has thoroughly tested the aircraft itself. Test campaigns aboard carriers are routine, this F/A-18E (left) being seen on Roosevelt in November 2002.

VX-23 'SALTY DOGS'

From 1945 the Naval Air Test Center at Patuxent River has been the principal US Navy test base. On 1 January 1992 the NATC became the Naval Air Warfare Center Aircraft Division (NAWC-AD), and its test operations were further divided into four squadrons covering strike/fighter, ASW, rotary-wing and test pilot training (USNTPS). On 1 May 2002 the three test squadrons at 'Pax' were given AIRTEVRON designations. The Naval Strike Aircraft Test Squadron (NSATS), to which various Hornet/Super Hornet models have been assigned, was rechristened VX-23. NSATS/VX-23 aircraft wear an 'SD' tailcode, standing for 'Strike Directorate'. The initials gave rise to the 'Salty Dogs' nickname.

pods." A total of 574 ATFLIRs is planned and will be compatible with -C/D models, as well. The pod, which measures about 72 in (1.83 m) and weighs approximately 400 lb (181 kg), mounts on the Super Hornet's left fuselage station, which means that additional stores may be carried on the wings.

MIDS-LVT

The F/A-18E/F is the lead aircraft platform for the new Multi-functional Information Distribution System – Low Volume Terminal (MIDS-LVT), a system designed to provide near-real time situational awareness for aircrews by integrating data from multiple information sources. MIDS provides an advanced, high-capacity, jam-resistant, digital communications link facilitating near real-time exchange of voice and data information. MIDS also allows the exchange of target information among multiple platforms using a network, such that a strike fighter with expended ordnance could pass targeting information to another shooter miles away. MIDS also provides Super Hornet crews with an overall integrated air picture that identifies friend and foe and displays velocities and headings. This obviously increases crew and section efficiency, as target pass-off times are significantly shortened.

Part of the Block II upgrade path, the system incorporates tactical air navigation functionality (replacing the ARC-118 TACAN system), which allows it also to serve as a navigational aid. Testing of the MIDS system commenced in 2002 and it completed OPEVAL later that autumn, going to sea with VFAs-14 and -41 during their Persian Gulf deployments in 2003. During OPEVAL, MIDS performed outstandingly and resulted in no 'blue-on-blue' engagements between aircraft using MIDS in 866 flight hours. Crews using the system have praised the substantial enhancement of situational awareness, saying it is superior to that provided by the TIDS/Link-16 aboard the F-14.

Joining the fleet

As the EMD and OPEVAL continued, plans commenced at NAS Lemoore, California, for the Super Hornet fleet readiness squadron, VFA-122. A former A-7 FRS, the 'Flying

Below: Bedecked in photo-calibration marks, a VX-23 F/A-18E takes the wire at Patuxent River. The squadron regularly uses the TC-7 catapult and Mk 7 arresting gear which equips Runway 14/32 at the base, enabling the unit to test carrier suitability of new aircraft or equipment in a tightly controlled environment before it goes to sea. The runways are also fitted with the same visual landing aids as found on carriers, and an Automatic Carrier Landing System (ACLS).

VFA-122 at Lemoore remains very busy as it converts both existing squadrons and new naval aviators. New squadrons are assigned their aircraft while the crews are still training. They take them away to their new assignment after having been certified 'safe to fly' by the FRS.

Below: The 'Bounty Hunters' exploited a brief window of opportunity to stage this formation of the outgoing VF-2 F-14D CAG-bird with the incoming VFA-2 F/A-18F CO's aircraft. Two-seat Super Hornets are replacing the F-14 in the long-range attack, fleet defence and reconnaissance roles.

Eagles' were officially reactivated on 1 October 1998 and stood up ceremonially on 15 January 1999, under the command of Commander Mark I. Fox; the FRS immediately began to establish the parameters for transitioning instructors, and later crews, to the new aircraft. To help smooth the overall transition, the Navy activated a Fleet Introduction Team at NAS Lemoore under the initial leadership of Commander Phil Tomkins, with its first order of business being the refurbishing of hangars and ready rooms necessary for the reactivation of VFA-122 in 1998.

VFA-122 – the Super Hornet FRS

VFA-122 still stands as the sole Super Hornet FRS, and is responsible for creating the Super Hornet training syllabus and for developing the tactics needed to employ the Super Hornet in accordance with its capabilities. The squadron began training future instructor pilots and WSOs in 1999 and received its first students in June 2000; new classes began every six weeks thereafter. With instructor pilot (IP), instructor WSO (I-WSO), and syllabus work completed, VFA-122 officially opened for business in June 2001, when four newly-winged pilots destined for VFA-115 arrived at VFA-122. The remainder of the squadron followed after completing its CVW-14 deployment in January 2001.

The Super Hornet FRS currently has 35 IPs and 10 I-WSOs. Approximately 485 enlisted maintenance personnel oversee the squadron's 37 jets and train future squadron maintenance personnel. Of VFA-122's allotment, 11 are single-seat Es and 26 are two-seat Fs. Eight of the latter are configured as twin-stick models to assist pilot training. To train new aircrews, VFA-122 uses its own aircraft plus those destined for new squadrons, which explains why many visitors to Lemoore see aircraft that appear to be squadron aircraft flown on 'Flying Eagle' sorties. These 'borrowed' aircraft are then released to new squadrons as soon as the unit is certified 'safe to fly'. So far, VFAs-115 (F/A-18E), -41 (F/A-18F), -14 (F/A-18E), -102 (F/A-18F), -2 (F/A-18F), and -137 (F/A-18E) have been certified and have joined the fleet. Former Tomcat squadrons VFAs-2, -14, -41 and -154 have permanently relocated to the Pacific Hornet base at NAS Lemoore; in November 2003, VFA-102 left Lemoore and replaced VF-154 in Atsugi, Japan.

VFA-2 'Bounty Hunters'

VF-2 (callsign BULLET) was established as the US Navy's second fleet Tomcat squadron in October 1972. During its long association with the F-14 it saw combat over Iraq (Desert Storm and Southern Watch). The last Tomcat cruise was in F-14Ds with CVW-2 aboard *Constellation* (CV 64). In early 2003 the unit transitioned to the F/A-18F – still assigned to CVW-2 ('NE' tailcode) but now for service aboard USS *Abraham Lincoln* (CVN 72), replacing CVW-14. The unit's aircraft retain the nose 'Langley stripe' – albeit in shades of grey on regular squadron machines – and a further representation is on the fin (including two stars to represent the squadron number) with a skull superimposed.

To minimise the effect on fleet operations, an effective plan was developed to transition squadrons to the -E/F between deployments as part of their normal work-up cycle. Using this system, as detailed below, full introduction of the -E/F can be accomplished within four years. Most instructional work is performed at NAS Lemoore, although weapons training detachments are made to Key West, Florida, and El Centro, California, as well as detachments to various fleet carriers as available for carrier qualifications. Most detachments consist of five or six aircraft and last about five days.

Like most other fleet readiness squadrons, VFA-122 operates a multi-tracked syllabus to train new Super Hornet crews. CAT 1 courses are tailored to students fresh from the training command and typically last about eight months. This syllabus has 10 phases, including familiarisation, formation flight, all-weather intercepts, section radar attack, basic fighter manoeuvres, fighter weapons, low-altitude tactical training, strike, strike fighter and carrier qualifications. The latter incorporates 10 day traps, six night traps, two day touch-and-goes and another two at night, as well as night-time refuelling missions. CAT 2 courses apply to former Tomcat and Hornet pilots. Some differences also exist for crews intended for the single and two-seat models. CAT 3 students are essentially Hornet pilots requalifying in the Super Hornet, and CAT 4 is for experienced F/A-18C pilots who are destined for the test pilot billet at either Patuxent River or China Lake.

Planned conversion schedule

Deliveries of new Super Hornets continue daily. Single-seat -E models largely replace older F/A-18C squadrons and two F-14 squadrons; two-place -Fs are replacing the remaining F-14 squadrons. As of mid-2003, sources indicate the following transition dates for 2003 through 2007:

Year	Squadron	Current aircraft	Super Hornet version
2003	VF-154	F-14A	F/A-18F
2004	VFA-22	F/A-18C	F/A-18E
	VF-154	F-14A	F/A-18F
2005	VFA-81	F/A-18C	F/A-18E
	VF-32	F-14B	F/A-18F
	VF-103	F-14B	F/A-18F
2006	VF-213	F-14D	F/A-18F
	VFA-86	F/A-18C	F/A-18E
	VF-211	F-14A	F/A-18F
2007	VF-11	F-14B	F/A-18E
	VF-143	F-14B	F/A-18F
	VF-31	F-14D	F/A-18F
	VFA-105	F/A-18C	F/A-18E
	VFA-146	F/A-18C	F/A-18E

Below left: VX-9 at China Lake continues to provide a vital bridge between the dedicated test units (VX-23 and VX-31) and the front line. Among its recent accomplishments was the evaluation of the JHMCS system at sea, confirming its suitability for use by embarked squadrons.

Below: On 5 September 2003 Boeing delivered the first Lot 26 Super Hornet with a redesigned forward fuselage. This has fewer parts, is cheaper to produce, and has a larger frame to allow it to take the APG-79 AESA radar.

Super Hornet armament

Stations and loadouts

The F/A-18E/F is equipped with 11 stations for the carriage of stores. On each wing is a wingtip Sidewinder missile launch rail and three underwing pylons. The outboard pylons can carry weapons in the 500-kg (1,100-lb) class, while the inboard and intermediate pylons can carry the larger stores, including fuel tanks. The fuselage has a centreline hardpoint for weapon, fuel tank or refuelling store carriage, either side of which is a hardpoint for the carriage of sensor pods or MRAAMS (AIM-7 or AIM-120). The number of hardpoints available allows a bewildering array of possible loadouts. By the use of twin racks on the two inner wing pylons the maximum number of weapons that can be carried is 14 (12 AIM-120s and two AIM-9s) for the air-to-air role, or 11 (Mk 82s or 83s, plus two AIM-9s) in the attack role.

When carrying asymmetric loads, the weights/moments on each wing have to be roughly equalled out. Here three Mk 83s are counterbalanced by a GBU-16 LGB and an AIM-120 AMRAAM.

Up to four GBU-31 JDAMs can be carried. A single AMRAAM is usually carried on the fuselage side pylon, but can be moved to the outer wing pylon if needed for balance purposes.

This 'five-wet' tanker (four 480-US gal/1817-litre tanks plus refuelling store) configuration leaves capacity for a baggage pod on the outer wing pylon and an AMRAAM for self-defence on the fuselage.

Air-to-air missiles

For the time being the AIM-9M is the standard wingtip missile, but it is being replaced by the Raytheon AIM-9X, an example of which is seen below on a VX-9 Super Hornet. The '9X' uses the motor and warhead from existing Sidewinders, but has a new imaging seeker and thrust-vectoring control. Although it has a small motor compared to other short-range air-to-air missiles, it has very low drag and its seeker allows engagements at beyond visual ranges.

The primary air-to-air weapon is the AIM-120C AMRAAM (above), which can be carried from any pylon apart from the centreline and wingtips. The four inner wing pylons can mount twin-rail launchers for AMRAAM. An air-to-air alternative to AIM-120 is the older AIM-7 Sparrow, of which up to eight can be carried. As the Super Hornet is tasked with fleet defence, it is a likely early candidate to receive any ramjet-powered AMRAAM follow-on, especially as the APG-79 AESA radar will greatly increase the Super Hornet's lethal range in the air-to-air role.

Air-to-surface

In the attack role the Super Hornet carries an array of unguided, near-precision and precision weapons. GPS-guided weapons such as the GBU-31 (2,000-lb/907-kg, of which four can be carried), GBU-32 (1,000-lb/454-kg, six) and AGM-154 JSOW (1,000-lb/454-kg, six) are widely used in guided attacks in all weathers. For greater precision laser-guided bombs are available from the Paveway II family, and the GBU-24 Paveway III is shortly to be cleared for service. Guided missile options include the AGM-65 Maverick (six) and the larger AGM-84H SLAM-ER stand-off weapon (four). Specialist weapons are the anti-ship AGM-84 Harpoon (four) and the AGM-88 HARM (six). Unguided munition options cover the range of Mk 80 series bombs and rocket pods.

The AGM-88 HARM (above and right) is used by the F/A-18E/F for attacking radars, although it is due to be replaced by the AARGM, which is designed to be better at attacking radars after they have shut down. For attacks against known radar site locations the AGM-154 JSOW (left) is often used, as this has proved far less likely to deviate from its target, with a consequent reduced risk of collateral damage.

Wings and fuel tanks
Based on the design of the original Hornet, the Super Hornet's wing is a multi-spar structure attaching to the fuselage by six main bolts. The interspar area forms an integral fuel tank, as does the interspar area of the twin fins. The remainder of the fuel is housed in a series of tanks in the upper fuselage. To the main wing structure is attached the trailing-edge flaps and leading-edge slats, and the outer folding panels. The latter also mounts control surfaces, and is attached to the main wing by a multiple ring/pin locking system. The actuating jacks are located in the inner panel near the trailing edge. The wingfold joint is covered by a porous fairing introduced to cure a wing-drop problem discovered during flight testing. This increases drag slightly, as does the toeing-out of the weapons pylons which was introduced to resolve stores separation problems.

Intakes
The intakes are of an RCS-re[duced design], although there is insufficient [space to] allow them to snake up and [over to a] great extent. Instead, there [is...] the engine. The intake sidew[alls...] boundary layer air is also eje[cted via] spillways in the upper side o[f the LEX] aft of the spoiler panels.

Radar
The first production Super Hornet lots are fitted with the Raytheon APG-73 radar, as installed in the later production F/A-18C/Ds. From 2007 the APG-79 AESA will become available as both a new-build and retrofit option. AESA will not, however, be available for aircraft built prior to Lot 26 as they do not have the redesigned forward fuselage necessary for AESA installation. They will retain APG-73 throughout their lives unless they undergo a structural modification.

Cockpit
The F/A-18E's pilot sits on a Martin-Baker SJU-17/A ejection seat, inherited from the later F/A-18C/D production models. As well as its state-of-the-art touch-screen displays, the cockpit is configured for the use of night-vision goggles and the JHMCS helmet-mounted sight and sensor-cueing system. Access is made via a ladder which extends from below the forward part of the LEX.

Loadout
This Super Ho[rnet...] comprising tw[o...] AGM-154 JSO[W...] AMRAAMs an[d the] internal M61 V[ulcan]

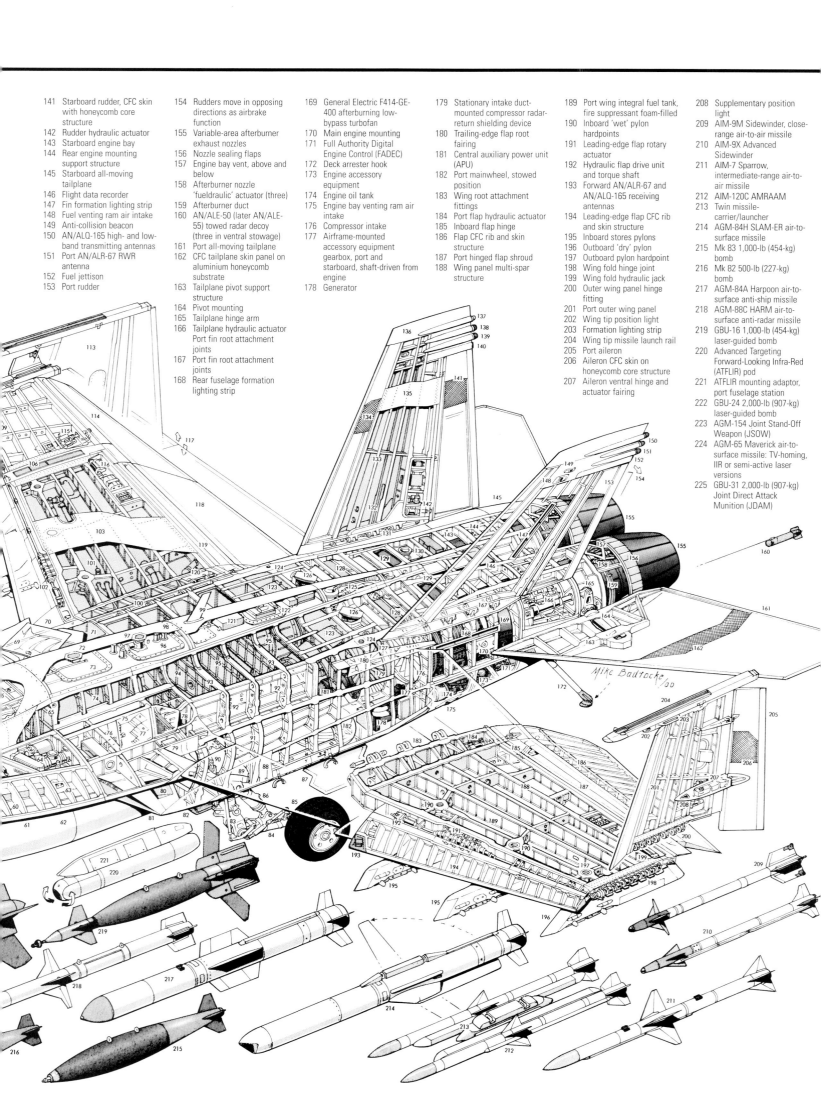

141 Starboard rudder, CFC skin with honeycomb core structure
142 Rudder hydraulic actuator
143 Starboard engine bay
144 Rear engine mounting support structure
145 Starboard all-moving tailplane
146 Flight data recorder
147 Fin formation lighting strip
148 Fuel venting ram air intake
149 Anti-collision beacon
150 AN/ALQ-165 high- and low-band transmitting antennas
151 Port AN/ALR-67 RWR antenna
152 Fuel jettison
153 Port rudder
154 Rudders move in opposing directions as airbrake function
155 Variable-area afterburner exhaust nozzles
156 Nozzle sealing flaps
157 Engine bay vent, above and below
158 Afterburner nozzle 'fueldraulic' actuator (three)
159 Afterburner duct
160 AN/ALE-50 (later AN/ALE-55) towed radar decoy (three in ventral stowage)
161 Port all-moving tailplane
162 CFC tailplane skin panel on aluminium honeycomb substrate
163 Tailplane pivot support structure
164 Pivot mounting
165 Tailplane hinge arm
166 Tailplane hydraulic actuator Port fin root attachment joints
167 Port fin root attachment joints
168 Rear fuselage formation lighting strip
169 General Electric F414-GE-400 afterburning low-bypass turbofan
170 Main engine mounting
171 Full Authority Digital Engine Control (FADEC)
172 Deck arrester hook
173 Engine accessory equipment
174 Engine oil tank
175 Engine bay venting ram air intake
176 Compressor intake
177 Airframe-mounted accessory equipment gearbox, port and starboard, shaft-driven from engine
178 Generator
179 Stationary intake duct-mounted compressor radar-return shielding device
180 Trailing-edge flap root fairing
181 Central auxiliary power unit (APU)
182 Port mainwheel, stowed position
183 Wing root attachment fittings
184 Port flap hydraulic actuator
185 Inboard flap hinge
186 Flap CFC rib and skin structure
187 Port hinged flap shroud
188 Wing panel multi-spar structure
189 Port wing integral fuel tank, fire suppressant foam-filled
190 Inboard 'wet' pylon hardpoints
191 Leading-edge flap rotary actuator
192 Hydraulic flap drive unit and torque shaft
193 Forward AN/ALR-67 and AN/ALQ-165 receiving antennas
194 Leading-edge flap CFC rib and skin structure
195 Inboard stores pylons
196 Outboard 'dry' pylon
197 Outboard pylon hardpoint
198 Wing fold hinge joint
199 Wing fold hydraulic jack
200 Outer wing panel hinge fitting
201 Port outer wing panel
202 Wing tip position light
203 Formation lighting strip
204 Wing tip missile launch rail
205 Port aileron
206 Aileron CFC skin on honeycomb core structure
207 Aileron ventral hinge and actuator fairing
208 Supplementary position light
209 AIM-9M Sidewinder, close-range air-to-air missile
210 AIM-9X Advanced Sidewinder
211 AIM-7 Sparrow, intermediate-range air-to-air missile
212 AIM-120C AMRAAM
213 Twin missile-carrier/launcher
214 AGM-84H SLAM-ER air-to-surface missile
215 Mk 83 1,000-lb (454-kg) bomb
216 Mk 82 500-lb (227-kg) bomb
217 AGM-84A Harpoon air-to-surface anti-ship missile
218 AGM-88C HARM air-to-surface anti-radar missile
219 GBU-16 1,000-lb (454-kg) laser-guided bomb
220 Advanced Targeting Forward-Looking Infra-Red (ATFLIR) pod
221 ATFLIR mounting adaptor, port fuselage station
222 GBU-24 2,000-lb (907-kg) laser-guided bomb
223 AGM-154 Joint Stand-Off Weapon (JSOW)
224 AGM-65 Maverick air-to-surface missile: TV-homing, IIR or semi-active laser versions
225 GBU-31 2,000-lb (907-kg) Joint Direct Attack Munition (JDAM)

Inside the F/A-18E/F

F414 engine

The F414 is a direct descendant of the F404 which powers the 'legacy' Hornet, offering considerable advances throughout the engine and more power. It has a larger fan for increased mass flow. The engine is a two-shaft powerplant built around a core which was similar to that developed for the F412 engine for the cancelled A-12, and features an afterburner section based on technology from the YF120 engine developed for the F-22/F-23. Upstream of the three-stage fan is a row of inlet struts and one of variable stators. The compressor has seven stages, while the low-pressure and high-pressure turbines each consist of a single stage.

Boeing F/A-18E cutaway

1. Composite radome
2. Radome open position for access
3. Raytheon AN/APG-73 multi-mode radar
4. Radome hinge
5. Scanner tracking mechanism
6. Radar mounting bulkhead
7. AN/APG-79 AESA radar for future integration
8. Low-band antenna
9. Radar equipment module
10. Electro-luminescent formation lighting strip
11. Cannon barrels
12. Cannon port and blast-diffuser vents
13. Flight refuelling probe, extended
14. Probe actuating link
15. Upper combined IFF interrogator antenna
16. M61A2 Vulcan 20-mm cannon
17. Cannon ammunition drum, 570 rounds
18. Incidence transmitter
19. Lower VHF/UHF/L-band antenna
20. Pitot head
21. Gun gas vents
22. Cockpit front pressure bulkhead
23. Nosewheel door
24. Ground power socket
25. Avionics ground cooling air fan and ducting
26. Rudder pedals
27. Instrument panel, full-colour multi-function CRT displays
28. Instrument panel shroud
29. Frameless windscreen panel
30. Head-up display (HUD)
31. Upward hingeing cockpit canopy
32. Martin-Baker NACES 'zero-zero' ejection seat
33. Starboard side console panel
34. Control column, digital fly-by-wire flight control system
35. Port side console with engine throttle levers, full HOTAS controls
36. Sloping seat-mounting bulkhead
37. Boarding step
38. Forward fuselage lateral equipment bays, three per side
39. Nosewheel leg pivot mounting
40. Landing light
41. Deck approach signal lights
42. Nosewheel steering unit
43. Catapult shuttle link
44. Twin nosewheels, forward retracting
45. Torque scissor links incorporating holdback fitting
46. Folding boarding ladder
47. Nosewheel retraction jack
48. AN/ALQ-165 EW transmitting antenna
49. Boarding ladder stowage
50. LEX equipment bay
51. Cockpit rear pressure bulkhead
52. Cockpit avionics equipment bay
53. Canopy rotary actuator
54. Starboard AN/ALQ-165 transmitting antenna
55. Canopy actuating strut
56. Canopy hinge point
57. No. 1 fuselage bag-type tank
58. Sloping bulkhead, structural provision for two-seat F/A-18F
59. EW receiver
60. LEX rib structure
61. Port leading-edge extension (LEX) chine member
62. 480-US gal (1817-litre) external fuel tank, centreline refuelling store as alternative
63. Port position light
64. Liquid cooling system equipment, reservoir, heat exchanger and ground running fan
65. Forward slinging point
66. Forward tank bay access panel
67. Starboard position light
68. Starboard LEX avionics equipment bay
69. Spoiler panel
70. LEX vent, operates in conjunction with leading-edge flap
71. Intake boundary layer spill duct
72. GPS antenna
73. No. 2 tank bay access panel
74. No. 2 bag-type fuel tank
75. Port spoiler
76. Spoiler hydraulic actuator
77. Boundary layer bleed air ducts
78. Bleed air spill duct
79. Port LEX vent
80. Perforated intake wall bleed air spill duct
81. Port fixed-geometry air intake
82. Mainwheel leg door
83. Main undercarriage leg strut
84. Trailing axle suspension
85. Port mainwheel
86. Shock absorber strut
87. Mainwheel door
88. LAU-116 missile carrier/launch unit
89. Mainwheel leg pivot mounting
90. Hydraulic retraction jack
91. Intake duct framing
92. Wing panel attachment joints
93. Machined titanium fuselage main bulkheads
94. No. 3 bag-type fuel tank
95. No. 4 bag-type fuel tank
96. No. 3 tank access panel
97. IFF antenna
98. Dorsal fairing access panels
99. Upper VHF/UHF/L-band antenna
100. Starboard wing panel bolted attachment joints
101. Starboard wing integral fuel tank
102. Leading-edge flap hydraulic drive unit and rotary actuator
103. Wing carbon-fibre composite (CFC) skin panelling
104. Starboard stores pylons, wing pylons canted 4° outboard
105. Leading-edge dogtooth
106. Wing-fold hinge fairing porous panel
107. Outboard leading-edge flap rotary actuator
108. Two-segment leading-edge flap
109. Outer wing panel dry bay
110. Wing tip position light
111. Formation light fairing
112. Wing tip missile installation
113. Starboard outer wing panel folded position
114. Drooping aileron
115. Aileron hydraulic actuator
116. Wing-fold hydraulic jack
117. Aileron and flap opposed movement as airbrake function
118. Starboard single-slotted trailing-edge flap
119. Hinged flap shroud
120. Flap hydraulic actuator
121. Dorsal equipment bay
122. No. 4 tank bay access panel
123. Ram air from intake duct for ECS
124. Rear fuselage slinging points
125. Environmental control system (ECS) equipment bay
126. ECS hinged auxiliary intake doors
127. Fuselage fuel vent tanks, port and starboard
128. Primary (starboard) and secondary (port) heat exchangers
129. Heat exchanger exhaust ducts
130. Engine pressure balance vent
131. Starboard fin bolted attachment joints
132. Fin integral vent tank
133. Multi-spar fin structure
134. Leading edge structure, CFC skin with honeycomb core
135. Fin CFC skin panelling
136. CFC fin tip fairing
137. Rear position light
138. Aft AN/ALQ-165 receiving antenna
139. AN/ALR-67 RWR antenna
140. Fuel jettison

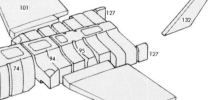

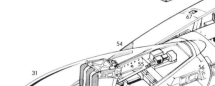

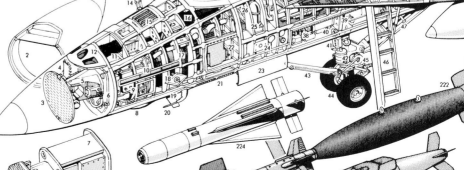

F/A-18E/F Super Hornet

AGM-88 HARM
HARM has been in service for 20 years, and remains the West's primary anti-radiation missile. The missile is 13 ft 8 in (4.17 m) long and weighs 796 lb (361 kg) at launch, of which 145 lb (66 kg) is a high-explosive blast/frag warhead. Range is up to 50 miles (80 km) for a high-altitude launch and the missile reaches Mach 2.9 during the fly-out. The latest versions feature much improved guidance systems to avoid potential fratricide and collateral damage, although the future of a GPS-aided version is uncertain.

Boeing F/A-18E Super Hornet
VFA-14 'Tophatters'
CVW-11, USS *Nimitz*
Operation Iraqi Freedom, 2003

Freshly converted to the Super Hornet from the Tomcat, VFA-14 embarked on *Nimitz* alongside VFA-41. With Operation Iraqi Freedom in full swing, the carrier headed for the war zone, arriving in April 2003 in time for Air Wing Eleven to see action during the last weeks of the military campaign. The carrier and its air wing remained on-station to provide air cover for US troops as they began the daunting task of bringing security to Iraq after the fall of Saddam Hussein.

ducing diamond shape,
room in the fuselage to
n to the engine face to any
s a radar blocker in front of
alls have a spill door, while
cted through V-shaped
f the wingroot fairing, just

AGM-154 JSOW
This weapon is a glide bomb guided by GPS/INS and fitted with pop-out wings to increase its stand-off range. The baseline AGM-154A version is a munitions dispenser, and carries 145 BLU-97 bomblets. It is 13 ft 5 in (4.1 m) long and weighs 1,067 lb (484 kg). As it nears the target the AGM-154A is placed automatically into a shallow dive over the target and, at the correct point, the payload covers are blown off and the sub-munitions ejected by an inflatable bladder. Each bomblet weighs 3.3 lb (1.5 kg), and contains a 10 oz (287 g) shaped-charge warhead. The bomblets have a retarding parachute and an extending nose tube. The latter is used to detonate the warhead at the optimum distance above the target.

LEX
The leading-edge extensions are multi-ribbed structures which attach to the fuselage. The forward tip of the LEX is strengthened to provide a boarding step. As well as mounting forward-hemisphere countermeasures antennas, the LEXs also provide capacity for some of the EW black boxes and ancillary equipment, including the liquid cooling system.

et carries a typical defence suppression loadout,
AGM-88 HARM anti-radar missiles and a pair of
s. The armament is completed by two AIM-120
two AIM-9 Sidewinders for self defence, plus the
Ican cannon.

Specifications – Super Hornet vs. Hornet

	F/A-18E	F/A-18C
Dimensions		
Wing span	44 ft 8½ in (13.62 m)	40 ft 5 in (12.32 m)
Wing span (folded)	30 ft 7¼ in (9.33 m)	27 ft 6 in (8.38 m)
Length	60 ft 3½ in (18.38 m)	56 ft 0 in (17.07 m)
Height	16 ft 0 in (4.88 m)	15 ft 3½ in (4.66 m)
Wing area	500 sq ft (46.45 m²)	400 sq ft (37.16 m²)
Weights		
Empty	30,500 lb (13835 kg)	29,619 lb (13435 kg)
Max. take-off	66,000 lb (29938 kg)	51,900 lb (23542 kg)
Carrier landing	42,900 lb (19459 kg)	33,000 lb (14969 kg)
Stores bringback	9,000 lb (4082 kg)	5,500 lb (2495 kg)
Internal fuel	14,460 lb (6559 kg)	10,860 lb (4926 kg)
External fuel	c. 16,290 lb (7390 kg)	c. 6,720 lb (3048 kg)
Performance		
Maximum speed	Mach 1.8	Mach 1.8
Approach speed	143.5 mph (231 km/h)	154 mph (248 km/h)
Service ceiling	50,000 ft (15240 m)	50,000 ft (15240 m)
Radius, attack	c. 760 miles (1220 km)	c. 540 miles (870 km)
CAP endurance at	1.8 hours	1 hour
230 miles (370 km) from carrier		
Powerplant		
Type	2 x F414-GE-400	2 x F404-GE-402
Afterburning thrust	22,000 lb (97.9 kN)	17,775 lb (79.1 kN)

Cockpit

Each cockpit (front, above left, and rear, above right) has four main screens. The larger screen is a Multi-Purpose Color Display (MPCD), while the two either side are monochrome. In the front cockpit the upper centre screen is the Active Matrix Liquid Crystal Display (AMLCD), which is touch-sensitive. AMLCDs have replaced the two side screens, while the MPCD is also due to be replaced in later production aircraft. A fifth screen in the front cockpit, below the left-hand display, shows engine and fuel information. Among the many displays which can be called up on the MPCD is a moving map (right). In this instance the aircraft (as represented by a simple symbol in the centre) is flying over Iraq, to the south of Baghdad. The front cockpit has a wide-angle head-up display (left) which displays flight and targeting data.

Undercarriage

The Super Hornet's undercarriage is similar in design to that of the F/A-18C/D, but is beefed up to cater for the higher weights. There is also more ground clearance to allow the aircraft to carry a large fuel tank on the centreline. The undercarriage retracts in the same way, swinging up and in so that stores can still be carried on the fuselage sides. When stowed the wheels lie flat beneath the intake trunking. The arrester hook fairing was identified as a key radar 'hot-spot', and has been treated with radar-absorbent material accordingly.

Defences

Rendering the Super Hornet more survivable in combat has been accomplished by the adoption of RCS reduction techniques and the use of advanced countermeasures. Although not a true 'stealth' aircraft, the Super Hornet has many RCS-reducing features, such as sawtooth edges on opening panels, application of RAM in key areas, and the adoption of radar 'blockers' in the intakes (left), which stop radar energy reaching the highly reflective fan face. The integrated electronic protection suite includes a towed radar decoy carried between the engines. In initial production aircraft this is the AN/ALE-50 (right), but this will be superseded in later aircraft by the AN/ALE-55. The AN/ALQ-214 system provides jamming, while a variety of mechanical countermeasures are available for decoying both IR- and radar-guided missiles away from the aircraft.

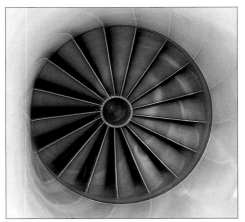
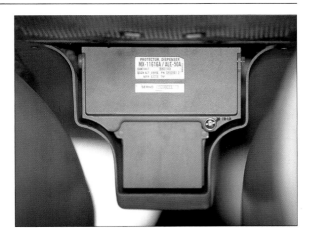

On 30 July 2003 the all-important trials with the APG-79 AESA radar began, undertaken on an F/A-18F assigned to VX-31 at China Lake. The radar is expected to be fitted to Block II production aircraft in 2006/7, and represents the most important systems upgrade for the Super Hornet. Among its many advantages, the APG-79 can operate virtually simultaneously in air-to-ground and air-to-air modes, while it offers longer range and greater resolution/target discrimination than the APG-73 currently fitted to the F/A-18E/F. AESA radar can only be fitted to aircraft with the new forward fuselage introduced at production Lot 26 in September 2003.

Much talk still surrounds the proposal to base a Super Hornet FRS on the East Coast. In 2002, the Navy conducted a feasibility study into locating the FRS at NAS Oceana, Virginia Beach. The two alternatives considered call either for basing six Super Hornet fleet squadrons and the FRS at Oceana, and locating four E/F squadrons at MCAS Cherry Point, North Carolina, or putting eight fleet squadrons and the FRS at Oceana (120 aircraft) and two squadrons at Cherry Point (24 aircraft). In September 2003, the Navy recommended the latter course and also the construction of a smaller outlying field at Washington County, North Carolina.

First blood

VFA-115 not only held the honour of the Super Hornet's first operational deployment, but also holds the honour of being the first Super Hornet squadron to see combat. VFA-115 initially flew missions over Afghanistan in support of Operation Enduring Freedom, but did not expend any ordnance. According to Lieutenant Stephen Walborn, a pilot with VFA-115, "The missions over Afghanistan were primarily of a 'show-the-flag' nature, but we were ready with ordnance to go where we were needed." Walborn said that the Super Hornets were typically armed with precision-guided munitions for these flights. "We tried to plan our flights around our bring-back capability, which is greater than that of the -C model, to ensure that we could recover what we left with." Walborn added, "[missions] essentially were planned around not dropping our ordnance." Many of the Afghanistan missions, Walborn noted, "were five to six hours long and involved tanking".

The 'Eagles' flew 214 combat missions during Operations Enduring Freedom and Southern Watch (the latter beginning in November 2002) and expended 22 JDAMs on a total of 14 targets during Southern Watch. VFA-115's Super Hornets also flew missions in support of Iraqi Freedom, beginning with sorties launched at the start of the war on 20 March 2003. Walborn said that the Iraqi Freedom missions were different from those flown during either Enduring Freedom or Southern Watch. "At this point we were planning to drop, so we didn't have to worry about bring-back."

Many of the Iraqi Freedom missions were flown with mixed loads of laser-guided precision bombs and GPS-guided JDAMs. Walborn: "The goal was to keep ourselves as flexible as possible. We had to be able to hit either a fixed or moving target, so we had to carry ordnance for both missions." Of course, the Super Hornets kept a basic armament for air-to-air contingencies, although none developed. The 'Eagles' flew 5,400 hours and made 2,463 arrested landings. Many of the sorties were made deep into Iraqi territory and certainly farther than its older C brethren flew. Although the Super Hornet squadron flew fewer overall combat missions than the C squadrons, it carried more ordnance, farther.

One of the most significant contributions by VFA-115 came in its tanking role. For these missions, the Es were launched in the so-called 'five wet' configuration, with four 480-US gal (1817-litre) tanks and one aerial refuelling store, plus a basic complement of air-to-air missiles. Pilots found the Super Hornet well suited to the tanking mission. "Except for tanking in bad weather, the Super Hornet handled the task well. We would essentially fly a slot position and hold while the aircraft manoeuvred onto the basket," Walborn explained. "One of the benefits of having the Super Hornet in the tanking role is its radar, something the S-3B people don't have. The radar helps with situational awareness and helps us better locate and guide our aircraft to us. It also helps us deconflict multiple aircraft that are heading in to refuel."

In a fully-configured load-out, the -E/F can give nearly twice the fuel of the S-3B. Equipped with four external tanks and a centreline A/A42R-1 aerial refuelling store, the Super Hornet can carry a total of 29,000 lb (13154 kg) of fuel. Lieutenant Walborn stated that typical operations saw a single Super Hornet "dragging a section in-country, tanking, then returning to the ship to refuel and meet the section en route back to the carrier". During the war most long-range tanking missions were flown by Super Hornets, while Vikings handled the task over the Gulf and overhead. "The Super Hornet let us send tanking assets into Iraq, which was something that we could not do with the S-3B. And that is primarily because of the range, and the Super Hornet's ability to self-protect." VFA-115 concluded its deployment on 1 May 2003, in a ceremony featuring President George W. Bush, which marked the end of major hostilities in Iraq. The squadron dropped more than 380,000 lb (172368 kg) of ordnance during Iraqi Freedom and received the Navy's Unit Commendation.

VFA-115's Super Hornets were a combination of early Lot models and were not equipped with MIDS, JHMCS or ATFLIR. Although the latter system was available in limited numbers – four LRIP ATFLIRs were sent to sea with CVW-14 – the squadron preferred to use the existing AAS-38B NiteHawk because of reliability problems in the early LRIP ATFLIR pods. According to one pilot, "It was essentially a trade-off between a vastly superior picture and reliability issues. We chose the degraded imagery as a price for ensuring that the instrument worked when we needed it." Walborn said that at one point four aircraft from the

Having performed sterling work in Iraqi Freedom, VFA-115 spent the summer of 2003 back at Lemoore. In late October/early November it was back at sea aboard USS John C. Stennis (CVN 74) for a COMPTUEX (Composite Training Unit Exercise) with the rest of CVW-14. In early 2004 the squadron will go to Fallon for a pre-cruise work-up, before deploying aboard Stennis. Air Wing Fourteen has been reassigned from Abraham Lincoln.

F/A-18E/F Super Hornet

Nimitz air wing (two each from VFA-14 and VFA-41) flew over to *Abraham Lincoln* in advance of the *Nimitz*'s arrival in the Persian Gulf to fly combat missions, and offered 'Eagles' crews a glimpse of their newer ATFLIRs and MIDS capabilities. "The ATFLIR images we were seeing were incredible compared to the NiteHawk and we were told that the reliability issues on the full-rate production models were minimal." VFA-115 is now fully equipped with new ATFLIRs, and Lieutenant Walborn indicated that the squadron's full-rate production pods are "wonderful" and are significantly improved in terms of reliability.

From a performance standpoint, Walborn praised the Super Hornet: "The aircraft is amazing and has performed as advertised." According to Walborn, the aircraft handles excellently around the boat, although he added that he initially found it challenging because the aircraft "can easily be overpowered on ball". Walborn believes that the new systems planned for the later Block aircraft, such as MIDS and the JHMCS, will greatly enhance the aircraft's overall situational awareness and lethality. "With JHMCS, one big advantage comes in the air-to-ground role. We waste a lot of time looking for targets or talking other aircraft onto a target. With JHMCS and MIDS, we can pass the information we have directly to other crews, resulting in a much faster weapons-on-target time." VFA-115 will deploy with MIDS and JHMCS in mid-2004.

EA-18G – replacement for the EA-6B

As early as the mid-1990s it had become apparent that a replacement would be needed for the EA-6B Prowler, which had been flying since the mid-1960s. Not only were Prowler airframes ageing, but even more hours were being accumulated as the aircraft assumed the jammer mission of the retired USAF EF-111 Raven. Moreover, operational demands from the numerous overseas commitments, such as Iraq's 'No-fly' Zones, Bosnia and Kosovo, began to take their toll; some Prowlers were showing fatigue stress on their wings, which limited their flight envelope.

While the Navy considered its options, Boeing (then McDonnell Douglas) began a government-funded six-month study of an electronic warfare version of the Super Hornet, referred to as the Command and Control Warfare (C^2W) variant. Following that initial study, Boeing continued concept development using its own funds. The initial concept called for incorporation of a single multi-band jamming pod to replace the Prowler's five ALQ-99 pods

VFA-137's CAG-bird rests at Lemoore during the squadron's conversion training period in 2003. It will deploy alongside VFA-2 in 2004. Current US Navy plans call for each air wing to have four Hornet squadrons: two with Cs, one with Es and one with Fs. In the next decade the F/A-18C is scheduled to be replaced by the Lockheed Martin F-35C Joint Strike Fighter.

VFA-137 'Kestrels'

Strike Fighter Squadron 137 was established as a new unit on 1 July 1985 to fly the F/A-18A Hornet, subsequently upgrading to the F/A-18C. Second of the F/A-18C squadrons to begin conversion to the Super Hornet, VFA-137 began its transition to the F/A-18E in 2003, and will partner VFA-2 in CVW-2 aboard *Abraham Lincoln*. Its badge comprises a stylised falcon's head superimposed on three aircraft pulling contrails.

Following VFA-137, the next two squadrons to convert will be VF-154 'Black Knights', which flew its F-14As back from Atsugi in September 2003 and stood up at Lemoore on 1 October, and VFA-22 'Fighting Redcocks' (F/A-18C).

Boeing fit-tested ALQ-99 pods on the first EMD F/A-18F at an early stage (below, with HARM and notional wingtip ALQ-218 pods), and in November 2001 flew the aircraft in an EA-18G configuration (above). Northrop Grumman is handling the integration of the EW suite, which is based on the EA-6B's ICAP-III system. As well as the ALQ-99 pods, a key component is the ALQ-218 receiver suite, while the MIDS/Link 16 will be enhanced with additional functions. While HARM is currently the baseline anti-radar weapon, the AARGM (Advanced Anti-Radiation Guided Missile) is being considered for delivery as part of the initial EA-18G system.

and relied on a crew of two, rather than four. The programme's $US2 billion cost quickly became an issue, largely due to the research and development associated with the new pod. As a result, the Navy asked Boeing to refocus its efforts on incorporation of the proposed ICAP-III technologies then planned for the Prowler. The resulting restructuring reduced overall programme costs by almost 60 per cent.

The initial C²W variant replaced the wingtip Sidewinders with wingtip multi-band receivers and added several low-band electronic surveillance antennas and SATCOM, and built on proposed F/A-18F technology such as MIDS and AESA. The aircraft later became known as the EA-18G 'Growler', a nickname that is not official; a competition is underway among the EA-6B community to select an official name for the EA-18G.

Following an extensive Advanced Electronic Attack Analysis of Alternatives that began in 2001, the Navy selected the EA-18G as the Prowler replacement in December 2002. Boeing serves as the prime contractor and Northrop Grumman acts as the principal subcontractor, responsible for integrating the electronic warfare suite. The EA-18G is a Block II F/A-18F, modified to accommodate the new features of the G, and retains all mission features of the -F. In fact, all -F models, beginning with Lot 30 aircraft, will be structurally provisioned to handle -G equipment, adding approximately 55 to 60 lb (25 to 27 kg) in structural changes. This production set-up will allow additional -F models to be converted to -G models if needed. However, the Navy has decided that once a Super Hornet has been converted to an EA-18G, it will not revert back to an F/A-18F configuration.

Boeing has conducted five flight tests since November 2001 using a modified -F (EMD F1) and hosted over 500 Prowler aircrews in its St Louis cockpit simulator. According to Paul Summers, Boeing's Program Manager for the EA-18G, the five flights "have systematically expanded the envelope with ALQ-99 pods, as well as tested for noise and vibration". Summers noted that the flight programme has achieved a maximum altitude of 30,020 ft (9150 m) and speeds of Mach 0.9. For these flights, an -F was configured with three instrument-loaded ALQ-99 pods and two 480-US gal (1817-litre) tanks. Summers says that the tests "produced very promising results and confirmed that we could do what we were saying with the aircraft". Boeing has also conducted extensive wind-tunnel tests and verified the Super Hornet's electromagnetic compatibility with fully-radiating ALQ-99 pods using anechoic chamber tests. Summers told *International Air Power Review*, "We tested the -F with fully radiating pods to make sure that the jamming did not interfere with the aircraft's electronics and flight systems, and the -F passed all tests."

'Growler' production

Production of 90 EA-18Gs is planned at a cost of approximately $US57 million per copy. This number is largely due to the fact that the Navy will be withdrawing from the Air Force EW mission beginning in 2010, and the fact that the Marines have yet to make their decision on a Prowler follow-on for their four electronic warfare squadrons. According to the Navy, the Marines should decide sometime in 2005-06 whether to pursue the EA-18G or an EW variant of the F-35. The Navy plans 10 squadrons of EA-18Gs, one for each of the carrier air wings and one FRS, although it has not yet decided how many EA-18Gs will outfit each squadron. The current number being considered is five per squadron. At this time, no EA-18Gs are planned for the Naval Air Reserve.

Boeing anticipates approval of funds late this year for entering the System Development and Demonstration phase in 2004 and for additional aircraft for the test programme. According to Paul Summers, four EMD E/F aircraft will be converted for air testing, plus two test production articles built as EA-18Gs – EA-1 and EA-2 – will join the test programme in FY08. Four additional production models will be added to the programme and will join the fleet following OPEVAL, which is scheduled for 2006; the EA-18G should achieve IOC in 2009. Low-rate initial production will commence in FY09 with 12 units, followed by 18 in FY10. Full-rate production will begin in FY11 with 22 aircraft, followed by 20 and 14 aircraft in succeeding years. These figures may be altered if the Marine Corps decides to adopt the EA-18G.

When production commences, the Block I EA-18G will be very similar to the -F, with MIDS, JHMCS and TAMMAC (Tactical Aircraft Moving Map Capability), and independent, missionised cockpits. The latter will enable independent pilot/ECMO sensor and weapons operations. F/A-18E/F/G front cockpits will be identical, and the aft cockpit will be common between the -F and -G with the exception that the -G will possess a master radiate switch. Early production models, although equipped to handle the AESA, will be fitted with the APG-73 radar and then retrofitted with the APG-79. AESA will bring even more capabilities to the -G and may be used to enhance the aircraft's long-range passive sensor and jamming capabilities. It will also allow integration with other onboard sensors and the ALQ-218 precision receiver system to enhance precision targeting.

The EA-18G will be 90 per cent common with -F models, retaining the full -F capability and its planned room for growth. Most of the changes will be related to software specific to the -G mission. The fully-equipped -G will weigh 1,300 lb (590 kg) more than the -F, but will not carry a gun. Boeing says that some consideration may eventually be given to a trade-off between -F capabilities and new growth room for the -G. The Navy is currently funding studies to evaluate both a single and dual multi-band pod system, which, if adopted, could free up an additional store for air-to-ground ordnance, making the EA-18G even more lethal.

One of the advantages of the EA-18G over the EA-6B is its ability to deliver a range of offensive weapons in addition to its EW and EA roles. Initial production EA-18Gs, referred to as Block I models, will carry HARMs to combat air-to-surface targets and AIM-120 AMRAAMs to thwart air threats. Follow-on store loadings will allow the Block II and III -Gs (currently not funded) to carry JSOW and JASSM, and may incorporate IDECM. Another benefit offered by the EA-18G is its ability to fly nearly identical profiles with the strike packages. Moreover, the commonality with other Super Hornet airframes in the carrier air wing should provide savings in maintenance costs.

Planned growth paths

A Block upgrade has been in the E/F roadmap from the programme's beginning, capitalising on the inherent growth space of the E/F. The Block I upgrade began in late 2001 and added new DMV-179 single-board mission computers to replace the AYK-14s, which had exhausted their memory, and the PMC-642 fibre cable network interface module. The new computers rely on COTS (commercial off-the-shelf) technology and provide significantly greater processing power and more memory, and use C ++ open architecture, thereby allowing easy upgrades as new technology evolves. Block I also saw the addition of a more advanced EW suite, additional weapons, advanced mission displays and JHMCS.

The second major upgrade, called Block II, is currently underway at Boeing, commencing with Lot 26 production. Block II replaces the two MFDs carried over from the -C/D models with an advanced display, and adds the advanced aft crew station with its larger, 8 x 10-in (20.3 x 25.4-cm) colour display and additional cockpit hand controllers. It also has provision for growth into AESA, and additional network-centric technology. Block II aircraft also incorporate MIDS technology and are equipped with ATFLIR. Using the added processing power of the new Block I computers, so-called 'smart' displays as used in the Block I models were replaced with 'dumb' displays that receive their data from the mission computers via a broadband high-speed databus.

Integration of the IDECM and decoupled cockpits is essential, as is placing priority on clearing all of the planned stores configurations. The decoupled cockpits, also called independent crew stations, allow F/A-18F crews to perform air-to-air and air-to-ground missions simultaneously. Scheduled for introduction in 2004, the decoupled system will enable crews independently to guide and control various weapons and onboard sensors. Modifications to the APG-73 will also follow (until AESA is delivered), incorporating RUG II SAR modes for generating highly-accurate ground maps. These modes are now only available on the F/A-18D used by the Marine Corps.

Block II provides the basis for further network-centric development (the referenced Block II+) and for the EA-18G. Boeing Chief Engineer Jim Young described the programme's goal as necking down to a few common platforms, thereby enhancing the ability to maintain the fleet at a realistic price. Super Hornets eventually will be fielded in three significant configurations: Block I, Block II and Block II-based EA-18G.

Brad Elward

With a smart new paint scheme, F/A-18F-1 conducted more EA-18G trials in 2002. Current Navy planning calls for 56 EA-18Gs to be purchased between FY06 and FY09 under the second multi-year procurement contract (MY2), and a further 34 under MY3. From Lot 30 all F/A-18Fs will be built as 'F-plus' aircraft, with the structural modifications necessary to turn them into EA-18Gs. This comes with a structural weight penalty of just 55 lb (25 kg). To cover a perceived shortfall in EW cover, it is possible that some Fs may be configured as interim EW platforms.

Top left: In 2002 F/A-18F-1 was given schemes representative of fleet EW squadrons. The green scheme (top) is of VAQ-209 'Star Warriors', while the red scheme is that of the Prowler FRS, VAQ-129 'Vikings'.

With a successful service entry and combat debut behind it, the Super Hornet has shaken off many of the criticisms which dogged the type's early years.

VARIANT FILE

US Military King Airs

The U-21 and C-12 family has served with all the branches of the US military except the Coast Guard, and while many of them were used in what became known as the operational support airlift role, others have been used as intelligence-gathering platforms for the US Army, creating some of the most unusual aircraft ever flown.

Founded in April 1932 in Wichita, Kansas, Beechcraft (more formally known as the Beech Aircraft Company) initially built a relatively small number of combat aircraft for the US military. However, it soon became one of the chief suppliers of trainers and utility aircraft to the services. Acquired by the Raytheon Company on 8 February 1980, Beech Aircraft was merged with Raytheon Corporate Jets in September 1994 and was renamed the Raytheon Aircraft Corporation.

U-21 Ute

The US Army took delivery of its first turbine-powered utility aircraft on 16 May 1967 when Beech Aircraft delivered the initial example of the King Air at its Wichita, Kansas, factory. Given the Beech model number 65-A90-1, the King Air was assigned the mission design series (MDS) designation U-21A. It was a hybrid aircraft that combined the fuselage of the earlier Beech Queen Air 65-80 with the wings, tail and undercarriage of the newer King Air 65-90, which first flew on 24 January 1964. The Army's initial contract with Beech covered the purchase of 88 U-21As at a cost of $US17.6 million. Prior to authorising the construction of the U-21A, however, the US Army Aviation Test Board at Fort Rucker, Alabama, had conducted a three-month evaluation of a modified model 65-80 Seminole that Beech had equipped with 500-shp (373-kW) United Aircraft of Canada (UAC) PT6A-6 turboprop engines under the designation NU-8F. In accordance with Army tradition it was named the Ute, honouring the Indian tribe from the Colorado/Utah area.

The U-21A's low wing had a span of 35 ft 6 in (10.82 m), an area of 280 sq ft (26.01 m²), and featured 7° of dihedral. Constructed from aluminium alloy using two spars, the wing was equipped with conventional ailerons, single-slotted trailing-edge flaps and pneumatic de-icing boots. The U-21A's semi-monocoque fuselage was built from lightweight aluminium alloy and was 35 ft 6 in (10.82 m) long. Its cabin provided seating for a two-person flight crew but, unlike the commercial King Air, the U-21A was not pressurised. In the utility role the Ute, which was equipped with a cargo door measuring 4 ft 5.5 in by 4 ft 3.5 in (1.36 by 1.31 m), was capable of carrying up to 10 combat troops, 3,000 lb (1361 kg) of cargo, three stretchers and three ambulatory patients or attendants, or six staff personnel. The tricycle landing gear featured single wheels on the nose and main landing gear struts, and was electrically operated. Intended for operations from prepared and semi-prepared runways, the pneumatic struts of the main landing gear retracted forward into semi-enclosed wells in the wing/engine nacelles, while the nose gear retracted aft into a fully-enclosed well.

The Ute was powered by a pair of 550-shp (410-kW) UAC PT6A-20 turboprops that drove constant-speed, three-bladed Hartzell propellers. Assigned the military designation T74-CP-700, the PT6A reverse-flow turboprop features a combination axial-centrifugal flow compressor comprising three axial and one centrifugal compressor stages. Two turbines, which operate on separate shafts, drive the accessory and propeller gearboxes. The engine

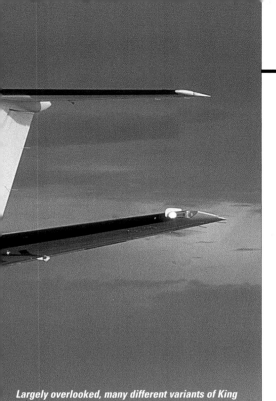

Largely overlooked, many different variants of King Airs have undertaken important roles in the US military for just short of four decades. Several different electronic warfare systems have been flown for the US Army (including the Left Foot in the RU-21E, left) while King Airs have served in the light transport or utility role with the Army, Air Force, Marines and Navy. This UC-12B is typical of the large fleet of Hurons flown in the operational support airlift role by the four services over the years.

Above: US Navy use of the U-21 was limited to a small number loaned by the Army to the NTPS at Patuxent River, Maryland, such as A-model 67-18096 '38'.

Below: Former JU-21A 67-18008 reverted to U-21A status after the Vietnam War, and served with the Army Reserve Command in the utility role.

is 5 ft 1 in (1.57 m) long, has a diameter of 1 ft 7 in (0.48 m) and weighs 289 lb (131 kg) dry.

Total production of the 65-A90-1 during 1966/67 was 141 aircraft, although the final 17 examples were equipped with a revised cockpit layout and designated U-21G. Nearly one-quarter of the fleet was eventually converted for a variety of electronic intelligence missions as EU/RU-21s. In addition, three new models were designed from the outset to operate as special electronic mission aircraft (SEMA).

VIP models

The first true King Airs to enter US Army service included five U-21Fs assigned to the Davison Aviation Command at Fort Belvoir, Virginia, in 1971. Developed from the King Air 100 series, the U-21F was an 'off-the-shelf' A100 version, which first flew on 20 March 1970. Its low wing, while similar to that of the model B90, featured fences on the upper portion of the outboard panels; it had a span of 45 ft 10.5 in (13.98 m), an area of 280 sq ft (26.01 m²), and 6° of dihedral. The wing was equipped with conventional ailerons, single-slotted trailing-edge flaps and pneumatic de-icing boots. It was constructed from light aluminium alloy using two spars, and the centre section had a constant chord whereas the outer panels were tapered. The horizontal and vertical stabilisers, which featured pneumatic de-icing boots, had also been redesigned and reached a length and height of 22 ft 4.6 in (6.82 m) and 15 ft 4.25 in (4.68 m), respectively. The pressurised fuselage was 39 ft 11.4 in (12.18 m) long and included a 4-ft 2-in (1.27-m) stretch. In addition, the U-21A's PT6A-20 engines were replaced by 680-shp (507-kW) PT6A-28 engines with four-bladed, reversible, constant-speed Hartzell propellers. Although capable of carrying up to 13 passengers, the U-21Fs were configured with VIP interiors and seating for eight.

Early SEMA models

Developed as a follow-on to the earlier RU-6 Beaver and RU-8 Seminole airborne radio direction-finding (ARDF) aircraft fielded in the early 1960s, the RU-21D was equipped with the AN/ARD-23 Laffing Eagle radio direction-finding set and was the first SEMA aircraft to be equipped with an inertial navigation system (INS). The RU-21D was initially deployed to Southeast Asia in 1968 and assigned to the 509th US Army Security Agency (USASA) Group at Davis Station, Tan Son Nhut Air Base, Saigon, Republic of Vietnam. Although a fleet of 34 RU-21Ds was planned, only 18 were ordered and the final 16 were delivered as RU-21Es and equipped with the AN/ARD-26 Left Foot ARDF.

Development of the AN/ARQ-28 Left Jab VHF signals intelligence (Sigint) ARDF system began in 1966 and was completed in 1970. Managed by the Army's Electronic Warfare Laboratory at Fort Monmouth, New Jersey, and designed by American Electronic Laboratories (AEL) Inc. on behalf of the USASA, the system was carried by three modified Utes given the designation JU-21A. Left Jab provided 360° direction-finding (DF) coverage and plotted the locations of hostile emitters by correlating data stored in an onboard computer with that provided by the aircraft's INS. The system was deployed to Southeast Asia in 1971 and assigned to the 138th Aviation Company (Radio Research) at Hue Phu Bai Air Base in the Republic of Vietnam. On 16 February 1973 JU-21A serial 67-18065 became the only U-21 lost to hostile file when it was shot down over North Vietnam, killing the five-man crew.

Although it never deployed to Vietnam – or anywhere, operationally – the AN/ARD-26 Left Foot radio direction-finding system was installed in 16 RU-21Es assigned the Beech model number 65-A90-4. Left Foot combined portions of the earlier Laffing Eagle system associated with the RU-21D and a new computer that automatically plotted target locations. The RU-21E retained the PT6A-20

A pair of JU-21As is seen in company with four of the earlier RU-21Ds. Both versions were used for radio direction-finding, and both were deployed to Vietnam. These two U-21 SEMA variants operated alone, while others were part of electronic warfare systems.

Variant File

The RU-21A, B and C (right) were all part of the Cefirm Leader direction-finding and electronic countermeasures system. The RU-21C formed the airborne jamming portion of Cefirm Leader. The series of Guardrail detecting and countermeasures systems have a long association with the King Air family, with early systems being carried by RU-21 Ute variants and later on RC-12 Hurons. The RU-21H (below) was part of Guardrail V (AN/USD-9), in service from 1978 until 1994.

Far right: U-21D 67-18110 was sold as N7154W to K & K Aviation in 1996, along with most of the U-21s.

Below: The first Super King Airs in US service were the three RU-21Js ordered in 1971. The aircraft were part of the Cefly Lancer Sigint system that, after a lengthy development programme, never entered operational service. The three aircraft were stripped of all their mission equipment following the end of the Cefly Lancer programme and used in the OSA role as C-12Ls.

(T74-CP-700) turboprops of the earlier U-21A, and its external configuration was similar to the RU-21D although with more external antennas.

McDonnell Douglas began development of the AN/ARD-22 Cefirm Leader direction-finding and countermeasures system in 1967. Also known as Crazy Dog, the Army's first co-ordinated multi-aircraft system was intended to perform three missions: signals intercept/direction-finding, master control and jamming. It encompassed nine aircraft and a large ground station, and missions were flown with two RU-21A DFs plus single RU-21B master control and RU-21C jamming aircraft.

The RU-21B and C models differed significantly from the RU-21A and were assigned the Beech model numbers 65-A90-2 and 65-A90-3, respectively. The aircraft were powered by the 620-shp (462-kW) PT6A-29 (military designation T74-CP-702), approved to operate at a take-off and landing weight of 10,900 lb (4944 kg), and were equipped with redesigned main landing gear struts. First used on Beech's Model 99 Airliner, the struts featured a dual-wheel configuration. The system underwent a lengthy development and gestation period and was fielded by the US Army Aviation Electronic Warfare Company at Fort Bliss/El Paso International Airport, Texas, in 1972. Problems with the DF portion of the system led to the eventual installation of more capable equipment, developed by Electromagnetic Systems Laboratories Inc. (ESL), on the RU-21As. The entire system was transferred to the US Army Reserve in 1977/78.

Guardrail

In July 1971, the National Security Agency issued a contract to ESL for the development of a remotely-operated airborne communications intelligence system in support of a so-called quick reaction capability. Unlike earlier Sigint systems, the Guardrail aircraft did not require onboard operators. The equipment was operated remotely by technicians in a ground-based integrated processing facility and data was transmitted between it and the aircraft (officially referred to as the airborne relay facility) via a secure datalink. Guardrail I's airborne sensors were installed in three modified US Army U-21Gs that deployed to Ramstein Air Base, West Germany, for the annual Reforger (Return of Forces to Germany) exercise in late 1971. Development of the system continued and a DF capability was added to the Guardrail II system, which was installed in six former RU-21E Left Jab aircraft that deployed to West Germany in July 1972. The subsequent incorporation of pre-planned product improvements resulted in the eventual operational deployment of the RU-21E Guardrail IIA with the 330th ASA Company at Ramstein in December 1972. A further equipment update that began in 1973 resulted in the Guardrail IV, which was also installed in modified RU-21Es. This system was deployed to the Republic of Korea in 1973 and assigned to the 146th ASA Company from 1974.

The sole USAF VC-6A, rests on the ramp at Andrews AFB, Maryland, during its service with the 89th MAW. The aircraft was used by President Lyndon Johnson to fly into his ranch in Texas while he was in office, before being used as a general hack until being retired in 1985.

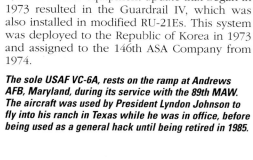

Right: C-12F2 86-60086 of OSACOM Det 42 is typical of the many Hurons used for operational support airlift (OSA) by the Army. Rationalisation of the US military fleet of OSA airframes saw the majority of the US Air Force's C-12C and Fs being turned over to the Army, where they replaced many of the earlier U-21s used in that role. Two different versions of the C-12F were ordered by the Army and they were differentiated by being designated as C-12F1s or F2s, while the US Air Force examples became C-12F3s.

Before the Huron was acquired for the USAF's OSA requirement, it was in service in the embassy support role, for which a total of 30 C-12As (upgraded as C-12Cs) and six C-12Ds was acquired. The majority still fly in that role, but the 412th TW at Edwards AFB, California, has a few on strength, including C-12C 76-1061 (below).

Development of a truly operational system began in 1976. However, even Guardrail V was intended only as an interim system that would bridge the gap until Cefly Lancer could be fielded. Known as the AN/USD-9 Special Purpose Detecting System, the Guardrail V equipment was installed in 21 U-21G and RU-21D/E/G airframes that were all subsequently given the common designation RU-21H. The new model was approved to operate at maximum take-off and landing weights of 10,200 lb (4627 kg) and 9,700 lb (4400 kg), respectively. The most significant identification feature associated with the RU-21H was the installation of wingtip-mounted pods that replaced many of the blade antennas installed on the earlier models. The RU-21H was initially fielded by newly-created aerial exploitation battalions in West Germany and Korea during 1978 and 1979.

Naval trainers

Seeking a replacement for its elderly fleet of TS-2A Tracker trainers, the US Navy established the VTAM(X) advanced multi-engine trainer aircraft programme. In 1976 the service announced that the Beech Aircraft Model H90 had been selected as the winner of the competition. Assigned the designation T-44A and named Pegasus, the aircraft flew for the first time in January 1977. Although similar in configuration to the U-21F, the T-44A was shorter at 35 ft 6 in (10.82 m) and featured a longer wing that spanned 50 ft 2.9 in (15.31 m). It was equipped with high-flotation landing gear as found on the earlier U-21s and was powered by two 550-shp (410-kW) PT6A-34B turboprops with three-bladed Hartzell propellers. The interior provided seating for an instructor and two student pilots, as well as seating for two passengers. The initial examples entered service with Training Squadron VT-31 at NAS Corpus Christi, Texas, on 4 April 1977. The Navy eventually acquired 61 T-44As, which provide multi-engine training for student naval aviators destined to fly the C-2, E-2, P-3 or C-130, as well as US Air Force C-130 students.

C-12 Huron

Initially referred to as the King Air A100-1, design of the Super King Air 200 began in October 1970 and the prototype first flew on 27 October 1972. The US Army ordered the first examples in 1971 when it purchased three under the designation RU-21J. In compliance with Army traditions, the aircraft was named Huron in honour of the Indian tribe that lived along the Great Lakes area of the northern United States. The RU-21Js were modified and equipped to test the AN/USQ-71 Cefly Lancer Sigint system, which was developed by AEL and intended as a replacement for Guardrail. At the conclusion of a lengthy test programme, the Sigint equipment was removed and the aircraft were reconfigured for the support role and given the designation C-12L.

Although initially assigned to the same MDS as the earlier King Airs, the RU-21J was a different aircraft, featuring a longer fuselage and redesigned wing and tail. Power was provided by two UAC PT6A-41 turboprops rated at 850 shp (634 kW), driving three-bladed Hartzell HC-B3TN-3/T10178 series propellers. Designed to operate from prepared runways, the Super King Air's semi-monocoque pressurised fuselage was constructed from lightweight aluminium alloy and measured 43 ft 10 in (13.26 m) in length. The cockpit provided dual controls and instruments for the pilot and co-pilot, who sat side-by-side. The main cabin was 16 ft 8 in (5.08) long (excluding the flight deck), with a maximum width and height of 4 ft 6 in (1.37 m) and 4 ft 9 in (1.45 m), respectively. The total volume was 393 cu ft (11.13 m^3), but a baggage compartment located aft of the boarding door added 53.5 cu ft (1.51 m^3) to the total volume. Like the earlier U-21 series, an integral airstair was installed on the port side of the fuselage aft of the wing. The Super King Air's low wing spanned 54 ft 6 in (16.61 m), and had an area of 303 sq ft (28.15 m^2) with 6° of dihedral. Equipped with conventional ailerons, single-slotted trailing-edge flaps and pneumatic de-icing boots, it was constructed from aluminium alloy and featured two spars. The centre section had a constant chord but the outer panels were tapered.

The Super King Air differed from the earlier A100 in being equipped with a swept, fixed-incidence 'T-type' horizontal stabiliser with conventional elevators. The swept vertical stabiliser had a conventional rudder and featured both dorsal and ventral fins. Pneumatic de-icing boots were installed on the horizontal stabiliser leading edge but not on the vertical stabiliser, as had been the case on the earlier U-21-series. The propellers and windshield were electrically de-iced and engine bleed air was used to de-ice the engine inlets. A retractable inertial separator installed in the engine inlet prevented the ingestion of ice, birds or other foreign objects.

The tricycle landing gear featured a single nosewheel and dual wheels were installed on the main landing gear struts. It was electrically operated and the main landing gear's pneumatic struts retracted forward into semi-enclosed wells in the wing/engine nacelles. The nose gear retracted aft into a fully-enclosed well. Bladder-type fuel cells located in the wings and engine nacelles were capable of carrying 549 US gal (2078 litres) or 3,733 lb (1693 kg) of fuel, of which just 5 US gal (19 litres) or 34 lb (15.4 kg) was unusable.

The Super King Air 200 was selected as the winner of the joint US Army/USAF UC-CX-X competition following a competitive fly-off, and 34 examples were ordered as the C-12A in August 1974 at a cost of $US20.6 million. The first examples were deployed by the Army at Fort Monroe, Virginia, in July 1975. It was intended to fulfil a range of missions including command and control, medical or casualty evacuation, liaison, administrative and embassy support. Seating for eight passengers was standard, but removing the seats in the main cabin allowed a variety of configurations.

US Navy use of the C-12 family was limited to the utility (OSA, as performed by the UC-12B, above right) and range support (RANSAC by the RC-12F, top, and RC-12M) roles, until the consolidation of US OSA assets resulted in surplus UC-12Bs becoming available for use as multi-engined trainers as TC-12Bs.

Above: The cockpit of the UC-12B dates back to the mid-1970s. Naval avionic upgrades have focussed on safety improvements, GATM compliance being the next step.

The US Army eventually took delivery of 60 C-12As and the US Air Force accepted 30 airframes. The latter were intended to support US embassies and military attachés worldwide. Although based on the model A200, the C-12A was equipped with 750-shp (559-kW) PT6A-38 engines and three-bladed Hartzell propellers. In 1978, the Army ordered 14 C-12Cs, which were powered by the uprated 850-shp (634-kW) PT6A-41 engines and equipped with a landing gear de-icing system. Surviving C-12As were redesignated C-12C after receiving the -41 engines and the brake de-icing system.

The small production batch of C-12Cs was followed by the first of 40 C-12Ds for the Army and six for the US Air Force. Based on the model A200CT, the C-12D differed from the C-12C in having high-flotation landing gear with larger, lower-pressure tyres that allowed the aircraft to operate from semi-prepared surfaces. It was also equipped with a brake de-icing system that utilised engine bleed air and had provision for wingtip-mounted 52-US gal (197-litre) auxiliary fuel tanks. An upward-hinging cargo door 4 ft 4 in by 4 ft 4 in (1.32 by 1.32 m) was installed in the port side of the fuselage aft of the wing. The aircraft retained the airstair, which became an integral portion of the cargo door. Of the 40 C-12Ds ordered by the Army between 1978 and 1984, 16 were modified as special electronic mission aircraft. The final six aircraft were equipped with 850-shp (634-kW) PT6A-42 engines, hydraulically-operated landing gear and other minor changes, and were referred to as C-12D2s; the earlier aircraft became C-12D1s.

In September 1983, the US Air Force selected both the Learjet 35 and the King Air B200C for use in the operational support airlift (OSA) role, as part of a five-year $US86.6 programme to replace the CT-39A Sabreliner. The first of 40 leased C-12Fs was delivered to the 375th Aeromedical Airlift Wing at Scott AFB, Illinois, in May 1984. Six additional C-12Fs were subsequently acquired for the Air National Guard later in 1984. Equipped with higher-powered PT6A-42 engines that drove four-bladed McCauley propellers, the C-12F also featured hydraulically-operated landing gear and retained the earlier C-12D's cargo door/airstair combination.

The Army purchased its first C-12Fs in 1985 when 11 examples were ordered. Unlike the USAF aircraft, these were based on the model A200CT and were powered by 850-shp (634-kW) PT6A-42 engines equipped with three-bladed Hartzell propellers. They were followed, beginning in 1986, by eight additional C-12Fs, based on the B200C but equipped with three-bladed McCauley propellers. The configuration differences between the USAF and US Army versions led to the use of the sub-designations C-12F1 and C-12F2 by the Army, whereas the Air Force examples were known as C-12F3.

Following a five-year gap, the Army ordered the first of 29 C-12Rs in 1992. Based on the B200C, the C-12R was similar to the US Air Force C-12F3 and was powered by the same -42 engine. It was, however, equipped with a different McCauley four-bladed propeller and its more advanced cockpit avionics included an electronic flight instrumentation system (EFIS), digital autopilot and GPS-equipped flight management system. The final C-12R was the last King Air built for any branch of the US armed forces.

Naval versions

The US Navy took delivery of 66 UC-12Bs between 1979 and 1982, the first example being delivered to NAS Patuxent River, Maryland, for testing on 15 September 1979. Based on the model A200C, the UC-12B featured a quick-change interior. It had the C-12D's cargo door, high-flotation landing gear, engines and propeller combination. The first operational UC-12B was assigned to VRC-30 at NAS North Island, California, in February 1980 and additional aircraft were subsequently assigned to Air Operations Departments at Naval and Marine Corps Air Stations.

The Navy ordered 12 examples of the model B200C in 1986, which were designated UC-12F. Similar in configuration to US Air Force C-12Fs, the aircraft were powered by the PT6A-42 but equipped with three-bladed Hartzell propellers. The final two UC-12Fs were modified for use as range surveillance aircraft (RANSAC) and equipped with the Litton AN/APS-140/504 sea search radar. The RC-12Fs, which are easily identified by the large antenna fairing carried on the centre of the lower fuselage, were assigned to the Pacific Missile Range Facility at Barking Sands, Kauai, Hawaii. Modified Fairchild Aircraft C-26D Metros previously operated by the US Air Force subsequently replaced them. The majority of the UC-12Fs were assigned to Air Operations Departments at Naval and Marine Corps Air Stations in Japan.

In 1987 the service ordered its final batch of King Airs, comprising 12 additional B200Cs. These aircraft differed slightly from the UC-12Fs and, accordingly, were designated UC-12Ms. The primary difference was the installation of McCauley three-bladed propellers, which were also used by the US Army C-12F2. The final two

Above: Few aircraft have carried such an extensive antenna array as the RC-12s of the different Guardrail systems. The RC-12K (this being 85-50149, Raytheon's testbed) is the airborne component of the Guardrail/Common Sensor System 4 (AN/URD-9D) and is seen carrying a larger number of aerials than fitted to the RC-12K when it entered service. The configuration seen here is similar to that of the RC-12N and P.

Right: The first SEMA C-12 was the RC-12D Improved Guardrail V, although it was preceded by the RU-21J version of the Super King Air. The RC-12D's advantages over the RU-21H included its higher ceiling, more internal space and an improved electronic payload.

UC-12Ms were also modified for the RANSAC mission and designated RC-12Ms. These aircraft were initially assigned to the Pacific Missile Test Center at NAS Point Mugu, California, in 1988 but were later transferred to Naval Station Roosevelt Roads, Puerto Rico. The majority of the UC-12Fs were assigned to duties in the Pacific, while the UC-12Ms were mostly assigned to facilities in Europe.

Airliners

Based on the Beech 1900 Airliner, which first flew on 3 September 1982, the C-12J was initially ordered by the US Air Force in March 1986, when six 1900Cs were purchased for the Air National Guard at a cost of $US20.8 million. The aircraft entered service in 1987 in the mission support role. Developed from the Super King Air 200, the 1900 differed in being stretched by 14 ft (4.3 m) and featured a cargo door measuring 4 ft 4 in by 4 ft 4 in (1.32 x 1.32 m). It retained the King Air's wing but was powered by a pair of 1,100-shp (820-kW) Pratt & Whitney Canada (PWC) PT6A-65B turboprops driving four-bladed Hartzell propellers. The aircraft is capable of carrying passengers or 5,300 lb (2404 kg) of cargo up to 1,570 miles (2527 km).

A single Beechcraft 1900D was delivered to the Army Material Command's Soldier and Biological Chemical Command at the Aberdeen Proving Grounds in Maryland during 1996 as part of a deal that saw 124 U-21s transferred to the civil register. Operating from Phillips AAF, it is primarily used to support the deployment of the command's Service Response Force, the Chemical Biological Rapid Response Team, and the Technical Escort Unit Response Team. It is also used to transport small quantities of chemical materiel in support of captive flight test programmes. The 1900D, which first flew on 1 March 1990, was 4 ft 8.5 in (1.43 m) longer than the earlier version and its cabin height was increased by 1 ft 2 in (0.36 m), although it retained the 19-passenger configuration. The aircraft was powered by two 1,279-shp (954-kW) PT6A-67D turboprops driving four-bladed Hartzell propellers. Its winglet-equipped wing was 3 ft 5.5 in (1.05 m) longer than its predecessor. Although it is a completely different aircraft, the 1900D was not assigned a new MDS designation and is referred to as a C-12J.

OSA fleet consolidation

Assigned to the 375th AAW, the US Air Force C-12Fs were flown by squadrons and detachments based at numerous US Air Force facilities within the continental US (ConUS) and in Germany and Japan. Training for the US Air Force crews was initially conducted by the 1375th Flying Training Squadron at Scott AFB, Illinois, but in July 1994 that responsibility was reassigned to the 45th Airlift Squadron at Keesler AFB, Mississippi. A reorganisation of the USAF's operational support airlift fleet resulted in the decision to transfer the C-12Fs to a training role in support of the Air Mobility Command tanker fleet. Although USAFE and PACAF retained their King Airs, the ConUS-based aircraft were transferred as companion trainers and assigned to air refuelling wings.

The mission, however, was short-lived, and on 1 October 1995 the US Air Force transferred the first of 44 C-12Fs and two C-12Js to the Army. The drawdown of the US Air Force C-12 fleet also resulted in the transfer of the training mission to the Army in 1995. Assigned to theatre aviation battalions, OSACOM regional flight centres and state flight detachments, and directly to major commands in Europe and Japan, the aircraft replaced U-21s and older C-12Cs. The US Air Force retained four C-12Js and pair of C-12Fs and continues to operate about 20 C-12C/Ds, which are based in foreign states and used in support of US embassy activities within their host nation. They are primarily tasked in support of the Defense Intelligence and Defense Security Cooperation Agencies. Although Air Force Materiel Command is responsible for the aircraft, aircrew training is now carried out by US Army, US Air Force and contractor personnel at Fort Rucker, Alabama, and nearby Dothan Regional Airport.

The influx of C-12Fs allowed the Army to retire its remaining U-21s, and in 1996 124 Utes were sold to K & K Aviation in Bridgewater, Virginia, for $US6.1 million. In November 1997, a single C-12L and 31 C-12Cs were transferred to law enforcement agencies nationwide, and K & K and its subsidiaries eventually purchased the remaining C-12Ls and several redundant C-12Cs as well. Today, fewer than 20 C-12Cs remain on the Army inventory, assigned to training and support duties. The last Utes were retired in 1999 when four U-21Fs assigned to the US Naval Test Pilot School at NAS Patuxent River, Maryland, were replaced by C-12Cs.

The same consolidation provided the Navy with enough surplus airframes to convert 20 redundant UC-12Bs as trainers. Now designated TC-12Bs, the aircraft augment the T-44A as multi-engined trainers and fill a void in capability created when the US Air Force shifted its turboprop training to the Navy in 1997. OSA mission scheduling for all ConUS-based aircraft is conducted by the Joint Operational Support Airlift Center at Scott AFB, Illinois. Training and maintenance missions are still scheduled by the local command.

Improved SEMA variants

Development of the Improved Guardrail V (IGRV) began in late 1981 when TRW, which had purchased ESL in 1976, received a contract for the first two, of four planned, systems.

Variant File

Right: The RC-12D Improved Guardrail V introduced a wide-band Interoperable Data Link (IDL) that required a pod to be hung under the port side of the nose and a fairing under the rear boom. This system allowed the transfer of data to members of different US services. The large blade aerial on the fuselage roof behind the TACAN antenna is a VHF/AM/FM command antenna. Some of the later RC-12 variants were windowless, while others, including the RC-12D, just has the windows blocked off. The IDL antenna on the port side of the RC-12N's nose (below) was balanced by a low-band dipole antenna on the starboard side. The RC-12N was part of the Guardrail/Common Sensor System 1 and featured a larger aerial array than the RC-12D. The aerials on the top of the fuselage consist of (front to back) a VHF-FM (SINCGARS) antenna, an upper low/mid-band vertical bent blade (with a similar example under the fuselage), a VHF/UHF communications antenna and a low band monopole. While the D had a rotating aft boom, the RC-12Ns had a pair of dipole antennas attached (below right). The P-band antenna (the black circle) is common to most of the Army's later RC-12s, while the three lumps around the rear window carry the (absent) CHAALS array.

Rather than the U-21 series, the airborne IGRV equipment was installed in the larger Beech C-12D Super King Air. The 1st and 2nd Military Intelligence Battalions (Aerial Exploitation) in West Germany fielded the two IGRV systems, which between them shared 13 RC-12Ds, during 1984 and 1985. Capable of operating at higher altitudes than the earlier RU-21Hs, the pressurised RC-12Ds featured a more reliable Delco Carousel IV-E inertial navigation system for more accurate target location. An advanced direction-finding system and a new interoperable datalink, which allowed more effective multi-service communications, were also incorporated. It retained the same powerplant and propellers as the earlier RU-21J and C-12D and was equipped with wingtip pods similar to those on the RU-21H. Rotating dipole antennas were mounted on a tail boom.

The RC-12D was equipped with defensive systems comprising the M-130 chaff/flare dispenser system, the AN/APR-39 radar signal detecting set and the AN/APR-44 radar warning set. Intended to defeat surface-to-air missiles, and first installed on the RU-21H, the M-130 system provided the aircraft with a single chaff dispenser in the aft portion of the right engine nacelle and a combination chaff/flare dispenser on the starboard aft fuselage side.

Two additional IGRV systems had been planned; however, the decision was made to upgrade them as Guardrail/Common Sensor (GR/CS) Systems 3 and 4, and in June 1984 TRW Avionics and Surveillance Group was awarded a contract for their development. Whereas IGRV was capable of intercepting communications signals, GR/CS added the ability to gather Elint including radar and non-communications signals. When AN/URD-9B GR/CS System 3 was fielded by the 3rd MIB(AE) in Korea in October 1988, the system was not equipped with the Advanced Quicklook Elint intercept classification and direction-finder capability or the Communications High Accuracy Airborne Location System (CHAALS), but both systems were retrofitted during 1996. System 3's airborne equipment was installed on six modified C-12D airframes, which were assigned the designation RC-12H because of the differences in capability between them and IGRV aircraft.

GR/CS System 4 was deployed to Germany with the 1st MI Bn (AE) in August 1991, replacing the Guardrail V system and its RC-12Ds. System 4 includes nine RC-12Ks, which are also based on the model A200CT. The RC-12Ks differ from the earlier aircraft in being equipped with structural modifications and rated for a take-off weight of 16,000 lb (7257 kg). RC-12Ks were powered by 1,100-shp (820-kW) PWC PT6A-67 turboprops and equipped with four-bladed McCauley propellers.

A contract for GR/CS Systems 1 and 2 was issued to TRW in August 1990 and the former was fielded to the 224th MI Bn (AE) at Hunter AAF, Savannah, Georgia, in August 1994. The 12 RC-12Ns, which are a component of System 1, replaced the earlier RU-21Hs that had equipped the battalion. Based on the C-12F, the RC-12N was the first King Air model to be equipped with an electronic flight instrumentation system and integrated aircraft survivability equipment. The aircraft retained the RC-12K powerplant and propeller but was rated for operations at higher weights than the earlier model. It was also the first Guardrail aircraft to incorporate a remote relay capability, which transmits data from the RC-12N to the IPL via a ground tethered satellite relay. That capability allows the aircraft to be forward deployed while the IPL is left at a location in the rear or even in the ConUS. The earlier Guardrail systems required that the IPL be forward deployed, near the RC-12, in order to receive data, which was transmitted via line-of-sight.

System 2, which includes 12 aircraft built to two different configurations, was fielded by the 15th MIB(AE) at Fort Hood, Texas, during 1999 and allowed the last IGRV system to be retired. The system includes nine RC-12Ps and three RC-12Qs. The RC-12P primarily operates in the intelligence-gathering role while the RC-12Q is designed to operate as a mothership. As such, it is equipped with a data correlator and a direct air-to-satellite relay (DASR) system. Rather than transmitting data to the IPL, the RC-12P sends information via datalink to the RC-12Q, which transmits the data to the IPL via line-of-sight or DASR. Both variants retained the capability to transmit via datalink, and the RC-12Q is also equipped to perform the intelli-

Above: All of the Army's RC-12 family have been painted in an overall grey scheme, with black antennas, and the RC-12Ns of the 224th MIB are no exception. Like all of the Guardrail series, the RC-12N acts as the airborne portion of the system. The RC-12N added the capability for long-range separation between the aircraft and the ground-based portion of the system.

Right: The latest RC-12 variant is the -12Q Direct Air-to-Satellite Relay, part of the Guardrail/Common Sensor System 2. Apart from the large bulge on its fuselage roof containing an AN/ARW-87 relay system, the RC-12Q is identical to the RC-12P, another component of System 2. The RC-12Q improves the Guardrail system's ability to disseminate information back to those that need to know.

gence-gathering mission. The RC-12Q is readily identified by the large radome on the top of the fuselage that houses the DASR antenna.

Crazyhorse

Beech/Raytheon modified 58 C-12s in support of SEMA programmes between 1976 and 1999. Guardrail was not, however, the only programme that utilised the C-12 airframe. Intended for service over Central America, the AN/URR-75 airborne intercept radio receiving set was developed by Sanders Associates, which is now a component of Lockheed Martin, for the National Security Agency. Known as Crazyhorse, the emitter locator system included a ground intercept facility and three modified C-12D aircraft. Designated RC-12G, the aircraft were similar in appearance and performance to the RC-12D IGRV. Unlike IGRV, however, the RC-12G could be flown as either a manned or unmanned platform and was equipped with two stations for onboard operators. Initially delivered to the US Army Echelon Above Corps Aviation Intelligence Company at Naval Air Engineering Center Lakehurst, New Jersey, in 1984, the system was deployed to Honduras in 1986 in support of US Army South and US Southern Command. In October 1989 the unit was reorganised as B Company, Military Intelligence Battalion (Low Intensity). The aircraft remained based at Soto Cano AB, Honduras, until transferred to Howard AFB, Panama, in 1993 and finally to Orlando International Airport, Florida, in May 1994. The entire system was eventually transferred to the US Army Reserve Command's 138th Military Intelligence Company in August 1994, with the RC-12Gs being retired in October 1998.

Upgrades

Forty C-12Fs assigned to OSACOM regional flight centres and state flight detachments were upgraded with the installation of the FDS-255 electronic flight instrumentation system, TCAS II traffic collision avoidance system and FMS-800 flight management system, making them functionally similar to the C-12R. Those C-12Fs modified with these systems were redesignated C-12Ts. The Army's 21 remaining C-12Fs are currently being upgraded with CNS (communication, navigation surveillance) equipment intended to make the aircraft compliant with global air traffic management (GATM) requirements. While similar to the modifications installed in the C-12T, these modifications also provide the aircraft with a terrain awareness warning system, cockpit voice (CVR) and flight data recorders (FDR), radios and navigation equipment. The GATM modifications will also be incorporated on the C-12T fleet and, as a result, the fleet will be updated to a single configuration known as C-12U. Already equipped with a 'glass' cockpit, the service's 29 C-12Rs will also receive a number of these systems and the modified aircraft will be known as C-12R1. The C-12U configuration received approval from the Federal Aviation Administration in early 2002 and the first examples were delivered to Fort Bliss, Texas.

The USAF is considering a similar modification for 27 of its King Airs but, unlike the Army, it intends to install the equipment on the earlier C-12C/D models along with the C-12F and J.

Modifications under way for the Navy's UC-12 fleet are part of its multi-phase Flight Safety Upgrade. Phase I provided the aircraft with GPS, CVR and FDR equipment and was completed in 1998. Phase II, which is currently under way, gives the aircraft a terrain awareness warning system, an enhanced ground proximity warning system, a new weather radar and a flight management system. These modifications have been incorporated on all RC/UC-12Ms, 16 UC-12Bs and a single UC-12F. The remaining UC-12Fs will be equipped with the Phase II FSU equipment by the end of 2005 and the remaining UC/TC-12B will follow by the end of 2007. A third phase, which is scheduled to commence in 2004, will provide all models except the TC-12Bs with the CNS equipment necessary to make the aircraft compliant with GATM requirements. Installation of this equipment will take from 2005 until 2010.

In addition to the avionics modifications, a number of the US Army aircraft have also been equipped with dual aft body strakes and ram air flow recovery systems, developed by Raisbeck Engineering. The strakes improve the aircraft's controllability and handling, climb and cruise performance, and provide a smoother and quieter ride. The system replaces the aircraft's original particle separator and ice protection system, and provides additional airflow to the engines. As a result, the engines are able to deliver their rated horsepower when operating in inclement weather. The system also allows the engines to operate at lower temperatures and so reduce fuel consumption.

Replacements

Although the US Army purchased its last C-12s in 1995, ongoing cockpit and structural upgrades will allow the transport versions to remain in service for several years to come. The Army, however, intends to develop a single aircraft and system to replace its mixed fleet of Elint/Sigint aircraft. There is currently only limited financial support for the Aerial Common Sensor programme, and the Guardrail aircraft will also be around for some time to come. Although the programme has been assigned a very low priority, the Navy has quietly begun looking for a replacement for its UC-12s. The King Air remains in production in Wichita for commercial and military customers, and the Model 350 is certainly considered a front-runner to replace the Navy's fleet.

Tom Kaminski

While only a single U-21 currently flies with the US military, the C-12 is still in widespread service. While the Army's Aerial Common Sensor (ACS) programme will replace the RC-12s with suitably equipped RC-20s (Gulfstream G450) or EMB-145s and the Navy has started looking for a UC-12 replacement, C-12 variants will remain in service for a long time to come.

US King Air Operators

The US Army is the largest military user of the Beech C-12/200 family. Its CECOM Flight Activity uses three RC-12Ds (including this example) for short-term test programmes.

UNITED STATES ARMY

The Army took delivery of its first SEMA-configured U-21 in 1968, and throughout its service during the Vietnam War the Ute proved to be an adaptable aircraft. At that time, the special electronic mission aircraft King Airs were assigned to a variety of units but most were attached to radio research aviation companies. After the war these aircraft were primarily assigned to the Army Security Agency Groups via ASA Aviation Companies. These ASA units eventually formed the basis for today's Military Intelligence Brigades, Battalions and Companies. The first RC-12s entered service in 1984. Through continued improvements to both aircraft and equipment, the King Air continues to serve as the Army's primary SEMA aircraft.

The US Army took delivery of its first utility variant of the U-21A Ute in 1967 and the initial C-12A followed in July 1975. Until 1992, when it consolidated its ConUS-based administrative/operational support aircraft and established the Operational Support Airlift Command, the Army's U-21/C-12 fleet was assigned to aviation companies, aviation detachments or installation flight detachments at facilities worldwide. Within the same timeframe, the aircraft based outside ConUS and those assigned to the US Army Reserve Command (USARC) flight detachments were similarly reorganised to form Theater Aviation Companies (TAC) and Battalions (TAB) and, eventually, all of the smaller detachments were eliminated. Although they were not relocated geographically, the aircraft assigned to the Army National Guard State Flight Detachments were attached to OSAC in 1994. Beginning in 1995, the US Air Force transferred most of its King Airs to the Army, allowing the latter service to retire all of its U-21s and C-12Cs from operational duties, leaving only the newer models in service. Today the Operational Support Airlift Command and theatre aviation units are responsible for the majority of the operational US Army's King Airs. A number of older examples continue to serve with test and training units.

SEMA assets

1st Military Intelligence Battalion (Aerial Exploitation)
Assigned to the V Corps at Campbell Barracks, Heidelberg, Germany, the 1st MIB(AE) is based at Wiesbaden AAF, Germany, and reports to the 205th Military Intelligence Brigade (Communications-Electronic Warfare and Intelligence) (MIBDE [CEWI]). The battalion's B Company – B/1st MIB(AE) – operated the RU-21H from 1984 to 1991, when it transitioned to the RC-12K. The company also operates a small number of RC-12D IGRVs.

3rd Military Intelligence Battalion (Aerial Exploitation)
Assigned to the US Army Intelligence & Security Command at Fort Belvoir, Virginia, Texas, the 3rd MIB(AE) is based at Desiderio AAF, Camp Humphreys, Seoul, Korea, and reports to the 501st MIBDE (CEWI). The battalion's B/3rd MIB(AE) operated the RU-21H from 1979 until 1988, when it transitioned to the RC-12H. The company is also assigned a small number of RC-12D IGRVs.

15th Military Intelligence Battalion (Aerial Exploitation)
Assigned to III Corps at Fort Hood, Texas, the 15th MIB(AE) is based at Robert Gray AAF on Fort Hood and reports to the 504th MIBDE (CEWI). Although B/15th MIB(AE) initially operated the JU-21A and RU-21D, it transitioned to the RU-21E in 1979 followed by the RU-21H in 1985. The unit flew the RU-21H in Operation Desert Storm. It transitioned to the RC-12D in 1991 and flew the IGRV version until it placed the RC-12P/Q in service during 1999.

204th Military Intelligence Battalion (Aerial Exploitation)
Assigned to the US Army Intelligence & Security Command at Fort Belvoir, Virginia, Texas, the 204th MIB(AE) is based at Biggs AAF on Fort Bliss, Texas, but reports to the 513th MIBDE (CEWI) at Fort Gordon, Texas. The battalion's D/204th MI Bn (ARL) operates a small number of C-12Us as support aircraft for the RC-7B Airborne Reconnaissance Low-Multimission (ARL-M) reconnaissance aircraft.

224th Military Intelligence Battalion (Aerial Exploitation)
Assigned to XVIII Airborne Corps, the 224th MIB(AE) is based at Hunter AAF, Savannah, Georgia, but reports to the 525th MIBDE (CEWI) at Fort Bragg, North Carolina. The battalion operated the RU-21H from 1981 to 1994 when it transitioned to the RC-12N.

The RC-12D IGRV has been replaced in the intelligence gathering role but is still used as a 'bounce trainer' and hack. 81-23542 served with the 15th MIB(AE) until replaced by the RC-12P/Q.

Training units

1-223rd Aviation Regiment (Training)
Although the training of C-12 flight crew is conducted by Flight Safety International under contract to the Army, the Headquarters/Headquarters Company of the 1-223rd AVN (Training) is responsible for the surveillance of contractor-operated training and conducts C-12 transition training for previously rated aviators. It is assigned a small number of C-12Ds and is based at Cairns AAF, Fort Rucker, Alabama. The battalion is assigned to the US Army Aviation Center as a component of the Aviation Training Brigade.

305th Military Intelligence Battalion (Training)
Assigned to the 111th Military Intelligence Brigade (Training) at Libby AAF, Fort Huachuca, Arizona, E/305th MIB (Training) is responsible for training all flight crew destined to fly the Guardrail variant and operates both the RC-12D IGRV and RC-12N GRCS models. The Brigade reports to the Training and Doctrine Command (TRADOC) at Fort Monroe, Virginia.

84-24376 was one of a handful of C-12Ds used by TRADOC to train C-12 crews.

Operational Support Airlift Agency

Operational Support Airlift Command
Headquartered at Davison Army Airfield, Fort Belvoir, Virginia, the Operational Support Airlift Agency (OSAA) is a field operating agency of the Department of the Army and reports to the National Guard Bureau. OSAA's Operational Support Airlift Command (OSACOM) provides high-priority, short-notice air transportation for passengers and cargo movement for the Army and other branches of the DoD. Operating a variety of fixed-wing types, OSACOM is the Army's largest single operator of the C-12 Huron and more than 50 examples are currently under its direct control. These aircraft are assigned to Regional Flight Centers (RFC) and State Flight Detachments (SFD). OSAA is also responsible for the Fixed Wing Army National Guard Aviation Training Site (FWAATS), which conducts training for all ARNG fixed-wing aviators.

C-12F 86-60085 is seen as flown by the New York SFD of the Army National Guard. The Latham-based unit has since had its C-12Fs upgraded to C-12T standard.

Unit	Location	Type
FWAATS	Benedum Airport, Bridgeport, West Virginia	C-12D/T
OSACOM – Davison AAF, Fort Belvoir, Virginia		
Fort Belvoir RFC	Davison AAF, Fort Belvoir, Virginia	C-12T
Fort Hood RFC	Robert Gray AAF, Fort Hood, Texas	C-12T
Fort Lewis RFC	Gray AAF, Fort Lewis, Washington	C-12T
Puerto Rico RFC	Isla Grande Airport, Puerto Rico	C-12T
Alaska RFC	Elmendorf AFB, Alaska	C-12F
Hawaii RFC	Hickam AFB, Hawaii	C-12D
Alabama SFD (OSACOM Det. 5)	Montgomery Regional Airport, Alabama	C-12T
Alaska SFD (OSACOM Det. 54)	Elmendorf AFB, Alaska	C-12T
Arizona SFD (OSACOM Det. 31)	Papago AAF, Phoenix, Arizona	C-12R
California SFD (OSACOM Det. 32)	Sacramento Mather Airport, California	C-12T
Connecticut SFD (OSACOM Det. 6)	Bradley IAP, Windsor Locks, Connecticut	C-12T
Delaware SFD (OSACOM Det. 7)	Newcastle County Airport, Delaware	C-12D
Florida SFD (OSACOM Det. 8)	St Augustine Airport, Florida	C-12T
Idaho SFD (OSACOM Det. 35)	Boise Air Terminal-Gowen Field, Idaho	C-12T
Illinois SFD (OSACOM Det. 36)	Decatur Airport, Illinois	C-12T
Indiana SFD (OSACOM Det. 10)	Indianapolis International Airport, Indiana	C-12T
Iowa SFD (OSACOM Det. 34)	Boone Municipal Airport, Des Moines, Iowa	C-12T
Kansas SFD (OSACOM Det. 37)	Forbes Field ANGB, Topeka, Kansas	C-12T
Kentucky SFD (OSACOM Det. 11)	Capital City Airport, Frankfort, Kentucky	C-12T
Louisiana SFD (OSACOM Det. 38)	Lakefront Airport, New Orleans, Louisiana	C-12T
Maine SFD (OSACOM Det. 14)	Bangor International Airport, Maine	C-12D
Maryland SFD (OSACOM Det. 13)	Phillips AHP, Aberdeen Pvg Gnd, Maryland	C-12T
Michigan SFD (OSACOM Det. 15)	Capital City Airport, Lansing, Michigan	C-12D
Minnesota SFD (OSACOM Det. 39)	St Paul Downtown Airport, Minnesota	C-12T
Mississippi SFD (OSACOM Det. 16)	Hawkins Field, Jackson, Mississippi	C-12R
Missouri SFD (OSACOM Det. 40)	Memorial Airport, Jefferson City, Missouri	C-12T
Montana SFD (OSACOM Det. 41)	Helena County Regional Airport, Montana	C-12R
Nebraska SFD (OSACOM Det. 43)	Lincoln Municipal Airport, Nebraska	C-12T
Nevada SFD (OSACOM Det. 45)	Reno-Stead Municipal Airport, Nevada	C-12T
New Hampshire SFD (OSACOM Det. 18)	Concord Municipal Airport, New Hampshire	C-12T
New Jersey SFD (OSACOM Det. 19)	Mercer County Airport, W. Trenton, New Jersey	C-12D
New Mexico SFD (OSACOM Det. 44)	Santa Fe County Municipal Airport, New Mexico	C-12R
New York SFD (OSACOM Det. 20)	Albany County Airport, Latham, New York	C-12T
North Dakota SFD (OSACOM Det. 42)	Bismarck Municipal Airport, North Dakota	C-12T
Oklahoma SFD (OSACOM Det. 46)	Westheimer Airport, Norman, Oklahoma	C-12T
Oregon SFD (OSACOM Det. 47)	McNary Field, Salem, Oregon	C-12T
Pennsylvania SFD (OSACOM Det. 22)	Muir AAF, Fort Indiantown Gap, Pennsylvania	C-12T
Puerto Rico SFD (OSACOM Det. 56)	Isla Grande Airport, San Juan, Puerto Rico	C-12T
Rhode Island SFD (OSACOM Det. 23)	Quonset State Airport, Rhode Island	C-12D
South Dakota SFD (OSACOM Det. 48)	Rapid City Regional Airport, South Dakota	C-12T
Tennessee SFD (OSACOM Det. 25)	Smyrna Airport, Tennessee	C-12T
Texas SFD (OSACOM Det. 49)	Austin-Bergstrom International Airport, Texas	C-12T
Utah SFD (OSACOM Det. 50)	Salt Lake City Municipal Airport, Utah	C-12T
Vermont SFD (OSACOM Det. 27)	Burlington International Airport, Vermont	C-12T
Virginia SFD (OSACOM Det. 26)	Richmond International Airport/Byrd Field, Virginia	C-12T
Washington SFD (OSACOM Det. 51)	Gray AAF, Fort Lewis, Washington	C-12R
West Virginia SFD (OSACOM Det. 28)	Wood County Airport, Parkersburg, West Virginia	C-12D
Wyoming SFD (OSACOM Det. 53)	Cheyenne Airport, Wyoming	C-12T

Theater Aviation

6-52nd Aviation Regiment (Theater Aviation)
The 6-52nd AVN (TAB) is headquartered at Los Alamitos AAF, California and is a component of the USARC. The battalion supports the Eighth US Army at Yongsang Barracks, Seoul, Republic of Korea, and is assigned to the 17th Theater Aviation Brigade. Its individual companies, which comprise both USARC and active component units, include three companies and detachments that operate the C-12F, C-12J and C-12R. During peacetime those units stationed within the ConUS support operational support missions.

Unit	Location	Type
A(-)/6-52nd AVN (TA)	Seoul K-16 AB, Sung Nam, ROK	C-12F
Det. 1 A/6-52nd AVN (TA)*	NAF Atsugi, Japan	C-12J
B(-)/6-52nd AVN (TA) (USARC)	ASF McCoy AAF, Fort McCoy, Wisc.	C-12R
Det. 1 B/6-52nd AVN (TA) (USARC)	ASF Godman AAF, Fort Knox, Ky	C-12R
C(-)/6-52nd AVN (TA) (USARC)	ASF Los Alamitos AAF, California	C-12R
Det. 1 C/6-52nd AVN (TA) (USARC)	Robert Gray AAF, Fort Hood, Texas	C-12F

*Det 1 A/6-52nd AVN supports the 78th Command Aviation Battalion (Provisional) at Camp Zama, Japan. The battalion is a component of US Army Japan.

1-214th Aviation Regiment (Theater Aviation)
Headquartered at Coleman Barracks, Mannheim, Germany, the 1-214th AVN (TAB) is assigned to the US Army Europe/Seventh US Army at Campbell Barracks, Heidelberg, Germany. The battalion's companies and their detachments are based in Germany and Italy. In addition to C-12Fs and a C-12J, the battalion is supported by USARC C-12Rs on temporary duty.

Unit	Location	Type
B/1-214th AVN (TAB)	Wiesbaden AAF, Germany	C-12F/R
E/1-214th AVN (TAB)	Stuttgart AAF, Echterdingen, Germany	C-12F/J
Det. 1 E/1-214th AVN (TAB)	Vicenza, AB, Italy	C-12F

2-228th Aviation Regiment (Theater Aviation)
A component of the USARC, the 2-228th AVN (TAB) is headquartered at Naval Air Station Joint Reserve Base Willow Grove, Pennsylvania, and assigned to the US Army Forces, Central Command/Third US Army at Fort McPherson, Georgia. The battalion's three companies operate both the C-12R and C-12D versions of the King Air in support of utility and OSA missions from six individual facilities.

Unit	Location	Type
A(-)/2-228th AVN (TA) (USARC)	NAS Willow Grove, JRB, Pennsylvania	C-12R
Det. 1 A/2-228th AVN (TA) USARC)	Johnstown-Cambria County Airport, Pennsylvania	C-12R
B(-)/2-228th AVN (TA) (USARC)	McCoy AAF, Fort McCoy, Wisconsin	C-12R
Det. 1 B/2-228th AVN (TA) (USARC)	Godman AAF, Fort Knox, Kentucky	C-12R
C/2-228th AVN (TA) (USARC)	Simmons AAF, Fort Bragg, North Carolina	C-12D
Det. 1 C/2-228th AVN (TA) (USARC)	Camp Doha, Kuwait	C-12D

Test, development and support units

US Army Aeromedical Center Air Ambulance Detachment
Known as Flatiron, the US Army Aeromedical Center Air Ambulance Detachment is based at Cairns AAF on Fort Rucker, Alabama. Flatiron operates a pair of C-12Ds, which are the only Army King Airs configured as air ambulances and are used to transfer patients and for operational support roles. The unit is a component of the US Army Medical Command at Fort Sam Houston, Texas, and had previously operated a U-21G.

US Army Communications-Electronics Command Flight Activity
Assigned to the US Army Communications-Electronics Command at Fort Monmouth, New Jersey, the CECOM Flight Activity operates a single C-12C and three RC-12Ds from Naval Air Engineering Station Lakehurst, New Jersey, as part of the US Army Materiel Command, Alexandria, Virginia. The aircraft are primarily used in support of short-term development projects. The C-12C has been used numerous times as a UAV surrogate. The unit was previously known as the Airborne Engineering Evaluation Support Branch.

US Army White Sands Missile Range
Units assigned to the WSMR, which is a component of the Test and Evaluation Command at Aberdeen Proving Grounds, Maryland, include the WSMR Air Operations Detachment and the Electronic Proving Grounds Aviation Detachment. These units operate, respectively, single examples of the C-12D and C-12C from Holloman AFB, New Mexico, and Libby AAF, Fort Huachuca, Arizona.

US Army Soldier Biological and Chemical Command Aviation Det
Based at Phillips AAF, Aberdeen Proving Grounds, Maryland, the SBCCOM AVN DET operates a single Beech 1900D airliner in support of the US Army Soldier and Biological Chemical Command as part of the US Army Materiel Command, Alexandria, Virginia.

US Army Aviation Technical Test Center
A component of the Test and Evaluation Command at Aberdeen Proving Grounds, Maryland, the Aviation Technical Test Center is based at Cairns AAF, Fort Rucker, Alabama. It operates examples of the RC-12D and JRC-12G in support of flight test programmes and had previously operated examples of the JU-21H and JC-12D.

C-12J 86-0079 is used by the Headquarters US European Command as a utility transport aircraft. It is one of two of the type transferred from the Air Force.

US King Air Operators

SEMA Units in Southeast Asia

224th AVN BN (Radio Research) (509th USASA Group)
Activated on 1 June 1966 and headquartered at Davis Station, Tan Son Nhut Air Base, Saigon, Republic of Vietnam, the 224th was assigned to the 509th US Army Security Agency Group (which was also known as the 509th Radio Research Group). The 224th was responsible for the entire U-21 special electronic mission aircraft force deployed in Southeast Asia. The battalion relocated to Long Thanh North in January 1970 but returned to Tan Son Nhut in August 1972. The 224th was deactivated in Oakland, California, on 3 March 1973. Its assigned units included:

Unit	Base	Aircraft
138th AVN CO (RR)	Da Nang AB/Phu Bai AB, RVN	JU-21A, RU-21D
144th AVN CO (RR)	Nha Trang AB, RVN	RU-21D
146th AVN CO (RR)	Tan Son Nhut AB/ Long Thanh North AB/ Can Tho AB, RVN	RU-21D

7th Radio Research Field Station
As activities in Vietnam wound down, the remaining SEMA assets were transferred to the 7th RRFS Flight Detachment at Udorn Royal Thai Air Force Base (RTAFB) in early 1973. The unit later relocated to U'Tapao RTAFB in July 1974 but was finally deactivated in May 1975. The detachment operated the JU-21A and RU-21D.

The three Left Jab JU-21A aircraft were based at Da Nang AB and later Phu Bai AB with the 138th AVN Co (RR) during the Vietnam conflict.

UNITED STATES AIR FORCE

The US Air Force's first C-12Cs were assigned to support US State Department embassies worldwide and the King Air continues to fulfil that mission. In 1984 the aircraft entered service in the operational support airlift role with the 375th Aeromedical Airlift Wing at Scott AFB, Illinois. The aircraft eventually saw service in the OSA role in Europe, the Pacific and within the ConUS. By the early 1990s the US Air Force had reassigned these aircraft as companion trainers in support of its aerial refuelling units. However, in 1995 the service transferred the majority of its aircraft to the US Army. Although the C-12F/J has been retired from its OSA duties within the ConUS, it continues to fulfil that mission in support of the Pacific Air Forces. C-12Js also support test operations in support of the Air Force Materiel Command.

3rd Wing
The 3rd Wing is headquartered at Elmendorf AFB, Alaska, and reports to the Pacific Air Forces at Hickam AFB, Hawaii, via the Eleventh Air Force, which is co-located with the wing in Fairbanks. A detachment of the Wing's 517th Airlift Squadron, equipped with the C-130H Hercules, flies examples of both the C-12F and C-12J in support of PACAF OSA requirements.

46th Test Group
Based at Holloman AFB, New Mexico, the 46th TESTG is assigned to the Air Armament Center, via the 46th TW Eglin at Eglin AFB, Florida. The AAC is a component of the Air Force Materiel Command at Wright-Patterson AFB, Ohio. The 46th TESTG's 546th FLTS operates a single C-12J in support of test projects.

51st Fighter Wing
Based at Osan Air Base, Republic of Korea, the 51st FW is a component of the Pacific Air Forces at Hickam AFB, Hawaii, and reports to the Seventh Air Force, which is also located at Osan. The Wing's 55th Airlift Flight operates a single C-12J in support of OSA duties.

412th Test Wing
Located at Edwards AFB, California, the 412th TW is assigned to the Air Force Flight Test Center, which is a component of the Air Force Materiel Command at Wright-Patterson AFB, Ohio. Assigned to the wing are a small number of C-12Cs operated by the 418th Flight Test Squadron. Besides supporting test projects, the aircraft are utilised by the USAF Test Pilot School, which is also a component of the AFFTC. The aircraft are also used by NASA's Dryden Flight Research Center, which is responsible for their maintenance and upkeep.

Air Mobility Command Air Operations Squadron (AMCAOS)
Although Flight Safety International conducts all flight instruction for US Air Force C-12C/D pilots at its Dothan, Alabama, facility, Detachment 5, AMCAOS oversees this training, stationed at Lowe Army Heliport on Fort Rucker, Alabama. The detachment is responsible for the curriculum-associated USAF C-12 initial qualification and initial instructor qualification courses. The unit's four evaluator pilots monitor contractor training and student performance, and conduct all end-of-course flight evaluations for C-12 pilots. The USAF Instrument Refresher Course is taught by Det. 5 instructors as part of the USAF Initial Qualification Course. More than 60 students are trained annually.

Air Force Security Assistance Center
Located at Wright-Patterson AFB, Ohio, AFSAC is responsible for the majority of the US Air Force's C-12C/D Hurons. These aircraft are stationed overseas in support of US embassies, the Defense Intelligence Agency and the Defense Security Cooperation Agency.

Location	Type	Location	Type
Buenos Aires, Argentina	C-12D	Abidjan, Ivory Coast	C-12C
Canberra, Australia	C-12C	Nairobi, Kenya	C-12D
La Paz, Bolivia	C-12D	Islamabad, Pakistan	C-12C
Bogotá, Colombia	C-12C	Manila, Philippines	C-12C
Cairo, Egypt	C-12C	Dhahran, Saudi Arabia	C-12C
Tegucigalpa, Honduras	C-12C	Riyadh, Saudi Arabia	C-12C
Budapest, Hungary	C-12D	Bangkok, Thailand	C-12C
Jakarta, Indonesia	C-12C	Ankara, Turkey	C-12C

Left: The 55th Airlift Flight of the 51st Fighter Wing based at Osan AB, Republic of Korea, operates a single C-12J. The aircraft is one of four that are still used by the Air Force.

Above: C-12A 73-1216 wears the smart scheme applied to the majority of the aircraft involved in the embassy support role. This aircraft was upgraded as a C-12C and assigned to Ankara, Turkey.

Companion Trainer Program

After removing the C-12F from service as an operational support aircraft in the early 1990s, the US Air Force assigned the aircraft to new duties as companion trainers. Under this programme the aircraft were used to provide additional flight hours to co-pilots rated on the KC-10 Extender and KC-135 Stratotanker. The C-12Fs were assigned to a total of seven Air Refueling Wings and a single Air Mobility Wing. The programme was terminated in the mid-1990s.

Unit	Base
19th ARW	Robins AFB, Georgia
22nd ARW/384th ARS	McConnell AFB, Kansas
43rd ARW/91st ARS	Malmstrom AFB, Montana
92nd ARW/43rd ARS	Fairchild AFB, Washington
305th AMW/32nd ARS	McGuire AFB, New Jersey
319th ARW/905th ARS	Grand Forks AFB, South Dakota
380th ARW	Plattsburgh AFB, New York
722nd ARW/9th ARS	March AFB, California

C-12F3 84-0161 served with the 380th ARW during the time of the Companion Trainer Program. All but two of the USAF's C-12Fs, were transferred to the Army from October 1995.

UC-12B BuNo. 161201 was based with the AOD of NAS Cubi Point in the Philippines. While Cubi Point is no more, the Huron continues to be flown by the majority of the AODs at the US Navy's air stations and facilities.

UNITED STATES NAVY

The majority of the US Navy's King Airs are operated by Air Operations Departments that are assigned to Naval Stations, Naval Air Stations or Naval Air Facilities. Mission scheduling for those UC-12s within the ConUS is conducted by the Joint Operational Support Aviation Command at Scott AFB, Illinois. The units outside ConUS have their missions scheduled by the Naval Aviation Logistics Office at NAS Joint Reserve Base New Orleans, Louisiana. The majority of the navy UC-12s remain in service, some as TC-12B trainers.

Commander US Naval Forces Europe
The Air Operations Departments (AOD) at Naval Air Facility Mildenhall, England, and Naval Station Rota, Spain, both operate the UC-12M in support of COMUSNAVEUR, which is headquartered in London, England. The former unit is assigned directly to COMUSNAVEUR while the latter reports to Commander Fleet Air Mediterranean at NSA Naples, Italy.

Commander US Naval Forces Central
Located at Naval Support Activity Manama, Bahrain, COMUSNAVFORCENT's King Airs operate throughout the US Central Command area of responsibility. The UC-12Ms are assigned to Flight Support Detachment and are based at Bahrain International Airport.

Commander US Naval Forces Southern Command
Based at NS Roosevelt Roads, Puerto Rico, COMUSNAVSO is responsible for that facility as well as NS Guantanamo Bay, Cuba. The AOD at the former operates both the UC-12M and RC-12M models, while that of the latter is assigned a UC-12B.

Commander Naval Air Force Reserve
Headquartered at Naval Support Activity New Orleans, Louisiana, COMNAVFORRES is responsible for five AODs, each operating the UC-12B. The units are located at NAS Atlanta, Georgia; NAS JRB Fort Worth, Texas; NAF Washington, Maryland; NAS JRB Willow Grove, Pennsylvania; and NAS JRB New Orleans, Louisiana.

Commander US Naval Pacific Fleet
COMPACFLT is located at Naval Station Pearl Harbor, Hawaii, and is responsible for AODs within the ConUS and in Japan. Those units at NAS Whidbey Island, Washington, NAS Fallon, Nevada, and NAF El Centro, California, operate the UC-12B. NAS North Island's AOD operates the UC-12B, UC-12F and the sole NC-12B. Reporting to COMPACFLT via the Commander Airborne Early Warning Wing Pacific at Naval Base Ventura County/Point Mugu, California, VRC-30 is also based at NAS North Island and serves as a fleet readiness squadron for the UC-12B and UC-12F, and operates a pair of former RC-12F range support aircraft. Three AODs are also located within Japan and report to COMPACFLT via Commander Fleet Air Western Pacific at NAF Atsugi, Japan. The units are attached to NAF Atsugi, NAF Misawa and NAF Kadena, and each operates the UC-12F.

Commander US Naval Atlantic Fleet
Headquartered at NS Norfolk, Virginia, COMLANTFLT is responsible for AODs at five facilities, only three of which operate fixed-wing aircraft. The AODs at NAS Jacksonville and NAS Key West, Florida both operate the UC-12B. Located at NS Norfolk, Virginia, Chambers Field operates as a satellite of NAS Oceana and its AOD operates both the UC-12B and UC-12M, and serves as a fleet readiness squadron for both types.

Current US Navy operators

Code	Unit	Location	Aircraft
7A	AOD	NAS Patuxent River, Maryland	UC-12B
7B	AOD	NAS Atlanta, Georgia	UC-12B
7C	AOD	Chambers Field, Naval Station Norfolk, Virginia	UC-12B/M
7D	AOD	NAS Fort Worth JRB, Texas	UC-12B
7E	AOD	NAS Jacksonville, Florida	UC-12B
7G	AOD	NAS Whidbey Island, Washington	UC-12B
7H	AOD	NAS Fallon, Nevada	UC-12B
7M	AOD	NAS North Island, California	UC-12B/F, NC-12B
7N	AOD	NAF Washington, Maryland	UC-12B
7Q	AOD	NAS Key West, Florida	UC-12B
7W	AOD	NAS Willow Grove JRB, Pennsylvania	UC-12B
7X	AOD	NAS New Orleans JRB, Louisiana	UC-12B
(8A)	AOD	NAF Atsugi, Japan	UC-12F
(8D)	AOD	NS Rota, Spain	UC-12M
8E	AOD	NS Roosevelt Roads, Puerto Rico	UC-12M, RC-12M
8F	AOD	NAS Guantanamo, Cuba	UC-12B
(8G)	AOD	NAF Mildenhall, UK	UC-12M
8H	AOD	NAF Kadena, Okinawa	UC-12F
(8K)	FSD	NSA Bahrain IAP, Bahrain	UC-12M
8M	AOD	NAF Misawa, Japan	UC-12F
8N	AOD	NAF El Centro, California	UC-12B
G	VT-31	NAS Corpus Christi, Texas	T-44A
G	VT-35	NAS Corpus Christi, Texas	TC-12B, UC-12B
RW	VRC-30	NAS North Island	UC-12B/F, RC-12F
n/a	USNTPS	NAS Patuxent River, Maryland	C-12C

Note: Codes shown in parentheses are assigned but are not normally carried on the aircraft. USNTPS aircraft carry individual two digit number codes.

The T-44A Pegasus replaced the final Trackers in US naval squadron service. Initially based at NAS Pensacola, Florida, the T-44 fleet is now at NAS Corpus Christi, Texas.

US King Air Operators

Above: US Army aircraft and helicopters are loaned to the USNTPS to increase the variety of types flown by the students. This C-12C is the latest member of the King Air family to serve with the unit. U-21As and U-21Fs have previously been used.

Above right: The TC-12B is a re-roled UC-12B used for the same task as the T-44A – that of training student aviators in the multi-engined flying techniques. USAF pilots destined for the C-130 fly the TC-12B/T-44A alongside naval students.

Commander US Naval Air Systems Command
Headquartered at NAS Patuxent River, Maryland, COMNAVAIRSYSCOM's King Air units include the air station's own AOD and the US Naval Test Pilot School. The former operates the UC-12B and the latter is assigned three C-12Cs that are actually owned by the US Army. The school had previously operated both the U-21A and U-21F versions of the Ute.

Chief Naval Air Training Command
Based at NAS Corpus Christi, Texas, Training Air Wing Four (TAW-4) reports to CNATRA at NAS Pensacola, Florida. The Wing's two squadrons, which include VT-31 and VT-35, operate the T-44A Pegasus and TC-12B Huron as advanced trainers. The latter unit also operates a UC-12B is support of OSA missions on behalf of the air station.

Former US Navy operators

Code	Location
7D	NAS Dallas, Texas
7F	NAS Brunswick, Maine
7J	NAS Alameda, California
7K	NAS Memphis, Tennessee
7P	NAWS China Lake, California
7R	NAS Oceana, Virginia
7S	NAS Lemoore, California
7T	Naval Air Test Center, NAS Patuxent River, Maryland
7T	NAS Moffett Field, California
7U	NAS Cecil Field, Florida
7V	NAS Glenview, Illinois
7Y	NAF Detroit, Michigan
7Z	NAS South Weymouth, Massachusetts
8B	NAS Cubi Point, Philippines
8C	NAS Sigonella, Italy
8J	NAF Agana, Guam
8U	NAF/NAS/NS Mayport, Florida
F	TAW-6 NAS Pensacola, Florida
n/a	Pacific Missile Test Center, NAS Point Mugu, California
n/a	Pacific Missile Range Facility, Barking Sands, Kauai, Hawaii

UNITED STATES MARINE CORPS

Within the US Marine Corps the UC-12 fleet is attached to individual Marine Corps Air Stations. Mission scheduling for those UC-12s within the ConUS is conducted by the Joint Operational Support Airlift Command at Scott AFB, Illinois. Units outside the ConUS have their missions scheduled by the Naval Aviation Logistics Office at NAS JRB New Orleans, Louisiana.

Marine Force Pacific
Headquartered at Marine Corps Base (MCB) Camp H.M. Butler, Hawaii, MARFORPAC is responsible for two Marine Air Wings (MAW): the 1st MAW at MCB Camp Smedley D. Butler, Okinawa, Japan, and the 3rd MAW at Marine Corps Air Station Miramar, California. The UC-12s assigned to each of these components are operated by the air station's Headquarters & Headquarters Squadron (H&HS). The units assigned to 1st MAW are based at MCAS Iwakuni and MCAS Futemma, Japan, and both squadrons operate the UC-12F. The H&HSs assigned to the 3rd MAW are stationed at MCAS Miramar, California, and MCAS Yuma, Arizona, and both operate the UC-12B.

UC-12Bs of the US Marine Corps (from the H&HS, MCAS Yuma, Arizona) and Navy (AOD, El Centro, California) are seen together in the late evening sun in July 2000. The majority of both Marine and Navy Hurons wear a basic white scheme, with a small number adopting an overall grey finish.

Below: UC-12B BuNo. 161324 was based at MCAS El Toro, California, being amongst the first UC-12s with the '5T' tailcode. Today '5T' is allocated to H&HS MCAS Miramar.

US Marine Corps operators

Code	Unit	Location	Aircraft
(5A)	MASD Andrews	NAF Washington, Maryland	UC-12B
5B	H&HS	MCAS Beaufort, South Carolina	UC-12B
(5C)	VMR-1	MCAS Cherry Point, North Carolina	UC-12B
5D	H&HS	MCAS New River, North Carolina	UC-12B
(5F)	H&HS	MCAS Futemma, Japan	UC-12F
(5G)	H&HS	MCAS Iwakuni, Japan	UC-12F
5T		MCAS El Toro, California	*
(5T)	H&HS	MCAS Miramar, California	UC-12B
5W		MCAS Kaneohe Bay, Hawaii	*
5Y	H&HS	MCAS Yuma, Arizona	UC-12B
EZ	MASD Belle Chasse	NAS New Orleans, Louisiana	UC-12B

* units no longer operate UC-12s.
Note: Codes shown in parentheses are assigned but are not normally carried on the aircraft

Marine Force Atlantic
The 2nd MAW is the air arm assigned to the MARFORLANT (Marine Force Atlantic), which is headquartered at NS Norfolk, Virginia. UC-12Bs assigned to two of the command's air stations – MCAS New River, North Carolina and MCAS Beaufort, South Carolina – are assigned to the H&HS, while at MCAS Cherry Point the UC-12Bs are assigned to Marine Transport Squadron One (VMR-1). Each of these units reports to the respective air station.

Marine Forces Reserve
Headquartered at Naval Support Activity New Orleans, Louisiana, the MARFORRES (Marine Forces Reserve) is responsible for the co-located 4th MAW. Two units within the Wing operate the UC-12B: the Marine Aviation Support Detachment (MASD) Belle Chasse at NAS New Orleans, Louisiana, and MASD Andrews at NAF Washington. Both of these units operate the UC-12B.

Above: The H&HS at MCAS Beaufort, South Carolina, uses UC-12Bs and HH-46Ds for utility and base SAR duties.

NATIONAL AIR & SPACE ADMINISTRATION
NASA operates a small number of King Air 200s as logistic support aircraft. Similar in configuration to the UC-12B/F and C-12F1, individual aircraft are assigned to the Wallops Flight Facility and Langley Research Center in Virginia. The Dryden Flight Research Center, located at Edwards AFB, California, has two King Airs assigned and shares three C-12Cs with the US Air Force's 412th Test Wing and the USAF Test Pilot School at Edwards.

Super King Air B200 N529NA is based at the Langley Research Center and used in the communications role. It is one of two that have been upgraded with PT6A-60A engines. Like most of their fleet, NASA's Beech 200/C-12s are all finished in the house colours of blue and white.

DEPARTMENT OF HOMELAND SECURITY
Bureau of Immigration and Customs Enforcement
The Office of Air and Marine Operations is headquartered in Washington D.C. and was known as the Air and Marine Interdiction Division when it was a component of the US Customs Service and part of the US Department of the Treasury. The organisation was transferred to the newly-established Department of Homeland Security on 1 March 2003. The division's mixed fleet of more than 130 aircraft includes C-12C King Airs that were previously operated by the US Army. Several aircraft were originally bailed to Customs by the Army but later transferred outright to USCS and now wear civil registrations. Additional aircraft were transferred when the Army phased out the majority of its C-12Cs. Several of OAMO's C-12s and/or B200s are equipped with sea search radar and are similar in configuration to the US Navy's RC-12F/M range control aircraft. These variants are referred to as C-12Ms. Although registered to the Customs National Aviation Center at Will Rogers World Airport in Oklahoma City, Oklahoma, the King Airs are assigned to air and marine branches and air units throughout the ConUS and its territories.

DEPARTMENT OF TRANSPORTATION
Federal Aviation Administration
Headquartered in Washington, D.C. are two former US Air Force C-12Cs that wear civil registrations, operated by the Federal Aviation Administration. Registered to the National Flight Program Oversight Office at Will Rogers World Airport in Oklahoma City, Oklahoma, the aircraft were transferred to the FAA during 2001 after being stored at the AMARC at Davis-Monthan AFB, Arizona. In addition to the former military airframes, the FAA operates 19 examples of the more powerful King Air 300 that are similar to the C-12S operated by the Drug Enforcement Agency. Also assigned are a single Model 200, five earlier King Air C90s and a pair of F90 King Airs.

N1560, a C-12C (ex 73-22258), is one of a number of Hurons first loaned and then transferred to the US Customs Service, today part of the Dept of Homeland Security.

Former USAF C-12C 73-1205 was transferred to the FAA as N11. It was operated alongside N12 (ex 73-1212) from Will Rogers World Airport, Oklahoma.

US GOVERNMENT CONTRACTORS
Science Applications International Corporation
SAIC's Aviation Integration and Test Division, located at Dothan Regional Airport, Alabama, operates individual examples of the U-21F and C-12C (serials 70-15908 and 78-23135) on behalf of the US Army Communication and Electronic Command. The aircraft were modified by the contractor on behalf of CECOM's Product Manager – Night Vision/Reconnaissance, Surveillance and Target Acquisition and are equipped with the General Atomics Lynx synthetic aperture radar, electro-optical and infra-red imaging equipment. The aircraft deployed to Bosnia during 2000 as part of a 90-day technology demonstration for US Army Europe.

Raytheon Aircraft Company
RC-12K serial 85-50149 is bailed to Raytheon and serves as a testbed for structural modifications associated with the Guardrail fleet. The aircraft is based at the contractor's facility in Wichita, Kansas.

Raytheon Range Systems
Raytheon Corporation's Range Systems Engineering Company operates two Beech 1900D airliners in support of US Navy test activities. The aircraft are based at Palm Beach International Airport, Florida, and provide a shuttle service between the main base of the Atlantic Underwater Test and Evaluation Center (AUTEC) in Florida and its detachment on Andros Island, Bahamas. Until recently, three aircraft supported the Army's Ronald Reagan Ballistic Missile Test Site on Kwajalein Atoll in the Republic of the Marshall Islands, flying shuttle missions between the site's facilities.

Flight Safety International
Located at Dothan Regional Airport, Alabama, FSI provides flight training for C-12 crew members in support of the US Army, Navy and US Air Force. The contractor currently operates 11 C-12C/Ds for this requirement.

US Military King Air Variants Part 1: U-21

YU-21 prototype

Beech Aircraft modified a single Model 65-80 by replacing its six-cylinder Lycoming reciprocating engines with a pair of 500-shp (373-kW) United Aircraft of Canada (later Pratt & Whitney of Canada) PT6A-6 turboprops. The modified aircraft (c/n LD-75), which first flew on 15 May 1963, was assigned the MDS designation NU-8F. In addition to receiving the new construction number LG-1, serial 63-12902 was later assigned the designation YU-21. In March 1964 it was delivered to Fort Rucker, Alabama, where the Army evaluated the aircraft. It was later operated by the US Army Test and Evaluation Command and eventually assigned to the US Army Transportation School at Fort Eustis, Virginia. The YU-21 was retired to the US Army Aviation Museum at Fort Rucker, Alabama, in August 1984 but was struck from the collection in 2001.

Serial	No.	Build No.
63-12902	1	LG-1

The first of a large family of military King Airs, the YU-21 was derived from the L-23 (later U-8) Seminole, of which 280 were delivered to the US Army. The Seminole itself was a military version of the Twin Bonanza, and (in its L-23F guise) the Queen Air 65.

U-21A Ute

Assigned the Beech model number 65-A90-1, the U-21A was powered by a pair of 550-shp (410-kW) PWC PT6A-20 turboprops that each drove a three-bladed, reversible, constant-speed Hartzell HC-B3TN-3/T10173 series propeller. Beech delivered 102 U-21As but at least 30 of the utility variants were modified as special electronic mission aircraft. The U-21A had an overall length of 35 ft 6 in (10.82 m), a wingspan of 45 ft 10.5 in (13.98 m) and a height of 14 ft 2.6 in (4.33 m). It was capable of a maximum take-off weight of 9,650 lb (4377 kg) and rated for a maximum landing weight of 9,168 lb (4159 kg). The U-21A carried 370 US gal (1400 litres) or 2405 lb (1091 kg) of fuel in wing and nacelle tanks, and was capable of a maximum speed of 208 kt (385 km/h) and a maximum altitude of 30,000 ft (9144 m).

The production version of the YU-21, the U-21A was a basic utility aircraft, serving mainly as a unit 'hack'. 67-18082 (above) was operated by the USAATCA in this attractive silver and white scheme, while 66-18033 (left, seen during its service with 1-158 AVN) wears the all-over grey scheme in vogue in the 1980s.

Serial	No.	Build No.
66-18000/18040	41	LM-01/41
67-18041/18076	36	LM-42/77
67-18078/18084	7	LM-78/84
67-18086	1	LM-85
67-18088	1	LM-86
67-18090/18092	3	LM-87/89
67-18094/18103	10	LM-90/99
67-18116/18118	3	LM-112/114

U-21A 67-18070 of Operational Support Airlift Command has the smart chocolate brown and white scheme used by most of the U-21s in Army utility service. The basic lines of the Beech 65-A90 transport models changed little over the years.

US military King Air variants

EU-21A

At least five U-21A aircraft, including serials 66-18000, 66-18013, 66-18027, 67-18055 and 67-18058, were modified for use as airborne radio relay platforms and were designated EU-21A. With the exception of the specialised mission equipment, the EU-21As retained the basic U-21A configuration and were not visually distinguishable from utility models. The equipment was not permanently installed and the aircraft were often used for utility missions. Folding canvas benches were installed in the EU-21As and the aircraft were not equipped with airline-type seats in the rear cabin.

The EU-21As were initially assigned to the 1st Signal Brigade Aviation Detachment during the Vietnam War, operating from Long Thanh North. The mission equipment was later removed and the aircraft reverted to the utility role, ending their careers as standard U-21As.

Externally there was no difference between the U-21A and the EU-21A, the radio relay equipment being contained within the cabin and able to be removed to allow the aircraft to resume the utility role. 66-18000 was one of at least five examples converted, ending its days with the USNTPS at Patuxent River.

JU-21A Left Jab

Three U-21As (serials 67-18063, 67-18065 and 67-18069) were equipped with the AN/ARQ-28 VHF SIGINT and radio DF system called Left Jab, which was developed by the Army's Electronic Warfare Laboratory at Fort Monmouth, New Jersey. The most visible modification made to the JU-21A was the installation of a retractable, oval-shaped pod for the direction-finding system on the lower fuselage. The aircraft carried a five-man crew of two flight crew and three system operators. One of these JU-21A aircraft (serial 67-18065) was shot down over Vietnam. After returning from southeast Asia, the surviving aircraft were assigned to the US Army Reserve and later deconfigured, reverting to U-21A standard.

One of the more unusual conversions of the U-21 family was the JU-21A fitted with the Left Jab SIGINT and direction-finding gear. A large retractable pod under the aircraft's centre section, the blade antennas on the rear fuselage underside and a large whip aerial on the cockpit roof were all features of the Left Jab aircraft. This example is 67-18063.

RU-21A Crazy Dog (Cefirm Leader)

RU-21A 67-18112 is seen as it originally appeared with the flat dipole antennaes on the wings, before they were replaced by wingtip pods. The aircraft is seen at St Louis, Missouri, McDonnell Douglas undertaking the installation programme.

Serial	No.	Build No.
67-18112/18115	4	LM-108/111

Beech RU-21A Cefirm Leader

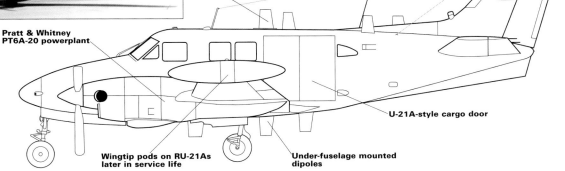

Four U-21As were modified as part of the Army's AN/ULQ-11 counter-measures jamming set and designated RU-21As. The integrated airborne intercept, direction-finding and jamming system known by the programme names Crazy Dog and later Cefirm Leader was carried by three different variants of the U-21A. The RU-21A special electronic mission aircraft were equipped with the AN/ARD-22 radio DF set and carried a four-man crew comprising two flight crew and two equipment operators. Based on the

Variant File

model 65-A90-1, the RU-21A retained the U-21's PT6A-20 turboprops and was approved for a maximum operating weight of 10,200 lb (4627 kg). It had four flat blade antennas installed along the cabin roof and four flat blade dipole antennas projecting from the upper and lower wings at the tips. The latter antennas were later replaced by wing pods similar to those carried by the RU-21H Guardrail V aircraft.

RU-21A 68-18813 shows off the upper surfaces of the variant after the wingtip pods had replaced the dipoles. The fuselage top has five distinct aerials, which are mirrored on the under sides of the fuselage by a further pair of aerials. In service, the RU-21A was flown by the US Army between 1972 and 1993, alongside the other two variants of the RU-21 that were part of Cefirm Leader.

RU-21B Cefirm Leader

Three RU-21Bs differed significantly from the standard U-21A and were assigned the Beech model number 65-A90-2. The aircraft were equipped with the AN/ALR-32 electronic countermeasures receiving set and formed the intercept component of the AN/ULQ-11 countermeasures jamming set known as Cefirm Leader. The RU-21B differed from the earlier variants of the Ute in being equipped with a new main landing gear strut that featured dual wheels and tyres as found on the Beech 99 and later models of the King Air 100. It had a slightly smaller fuel capacity and was powered by two 620-shp (462-kW) UAC PT6A-29 engines that drove three-bladed Hartzell propellers, and was rated for a maximum speed of 208 kt (385 km/h). The aircraft was capable of both taking off and landing at a maximum weight of 10,900 lb (4944 kg), and its five-member crew comprised two flight crew and three equipment operators. The RU-21B carried 396 US gal (1499 litres) or 2,574 lb (1168 kg) of fuel.

Serial	No.	Build No.
67-18077	1	LS-1
67-18087	1	LS-2
67-18093	1	LS-3

The RU-21B was the intercept platform for the Cefirm Leader system, with similar fuselage aerials but lacking the wing-mounted dipoles or wingtip pods. The variant introduced a new twin-wheel main undercarriage. RU-21Bs flew with the US Army EW Co based at El Paso International Airport near Fort Bliss, Texas, alongside the RU-21A and Cs.

RU-21C Cefirm Leader

Two RU-21Cs, similar in configuration to the RU-21B, were assigned the Beech model number 65-A90-3. The aircraft were equipped with the AN/ALT-29 countermeasures transmitting set and served as the jamming component of the AN/ULQ-11 countermeasures jamming set known as Cefirm Leader. Like the RU-21B, the jamming aircraft's main landing gear strut was equipped with dual wheels and tyres, and it was powered by the same PT6A-29 turboprops and Hartzell propellers. Although its external configuration differed slightly from the RU-21B, the aircraft was rated for operations at the same speeds and weights. Its four-member crew comprised two flight crew and two equipment operators.

The RU-21C was similar to the RU-21B but differed in the role that it undertook, although both aircraft were part of the AN/ULQ-11 Cefirm Leader system. Initially both the RU-21B and C had the squat round antennas above the cockpit, while the wingtip bulge and downward aerial were added to the C post-1968. 67-18089 is seen while serving with USAECOM.

Serial	No.	Build No.
67-18085, 18089	2	LT-1/2

RU-21D Laffing Eagle

The RU-21D was the platform used for the Laffing Eagle (AN/ARD-23) airborne radio direction-finding system. Although 34 RU-21Ds were ordered, only 18 were completed in this configuration. The RU-21D was also based on the U-21A and fell into the 65-A90-1 design series. The remaining 16 RU-21Ds were delivered as RU-21Es under model number 65-A90-4. These aircraft were equipped with uprated engines and an advanced direction-finding (DF) system called Left Foot (AN/ARD-26). In addition to the flight crew, the RU-21D carried two system operators for the separate intercept and DF systems. Laffing Eagle was subjected to a rigorous test programme but was never fielded, and the aircraft were further modified to Guardrail IIA and IV configurations. Like the RU-21C, this model was powered by the PT6A-28 or -34 and its performance mirrored that version of

Dipole antennas re-appeared on the RU-21D. Prior to delivery 67-18104 was painted in the overall brown Army scheme with its c/n on the nose u/c door.

US military King Air variants

U-21D Ute

A number of RU-21Ds were deconfigured for the operational support role as U-21Ds. Examples include 67-18104, 18106, 18108, 18110, 18120, 18122 and 18128, with some of the other 18 RU-21Ds also becoming U-21Ds.

Right: Devoid of the RU-21D's electronic equipment, the U-21Ds were distinguished from other utility models by the blade aerials on top of the fuselage, reflecting their previous use. 67-18108 was assigned to the US Army Air Traffic Control Activity (USAATCA).

RU-21D 67-18110 wore a colourful red and white finish during its time with the USAATCA. The dipole antennas on both the wings and tailplane have been removed and the aircraft was later reclassified as a U-21D. Inside, the AN/ARD-23 Laffing Eagle direction-finding equipment had long been removed.

the Ute. Visually-distinguishable modifications made to the RU-21D included the installation of a series of flat blade dipole antennas that extended vertically from the upper and lower surfaces of the wings, and horizontal stabilisers. Three examples were converted as RU-21Hs while others were deconfigured and finished their careers flying operational support missions as U-21Ds with various Army organisations.

Serial	No.	Build No.
67-18104/18111	8	LM-100/107
67-18119/18128	10	LM-115/124

RU-21E Left Foot/Guardrail II/IIA/IV

Although assigned the model number 65-A90-4, the RU-21E retained the PT6A-20 series engine and its performance and specifications were the same as the earlier U-21A model. Intended from the start as a SEMA platform, the RU-21E was equipped with the advanced AN/ARD-26 radio direction finding system known as Left Foot. The RU-21E's four-man crew comprised of a pilot, co-pilot and two systems operators. Although the system was more advanced than the earlier RU-21D/Laffing Eagle, it was never operationally fielded. Twelve examples were eventually modified to Guardrail IIA and IV configurations but retained the earlier designation. All 16 RU-21Es were eventually converted to Guardrail V configuration and redesignated RU-21Hs. Visually, the RU-21E configuration was similar to that of the earlier RU-21D and included the installation of flat blade dipole antennas that were similar in configuration to those carried by the earlier aircraft and extended upward and downward from the wings and horizontal stabilisers and upwards from the fuselage.

Guardrail II/IIA conversions comprised of serials 70-17876, 70-15880 and 70-15883 through to 70-15886. Those airframes modified to Guardrail IV configuration included serials 70-15875, 70-15877, 70-15878, 70-15881, 70-15882 and 70-15888.

Serial	No.	Build No.
70-15875/15890	16	LU-01/16

Above: A line up of RU-21E Guardrail IIs is seen in the early 1970s. Six former Left Jab aircraft were modified as such, becoming the airborne relay facility of the system. Guardrail II was improved as IIA deployed with the 330th ASA Co.

Below: The RU-21E Guardrail IV was deployed to South Korea from 1973. Externally the three RU-21E sub-versions were all identical, featuring a large array of dipole antennas strategically located. Guardrail required no airborne operator.

Variants File

U-21F Ute

In 1970, the Army procured five standard Beech A100 King Airs under the designation U-21F. Equipped with the uprated 680-shp (507-kW) PT6A-28 engines, they also featured four-bladed, reversible, constant-speed Hartzell HC-B4TN-3/T10173 series propellers and executive interiors.

When they entered the Army inventory in 1971, the U-21Fs became the Army's first pressurised aircraft. They were initially assigned to the US Army Davison Aviation Command at Davison Army Airfield, Fort Belvoir, Virginia, and later the Operational Support Airlift Command. Although OSACOM retired the type in 1996, four examples were transferred to the US Naval Test Pilot School at NAS Patuxent River, Maryland, and served the school until they retired in March 1999. The fifth example remains on the inventory but is operated as an airborne testbed by Science Applications International Corporation's Aviation Integration and Test Division. Based at Dothan Municipal Airport, Alabama, the King Air supports a number of projects including tests of the General Atomics LYNX synthetic aperture radar.

The U-21F can operate at maximum take-off and landing weights of 11,500 lb (5216 kg) and 11,210 lb (5085 kg), respectively. Capable of carrying a maximum fuel load of 388 US gal (1469 litres) or 2,522 lb (1144 kg), it is rated for a maximum speed of 226 kt (419 km/h) and maximum altitude of 31,000 ft (9449 m). In addition to five U-21Fs, the Army placed a single example of the model B100 in service.

The aircraft had been acquired under the Confiscated/Excessed Aircraft Program and was operated as a non-standard aircraft using its commercial model number. The B100 differed from the A100/U-21F primarily in being powered by two 715-shp (533-kW) Garrett AiResearch TPE331-6-252B turboprop engines and rated for a take-off weight of 11,800 lb (5352 kg).

Type	Serial	No.	Build No.
U-21F	70-15908/15912	5	B95/99
B100	90-0060	1	BE-67

The U-21F was radically different from the other aircraft procured using the U-21 designation, being larger and more powerful. 70-15912 is seen serving with USADAC, before going to the USNTPS.

Beech U-21F Ute
- Hartzell HC-B4TN-3/T10173 propellers
- 35 ft 6 in (10.8 m) long fuselage for 13 passengers
- Swept vertical tail surfaces, as per other U-21 family members
- Pratt & Whitney PT6A-28
- Standard passenger door with airstairs

U-21G Ute

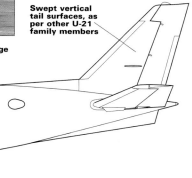

The final 17 examples of the Ute, although delivered under the model number 65-A90-1, were designated U-21G. Dimensions, weight and performance for the U-21G were shared with the U-21A and the chief difference was the configuration and layout of the cockpit avionics. The aircraft were primarily used in the utility role, although nine were eventually converted for SEMA duties.

Serial	No.	Build No.
70-15891/15907	17	LM125/141

The U-21G was the final new-built utility version of the Ute, differing from the original U-21A in cockpit layout. 70-15900 is seen while serving with the 1-228 AVN.

RU-21G Guardrail I

Three U-21Gs were equipped with the Guardrail I airborne communications intelligence system developed by ESL during 1971. Modifications included the installation of multiple antennas on the fuselage along with large sensor pods on the wingtips. Unlike earlier SEMA platforms, the Guardrail system was operated remotely from a ground-based integrated processing facility and the RU-21G normally operated with a two-man flight crew. Although the aircraft are often referred to as RU-21Gs, it remains unclear whether this designation was ever made official. All three examples, comprising serials 70-15893, 70-15898/15899, were later modified to RU-21H configuration.

The large antennas on the back of the RU-21G made it easy to distinguish from the other U-21s. They were used as development aircraft for the Guardrail system and, although not entering operational service, they were detached to West Germany for Reforger I in 1971.

RU-21H Guardrail V

Beginning in 1976, 21 RU-21D/E/G and U-21G airframes were equipped with the AN/ARD-9 Guardrail V system and received the designation RU-21H. These aircraft comprised RU-21D serials 67-18105, 67-18111 and 67-18119, RU-21E serials 70-15876, 70-15879/15880, 70-15883/15887 and 70-15889, RU-21G serials 70-15893, 70-15898/15899, and U-21G serials 70-15891, 15893, 15895 and 15902/15904. Once modified, the RU-21H was certified to operate at a maximum take-off weight of 10,200 lb (4627 kg) and could land at weights up to 9,700 lb (4400 kg). The most obvious visible difference between the RU-21H and earlier Guardrail models was the installation of two large pods on the wingtips that replaced the various configurations of dipole antennas. The RU-21H had provision for AN/ALQ-36 and ALQ-44 radar warning equipment and was equipped with M-130 chaff/flare dispensers. It also had more powerful electrical generators.

Right: RU-21H 70-15880 displays the large wingtip pods and mid-wing mounted dipoles of the variant. All the RU-21Hs were converted from earlier types, with 880 previously being a RU-21E.

70-15898, along with the other two RU-21G Guardrail Is, was modified to RU-21H standard with a new aerial set and an overall grey scheme. The RU-21H was the longest serving U-21 Guardrail version, being retired in 1994.

U-21H Ute

Following replacement of the RU-21Hs, Army planning called for modification of 36 RU-21E/H Ute aircraft to a utility variant known as the U-21H. Raytheon Aerospace carried out the modifications at its Craig Field facility in Selma, Alabama. According to US Army records, 23 aircraft were reconfigured before the transfer of the US Air Force C-12F fleet made the U-21 series redundant. All work on additional examples was halted. Known U-21H airframes include serials 67-18105, 18111, 18119, 70-15875/15876, 70-15878/15887, 70-15889, 70-15891, 70-15893/15895, 70-15898/15899 and 70-15902/15904.

U-21H 70-15875 of the 1-185 AVN was one of a handful that entered service with the US Army. They were RU-21Hs stripped of aerials for the utility role and had a short career before they were replaced by surplus US Air Force C-12s.

JU-21H

Originally delivered as RU-21E Left Foot aircraft, serials 70-15877 and 70-15888 ended their service careers as JU-21H test aircraft. Prior to this conversion, the latter airframe was modified to Guardrail IV configuration. After being deconfigured it was assigned to the US Army Aviation Technical Test Center at Fort Rucker, Alabama, replacing serial 66-18009, and was the last U-21 in US Army service. Following its retirement on 4 January 2000, the Ute was placed on display at the US Army Aviation Museum at Fort Rucker, Alabama.

70-15877 was based at the White Sands Missile Range. It was one of two JU-21H testbeds used by the US Army.

RU-21J Cefly Lancer

The RU-21J was the first of the Beech 200 Super King Air models to enter US military service. RU-21J serial 71-21060 (above, seen in July 1976), was operated by the US Army Airborne Engineering Evaluation Support Branch at NAS Lakehurst, New Jersey.

The Army purchased three examples of the model A100-1 in 1971 and the aircraft entered service as the RU-21J. It was intended to serve as a platform for the airborne portion of the AN/USQ-71 Cefly Lancer sigint system developed by American Electronic Laboratories (AEL). The RU-21J was powered by two 850-shp (634-kW) PT6A-41 engines driving three-bladed Hartzell propellers. It had an overall length of 43 ft 9 in (13.34 m), a wingspan of 54 ft 6 in (16.61 m) and a height of 15 ft (4.57 m). Capable of a maximum speed of 260 kt (482 km/h) and a maximum altitude of 31,000 ft (9449 m), the aircraft was rated for a maximum take-off and landing weight of 12,500 lb (5670 kg). The RU-21J could carry 549 US gal (2078 litres) or 3,569 lb (1619 kg) of fuel in nacelle and wing tanks.

The Cefly Lancer-equipped RU-21Js were delivered in 1974 but the mission equipment was not placed in service operationally. The aircraft, which were formally delivered to the Army by AEL on 1 August 1977, differed from standard Super King Airs in being equipped with uprated alternators, a fuel dump system and multiple external antennas. Additionally, a datalink system was installed in place of the nose-mounted weather radar. Intended to provide greater direction-finding coverage than the Guardrail system, numerous technical problems resulted in programme cancellation.

Subsequently configured for operational support missions, the RU-21Js were assigned the designation C-12L and were eventually retired in 1997. All three examples are currently on the civil register.

Serial	No.	Build No.
71-21058/21060	3	BB-3/5

COMBAT COLOURS

Escuadrilla Aeronaval de Exploración

Argentina's Lockheed patrollers

Today the escuadrilla flies six P-3B TACNAVMOD Orions. The unit's formal albatross badge is carried on the nose, while the informal badge of a fox, ship, submarine and the Southern Cross is worn on the fin.

In 1959 the Argentine navy began to use landplanes for maritime patrol. Since then, three generations of Lockheed patrol aircraft have been employed by the Escuadrilla Aeronaval de Exploración. Following the 1982 South Atlantic war, Argentina's highest award – 'Honor al Valor en Combate' – was bestowed on the unit.

The origins of the Escuadrilla de Exploración go back to the times of the big flying-boats. The first in Argentine service were eight Curtiss FSLs purchased in 1921, followed in 1922 by four Dornier G-52 Wals and six Supermarine Southamptons in 1929. Then came six Consolidated P2Y-3A Rangers, which on 22 January 1940 made an epic flight over the Malvinas/Falklands Islands to carry out the order of President Castillo to search for the British cruiser HMS *Ajax* in the wake of the Battle of the Río de la Plata (River Plate). They did not find the ship.

In 1946 six Grumman JRF-6Bs and G-21As (Goose) were introduced and were used by the Prefectura Nacional Marítima (Coast Guard) on the first operations of the Servicio de Aviación. That year also saw the arrival of 17 Consolidated PBY-5A Catalinas. Between 1954 and 1958 eight Martin PBM-5A Mariners entered the inventory. These aircraft formed the Flotilla de Exploración, together with the recently-arrived Lockheed P2V-5 (P-2E) Neptune.

The Neptunes originally had been delivered to the Royal Air Force's Coastal Command, which by the mid-1950s had decided to replace its Neptune MR.Mk 1s (as they were called in the UK) with the new Avro Shackleton. The Neptunes were offered to Argentina and Brazil, among other countries. Although Argentina originally intended to buy 15, in the end only eight were obtained, for $US206,000 each. All had suffered accidents during their careers, but they were repaired.

Having been turned down for Orions in the 1970s, and with a request for six Breguet Atlantics overturned after the 1982 South Atlantic war, Argentina turned to Electras procured from the second-hand civil market for its MPA requirements. This group outside the escuadrilla's hangar at Trelew includes one used as a spares source (nearest aircraft on left), the Electra Wave Sigint platform (centre aircraft on left) and the single machine converted to tanker status (right).

Escuadrilla Aeronaval de Exploración

Two views show 2-P-101, one of the first batch of ex-RAF P2V-5s. It was the only one of the eight aircraft to retain its original nose turret. At right it can be seen starting up, framed by the rear fuselage of a Catalina which served alongside both Neptunes and Mariners in the Flotilla de Exploración.

Before delivery to Argentine naval aviation, the Neptunes were modified by Aviolanda Maatshapijj voor Vliegtuigouw N.V. in Holland with the most modern components of the time. One aircraft, 2-P-101, kept the three original turrets: one in the nose, one in the tail (Emerson Aero 9B and 11A with two 20-mm/0.787-in guns, respectively) and one dorsal Martin 250CE328 with two 12.7-mm (0.5-in) machine-guns. 2-P-108 had an observation position fitted in the nose in place of the turret. The other six had the nose like -108 and an AN/ASQ-8 magnetic anomaly detector (MAD) in the tail, retaining only the dorsal turret. The aircraft had an excellent electronics system comprising two search radars, one in the ventral radome (AN/APS-20AC) and the other (AN/APS-31) in the right wingtip fuel tank. Also in the latter was the searchlight (AN/AVQ-2). The engines were two Wright R-3350-30W Turbo Compounds of 3,250 hp (2425 kW). In their bomb bays the Neptunes could carry a complete variety of anti-surface and anti-submarine weapons such as bombs, torpedoes and mines, to a maximum of 7,937 lb (3600 kg). In 16 stations under the wings they could carry up to 1,984 lb (900 kg) of weapons such as 5-in (12.7-cm) HVAR rockets, sonobuoys, smoke markers and depth charges.

The P2V-5s joined the Flotilla de Exploración, receiving the serials 0409 to 0415 and the callsigns 2-P-30 to 2-P-37, later changed to 2-P-101 to -108. The aircraft retained the 'P' for patrol in their callsign; naval aviation did not adopt 'E' for exploration, although it was the main role of the Neptune.

The first aircraft arrived in Argentina on 19 September 1958, flown by British crews, and the next day they were officially incorporated into the Comando de Aviación Naval Argentina (COAN). The crews had gone to the United States for training, and the Escuadrilla Aeronaval de Exploración was created on 21 June 1959 at Base Aeronaval Comandante Espora, in southern Buenos Aires province.

Combat and searches

When unidentified submarines were detected in the Golfo Nuevo along the coast of Patagonia, the Argentine Neptunes had their first real anti-submarine action between 20 and 27 October 1959, launching the few weapons they had at the time. The Martin Mariners, Catalinas and the whole Flota de Mar also participated. The situation was repeated in November, by which time naval aviation had received weapons from the United States. The most important incident was on 30 January 1960 when, during a navigation training sortie over the Golfo Nuevo with the destroyer ARA *Cervantes* (D-1) and the patrol ship ARA *Murature* (P-20), the patrol ship ARA *King* (P-21) made sonar contact with a submarine. An anti-submarine operation was immediately mounted with the ships of the Flota de Mar, and the Martin PBM-5M Mariners and Vought F4U-5N Corsairs of naval aviation.

On 2 February, Neptune P2V-5 0409/2-P-102 (*Golondrina*) arrived with other aircraft. On 6 February, 0414/2-P-107 made a sunrise bombing run against a submarine, after which it was detected sailing more slowly and surfacing more frequently in order to repair damage. On 10 February, two Avro Lincoln B.Mk 1 bombers of the Fuerza Aérea Argentina arrived, configured for long-range maritime surveillance. These bombers, which dated from 1947, were inadequate for this kind of mission, but for some time they were Argentina's only long-range aircraft equipped with radar and able to carry a heavy warload.

On 11 February, 2-P-105 arrived in the area and performed a 12-hour ASW surveillance flight, seven hours at night, but without results. Six days later 2-P-107, after receiving a positive return from the MAD, attacked the submarines with new Mk 43 torpedoes, which had been sent quickly from the United States. That flight lasted 15 hours, the longest of all these sorties.

Finally, after the submarines were corralled for one month inside the Golfo Nuevo, two of them were seen on the surface, one repairing damage. After this sighting and subsequent attacks, all contact was lost. The Argentine Navy concluded that the submarines had been from the Soviet Union, testing the response capacity of US allies.

That year, the aircraft were overhauled because of corrosion and operational wear. 1960 also marked the beginning of UNITAS exercises with the navies of the United States, Brazil and Uruguay; these exercises were mostly anti-submarine in nature and the Neptunes participated heavily.

On 12 June 1961, the unit made its first maritime traffic control flight, which became one of its most memorable. Commanded by Capitán de Corbeta Siro de Martini and with Capitán de Corbeta García Bosch as co-pilot, aircraft 0410/2-P-103 took off from Comandante Espora, and when it was 100 miles (161 km) east of the city of Puerto Deseado it turned southeast. The Neptune descended to 2,500 ft (762 m) and headed for the Malvinas/Falklands Islands, which it overflew at very low altitude. Reaching Puerto Argentino/Port Stanley via Isla Soledad/East Falkland, it increased altitude to take pictures before returning to the city of Río Gallegos, where it landed three hours later. One month later 0409/2-P-102, commanded by Capitán Canel, repeated the trip.

Another memorable flight was made by 0410/2-P-103 between 18.15 on 14 August and 16.05 on the following day, when the aircraft

Neptune – the first generation

Eager to acquire land-based long-range ASW patrol aircraft, Argentina was supplied with eight ex-RAF Neptune MR.Mk 1s (P2V-5s), which had been originally funded by the US under MDAP. They were used to equip the newly established Escuadrilla Aeronaval de Exploración, part of Escuadra Aeronaval N° 2 at BAN Comandante Espora (Bahía Blanca).

Below: The left-hand of these two aircraft (2-P-37, later 2-P-108) was the only aircraft to have the observation nose installed but not the MAD.

Above: Seen wearing its initial codes of 2-P-34, this aircraft has the observation nose and a MAD boom, but retains the dorsal turret.

The last of the first-batch Neptunes continued in service until 1973, although the fleet lost the redundant dorsal turret at an early stage. The Neptunes were involved in some notable incidents, including actions against Soviet submarines in the Golfo Nuevo incident in 1960, and participation in the 1963 internal fighting.

took off from Espora, headed to Isla de los Estados at the extreme south of Argentina and from there, flying over the Andes mountains, reached Salta at the extreme north, where the aircraft ended its journey.

On 2 April 1963, during a confrontation between two parts of the armed forces called the *azules* (blues) and the *colorados* (reds), the Neptunes attacked troops of the Regimiento de Infantería 3 Motorizada of La Tablada, in Buenos Aires province, destroying a jeep and a mobile communications centre. After this, and because their side had been defeated, 0410/2-P-103 and 0415/2-P-108 sought refuge in Uruguay, together with some aircraft of the Fuerza Aérea Argentina.

By this time, the Neptunes were old and suffering poor operational readiness, so the purchase of a P2V-5FE (EP-2E) was decided upon in 1965. It was the first Argentine Neptune to be equipped with two Westinghouse J34-WE-34 turbojets, of 3,250 lb (14.45 kN) thrust, mounted under the wings. The aircraft was serialled 0541/2-P-101.

Jet-augmented Neptune

Small jets were added under the wings of Neptunes from the P2V-5S (SP-2E) onwards to offset the continuing weight growth of onboard equipment. The first jet-equipped Neptune, an EP-2E, was delivered in 1965, with another three SP-2Es (below) following in 1970/72. They were retired around 1977 when another batch of four SP-2Hs (left) was received in place of the desired P-3 Orions. This last batch served for five years, and by the time of the 1982 South Atlantic war only two were operational. Despite their general obsolescence and the poor serviceability of their equipment, the two performed admirably during the conflict, but were retired immediately after.

On 7 November 1965 the first fatal Argentine P2V accident occurred, when 0414/2-P-107 flew into Josafat mountain in Brazil with the loss of 11 crew members. The aircraft had been returning to Canoas Air Base in Porto Alegre after a night mission during UNITAS VI exercise.

In May 1970 a P2V-5FS (SP-2E), serialled 0644/2-P-103, was delivered. This aircraft was equipped with the Julie-Jezebel sonobuoy search system, so had a crew of 12 instead of the original nine. The electronics systems of the SP-2E were more sophisticated than those of other versions and it was equipped with an AN/APS-20BE radar and AN/APN-122 Doppler. The installation of 166-Imp gal (757-litre) fuel tanks on the wingtips extended its radius of action.

December 1971 saw the final flight of 0409/2-P-102, which was the last of the first series Neptunes. It was the only P2V-5 (P-2E) to receive a grey-and-white paint scheme in place of dark blue and white. To augment the small fleet, two additional SP-2Es were received in 1972.

Over their operational career, the Neptunes made hundreds of flights over the Antarctic. Most important were the glaciological survey of maritime routes between the Antarctic bases and South America. Other flights were for support, such as dropping food and mail. Search and rescue missions were undertaken for lost ships and aircraft.

During a glaciological flight in bad weather over the South Shetland Islands, 0644/2-P-103 crashed on Livingston Island on 15 September 1976 and was completely destroyed. A search and rescue operation was launched immediately, but sadly all 11 occupants – 10 crew members and one civilian – had been killed.

By the mid-1970s plans were under way to replace the Neptunes with five Lockheed P-3B Orions, to be received from 1977. The US government, however, did not authorise the sale, so the Argentine Navy decided to buy four used SP-2H (P2V-7S) aircraft. Argentina also entered negotiations with France to buy five Breguet Atlantic Is, with deliveries from 1982.

In 1977 the four SP-2Hs arrived from VP-67 of the US Navy. They were serialled 0706/2-P-110, 0707/2-P-111, 0708/2-P-112 and 0718/2-P-114. (The number -113 was not used, because it is the tradition of the Argentine Navy not to use the number 13.) That year, the three surviving SP-2Es were discharged. The SP-2H was the last Neptune variant built and featured improved Wright R-3350-32W Turbo Compound engines of 3,500 hp (2611 kW) with two auxiliary Westinghouse J34-WE-36 jet

Escuadrilla Aeronaval de Exploración

2-P-112 was the SP-2H which guided two Super Etendards to attack British warships on 4 May to the south of the Falklands, resulting in the destruction of HMS Sheffield. It was the last Neptune to be retired, its withdrawal in August 1982 temporarily leaving the COAN without any long-range maritime radar/ASW patrol capability

engines. The main external difference was the canopy design.

A border conflict with Chile meant that by the end of 1978 the Neptunes had been deployed to Base Aeronaval Almirante Quijada at Río Grande (in Tierra del Fuego province). There, they received a green-and-blue camouflage and made reconnaissance missions over Chilean waters and the channels of Tierra del Fuego.

South Atlantic war

The onset of war with the UK took the escuadrilla by surprise and it had only two Neptunes in service (0707/2-P-111 and 0708/2-P-112), plus the two Beechcraft B-200 Super King Airs (2-G-47 and 2-G-48) that had arrived in 1979. Operations started on 23 March 1982 when 2-P-112 flew to Base Aeronaval Almirante Quijada and from there started anti-surface and anti-submarine exploration flights, in addition to search and rescue missions. After the Argentines landed on the islands on 2 April, the Neptunes made flights around the area and co-operated with Air Force aircraft, guiding them to targets and training the Air Force pilots in naval operations. Naval aviation crews complemented those of the Air Force's KC-130H Hercules on search missions and advised them on the tactics of maritime surveillance. They also assisted in recognising and identifying radar contacts.

Guiding exercises were made with the Super Etendards. The aircraft undertook anti-submarine support flights for the ELMA (Empresa Líneas Marítimas Argentinas/Argentine Maritime Lines Company) ships *Formosa* and *Río Carcarañá* as they crossed to the islands.

On 1 May, after a British attack, maritime surveillance flights were made by 0708/2-P-112, armed with two Mk 44 torpedoes because it was known British submarines were in the area. This information was proven when the British nuclear submarine HMS *Conqueror* torpedoed and sank the Argentine cruiser ARA *General Belgrano* (C-4).

Minutes after, the destroyer ARA *Piedrabuena* (D-29) radioed a message that the cruiser was adrift and sinking. All the ships and aircraft in that area immediately began a search and rescue mission to save the survivors. The Neptunes found the survivors and guided the ships to the rescue. After life rafts had been found, contact was passed to the B-200 Super King Air and Lockheed L-188 Electra, leaving the SP-2H free for surveillance.

During the war, the aircraft suffered from excessive oil consumption, a situation that was particularly problematic because they had to fly to Espora for maintenance. Nonetheless, due to the efforts of the mechanics, they were in service almost until the end of the war.

At 05.07 on 4 May, 2-P-112 took off from Río Grande on a mission ordered by the Comando de la Aviación Naval, to undertake maritime surveillance and assist in the arrival of three Air Force C-130s at the islands. During the flight the aircraft used electronic surveillance continuously and the radar occasionally; there were some problems, but the flight continued. At 07.10 the radar indicated a contact, at which point it was shut down. With the ESM, the crew detected the signal of the Marconi 965-type radar of a Type 42 destroyer, which itself had been alerted by the radar emission of the Neptune. Argentina has two ships of this class, so naval aviation was completely familiar with the capacity and characteristics of this kind of radar.

The contact was communicated to the Comando de Aviación Naval, which evaluated the situation and decided to prepare two Super Etendards for an anti-shipping mission. Each was armed with one Aérospatiale AM39 Exocet missile.

Meanwhile, the flight of 2-P-112 continued, maintaining contact with the target and sending its position to the Super Etendards as they approached. At 08.43 a new radar emission was made and they had three contacts, although radar emissions were detected with the ESM. At

Electras were initially procured to provide the Argentine navy with an organic transport capability, replacing C-54s. As well as operational taskings, they operated an airline-style service in the south. The three-aircraft fleet acquired individual names, worn on the nose.

EA51's three Electras were the COAN's main transport asset, used for carrying both passengers and cargo. Their main routes were between Buenos Aires and Patagonia, including the base at Río Grande.

Right: After the South Atlantic war 5-T-2 was converted to tanker status for trials, including the carriage of extra tanks in the cabin. It was subsequently further modified to Explorador standard.

Electra transports of EA51

Although the Electra was also flown by the air arms of Bolivia, Ecuador, Honduras, Mexico and Panama, Argentina was the biggest and best-known military operator of Lockheed's classic turboprop. The first batch of three Electras was procured in 1973 to perform in the transport role, acquired from McCulloch International Airlines, which overhauled the aircraft and modified them to L-188PF standard with rear cargo doors. Issued to 1 Escuadrilla de Sostén Logístico Móvil (EA51), the three Electras were supported by a single aircraft bought for spares. In 1982 the aircraft were used, alongside three F28s of 2 Escuadrilla de Sostén Logístico Móvil, for transport in support of the Malvinas campaign, including several hazardous supply/evacuation flights into Puerto Argentino/Port Stanley. An Electra was the last Argentine aircraft to operate from the airport, flying a resupply mission in the early hours of 14 June. They were also used for visual maritime patrols, augmenting the Neptunes of the Escuadrilla Aeronaval de Exploración. Throughout the conflict they performed their missions with no special equipment, and retained an airliner-style colour scheme. Ten years after the war, two of the aircraft were transferred to the maritime patrol escuadrilla to consolidate all Electra operations in one unit.

09.25 the Neptune's radar stopped working, prompting much effort by the operators to fix it in order to continue guiding the Super Etendards. At 10.30, 2-P-112 was able to make another emission and this time received three contacts at 60 miles (96 km), concluding they were two medium-sized ships of destroyer type and a large one such as an aircraft-carrier. This radar emission was made much closer to the ships – inside their radar range, in fact – and they could be painted with a marker on the radar. The Neptune's radar was shut down immediately after the emission and the aircraft descended and turned back, to be outside the range of the British radars.

The position was transmitted to the Super Etendards approaching the zone and at 11.14 those aircraft launched their missiles, destroying the British Type 42 destroyer HMS *Sheffield*.

Meanwhile, 0708/2-P-112 returned to its base and landed at 12.04, after more than seven hours flying. The aircraft did not have enough fuel to divert to an alternative airfield, so the practice was to use a corrected fuel mixture and modified jet engines in order to extend the range.

That afternoon, 2-P-111 set off to determine the condition of the attacked ships. It approached to within 40 miles (65 km) of an enemy ship, which ordered a combat air patrol of Sea Harriers to shoot it down. The communication was intercepted at Rio Grande and they informed the Neptune, which was able to escape, covered by two Air Force IAI Daggers.

On 7 May Electra 5-T-1 located five grey-painted ships and 2-P-111 immediately was sent to investigate; at the same time, the hospital ship ARA *Bahía Paraíso*, sailing near that area, requested air cover. It transpired that the ships were Soviet fishing vessels.

Neptune flights continued until 12 May, when 2-P-111 was sent to Espora suffering from a number of mechanical problems, and did not return. 2-P-112 went to Espora for the same reason on 15 May; after a partial fix it returned to Río Grande on 26 May, but was back in Espora by 4 June. Between 5 and 7 May, 2-P-111 made some flights to practise mine dropping, but these were not operational sorties. On 8 June the Neptunes withdrew from service almost entirely. On 28 July 2-P-111 flew for the last time, followed on 30 August by 2-P-112, closing the Neptune chapter with the Escuadrilla Aeronaval de Exploración of the Comando de Aviación Naval Argentina.

Electras step in

In 1973, given the necessity to replace the old Douglas DC-4/C-54s and the small budget with which to do it, three second-hand Lockheed L-188PF Electras were purchased and refurbished by McCulloch Airlines. A fourth, an L-188A, was bought from Sociedad Aeronáutica Medellín of Colombia to use for spares.

The four aircraft arrived in 1974 and the three L-188PFs began operations with the Primera Escuadrilla Aeronaval de Sostén Logístico Móvil (EA51), together with the last C-54 (serial 0297, callsign 5-T-40) remaining from seven the COAN operated. The new aircraft received callsigns 5-T-1 to 5-T-3.

From the beginning, the Electras performed transport missions and operated a passenger line through Patagonia. They resided at Base Aeronaval Ezeiza, in the suburbs of Buenos Aires. During the 1978 conflict with Chile they undertook logistical missions, supporting units deployed in the south.

Electras in the South Atlantic War

Just after the 2 April 1982 Argentine landing on the Malvinas/Falklands Islands, the Electras moved south to support naval infantry and naval aviation. They made a number of logistical flights, most of them to Puerto Argentino/Port Stanley, and some maritime surveillance flights. On 29 May 5-T-2 landed at Puerto Argentino airport under enemy fire, transporting ammunition, clothes and other material, and evacuated the crew of a sunken fishing vessel. During the conflict, the Electras flew

6-P-101 was from the second batch, one of three L-188As bought from Evergreen and overhauled before delivery. A fourth flying aircraft came from TAN of Honduras, while a spares recovery airframe was purchased from WKB Aviation at Fort Lauderdale. Prior to Explorador conversion, the aircraft flew patrols without any special equipment, and with this civil-style scheme.

Escuadrilla Aeronaval de Exploración

Electra Explorador

Sensor operators are seen at their work-stations in an Explorador (above), the yellow screen being the APS-70 radar display. Rear fuselage detail (below) shows the bulged port aft observation window and stores-dropping chute underneath.

To replace the Neptunes, the last of which retired in 1982, the COAN produced three maritime patrol Electras by conversion: two from the second batch and one from the original transport aircraft batch. The most obvious feature was the addition of an APS-705 search radar in a ventral radome, providing 360° coverage. Additionally, the aircraft were fitted with sonobuoy tubes, which could also be used for dropping maker flares, and a SAR package dispenser. Omega equipment was installed for accurate overwater navigation.

The first two aircraft joined the Escuadrilla Aeronaval de Exploración at Base Aeronaval Comandante Espora by the end of 1982. The first, 0792/2-P-101 then 6-P-66, was written off for use as a spares source and remained at Espora. The second was serialled 0791/6-P-102. 0789/6-P-101, the third Electra, landed at Espora on 1 January 1983. It stayed only briefly at Espora, because on 11 February the unit moved to Base Aeronaval Almirante Zar (BAAZ), close to Trelew City in Chubut province. On 20 December 1982 the Escuadrilla Aeronaval de Exploración had performed a flypast of the new base; from 14 April, it became part of BAAZ's Escuadra Aeronaval No. 6 (EA6E). These aircraft, equipped with only basic equipment, were to begin maritime traffic control missions over the Argentine Sea (the waters off the coast of Patagonia). Around the same time, three Beechcraft B-80 Queen Airs (6-G-82, 6-G-83 and 6-G-84) were received for light transport and liaison, soon followed by one Pilatus PC-6 Turbo Porter (6-G-2).

On 26 August 1983 the fourth Electra, 6-P-103, arrived at Trelew from the United States, and the final one, 6-P-104, landed on 25 September. On 30 May 1984, they carried donations for the victims of a flood at Resistencia in Chaco province. They made four flights over Antarctica: 6-P-101 on 9 July,

1,318 hours in 404 flights of support, search and rescue, and maritime surveillance.

After the war

An arms embargo was put in place by Great Britain and the United States after the war, leading Argentina to decide to buy five additional Lockheed Electras on the civil market.

One of the four operational second-batch Electras was used for transport duties, with a forward cargo door. It was withdrawn from use in 1997, along with the two surviving Exploradors.

The COAN's patrol assets of the 1980s are represented by this formation of an Electra leading an S-2E Tracker and King Air 200. The Trackers were assigned to the carrier 25 de Mayo, but routinely operated from shore bases, including wartime missions.

6-P-103 on 12 October, and 6-P-103 and 104 on 16 December. On 10 October 1984 6-P-102 flew for the first time at Espora, and went to Trelew on 14 November.

On 11 October 1989 the first maritime traffic control flight was made with Electra L-188A 5-T-3 of Primera Escuadrilla Aeronaval de Sostén Logístico Móvil of Ezeiza, with pilots of this unit but an EA6E mission crew.

The COAN decided to make extensive modifications to these machines, allowing them to perform their missions more effectively. Two aircraft initially were selected (6-P-101 and 6-P-102) and went to Taller Aeronaval Central (TAC, Central Naval Aviation Workshop) at Espora, where work began to install an APS-705 radar and other systems. This radar, similar to the one used by the ASH-3D Sea King, has 360° coverage and a range of 100 miles (160 km), with its antenna in a ventral radome. The radar was connected to a COTAC indigenous computer that enabled tactical operations. The meteorological radar on the nose was replaced by a more modern RCA Primus 400. Other electronic equipment was installed, such as a VLF Omega which was indispensable for overwater navigation, and an installation for dropping survival equipment during SAR missions. Finally, the Electras received four tubes for flare and sonobuoy launch, and observation windows on both sides of the fuselage. The aircraft were then known as Electra Explorador. 6-P-103 was never modified and continued in use as a cargo aircraft.

The limited budget meant that modifications were done slowly, including the Wave project which covered the modification of 6-P-104 for electronic warfare missions (Sigint and Elint). Work was done at Taller Aeronaval Central in Espora and Arsenal Aeronaval N° 5 at Ezeiza, with Israeli support. On 15 February 1990, 6-P-104 returned to BAAZ after modification.

Meanwhile, 5-T-1 (called *Antártida Argentina*), 5-T-2 (*Ushuaia*) and 5-T-3 (*Río Grande*) continued their duties at EA51 as transports. In the early 1990s 5-T-3 had suffered structural damage during a very turbulent flight, and in 1995 was written off; it now forms part

Electra Wave

Israel Aircraft Industries provided equipment and assistance to the COAN when it converted 6-P-104 to Wave configuration for the Sigint mission. As well as a 'farm' of antennas under the fuselage, presumably for Comint, the Wave was fitted with large receivers in a 'turret' above the flight deck, facing outwards at 45° to provide full forward hemisphere coverage. A large side-looking array of Elint antennas in bulged fairings was mounted either side of the rear fuselage. This rare inflight photograph (left) shows the Wave during the trials programme: later the sensor fairings were painted to match the aircraft's colour (below). It was the COAN's last flying Electra, serving with the escuadrilla until 2001.

Coastal patrol and support

As well as the main mission equipment, the Escuadrilla Aeronaval de Exploración has also operated a small number of support types, used mainly for liaison and light transport tasks, but also for coastal patrol duties. Two Beech King Air 200s and a single Pilatus PC-6 Turbo-Porter were on strength from 1979 to 1982, the King Airs being used during the South Atlantic war for patrol and SAR cover duties. They were subsequently reassigned to other units. Later, in the late 1980s, three Queen Airs were allocated to the escuadrilla. One was lost in a landing accident in 1990 and the other two were retired and sold in 1992.

Four PC-6B/H2 Turbo-Porters were procured by the COAN for general duties, including Antarctic support. When assigned to Escuadra Aeronaval Nº 6 the aircraft wore standard COAN scheme (above) with sharkmouth, later replaced by camouflage (left) when assigned to Escuadra Nº 4.

Below left: This Super King Air 200 was one of those assigned to the escuadrilla, but wears the codes of Escuadra Aeronaval Nº 4 after reassignment. Under the Cormoran programme the COAN converted some King Airs in the late 1990s with RDR-1500 radar to perform maritime patrol missions, assigned to Escuadra Aeronaval Nº 6 alongside the P-3 Orion fleet.

Below: One of the escuadrilla's B80 Queen Airs is seen in the hangar at Trelew. It wears the standard COAN overwater scheme adopted in the 1980s.

of a bar in the suburbs of Buenos Aires. 5-T-2 was converted into a tanker, receiving a refuelling hose drum unit on the fuselage and seven tanks in the cabin. This system allowed only one refuelling at a time, so its usefulness was limited and it was employed only for tests.

At the beginning of 1992, 5-T-1 and -2 were transferred to EA6E, and received the callsigns 6-P-105 and -106, respectively.

6-P-101 was lost in a non-fatal accident at Trelew in 1989, so 6-P-106 was modified to Explorador standard although it retained its tanker abilities. 6-P-103 and 6-P-105 continued in passenger and cargo transport roles. In 1991 the two remaining B80s were sold (6-G-84 had been lost after an inflight fire on 7 March 1990, but was able to land at Almirante Zar base and the crew escaped).

The locally-modified Electra Exploradors made a number of flights, and sometimes were intercepted by Royal Air Force Phantoms and Tornados on the boundary of the Argentine Sea and the exclusion zone proclaimed by the British surrounding the Malvinas/Falklands Islands. The first encounter was on 25 July 1983 186 miles (300 km) east of Cabo Vírgenes, when two Phantoms intercepted 6-P-102.

On 26 March 1986, 6-P-101 detected a submarine operating 43 miles (70 km) off the coast of Mar del Plata City, but it was impossible to determine the nationality.

In the 1990s, due to the increase in activity by foreign fishing ships in the Argentine Economic Exclusion Zone, the escuadrilla and its Electras were requested for duties. Unfortunately, the numbers and the equipment were not sufficient to deal with the scale of the incursions. Compounding the pressure was the fact that Argentina is responsible for conducting search and rescue in South Atlantic international waters.

The aircraft's length of time in service began to be a problem, exacerbated by the difficulty in finding spares because there are few operational Electras in the world. After evaluating the possibility of acquiring surplus, good-condition Lockheed P-3B Orions from the US Navy, the Electras were deactivated with only 6-P-104 remaining operational. The last Electra Explorador in service, 6-P-106, was taken to the Museo de Aviación Naval at Espora, also home to the fuselage of 6-P-66. 6-P-104 ceased flying in 2001 and was also assigned to the museum. EA6E plans to use part of the Wave electronics suite on one of the Orions.

Time for the Orion

The Electra acted as a good transition to the Orion. The process was drawn out, but allowed the crews to train on the new equipment and the last ferry flights from the United States were made by Argentine crews. The purchase of the aircraft, made possible by the support of the Secretaría de Agricultura, Ganadería, Pesca y Alimentación, gave new life to the escuadrilla, which was now able to undertake ASW and AS missions within a radius of 1,500 nm (2778 km) and lasting up to 12 hours. Six aircraft are part of the Escuadrilla Aeronaval de Exploración and have a long service life ahead, while another was bought for spares recovery and was dismantled in the United States, the airframe never going to Argentina.

The Orions are equipped with an APS-80 radar with a range of more than 120 miles (193 km); although it can detect objects it cannot identify them, which requires trained operators. The radar has two antennas, one in the nose and another in the tail in front of the MAD. This radar scans 360° and is associated with an ASW-124 computer. All the electronic equipment adds 2 tons to the aircraft's weight.

A torpedo bay has two hydraulically-activated doors for bombs, torpedoes, depth charges and mines. Sonobuoys, flares and an installation to drop survival equipment for SAR missions are also carried.

Combat Colours

During patrols over the South Atlantic the escuadrilla's aircraft were occasionally intercepted by RAF fighters (Phantoms until July 1992) flying from Mount Pleasant in the Falklands. Here a Tornado F.Mk 3 closes in on an Electra. British forces established a 200-mile (322-km) radius Total Exclusion Zone around the islands during the conflict, but the policed area was subsequently reduced to the 150-mile (241-km) Falkland Interim Conservancy Zone (FICZ), primarily intended to protect fishery interests but also used as an effective outer limit for air defence purposes.

The aircraft has two beds for the crew and a well-equipped galley, plus parachutes, and a ladder extending from the fuselage. A curtain separates the observation posts, so the observer in darkened conditions can retain night vision. A hangover from the Cold War, there is also a security box for secret orders.

The first important operation of the Orion was a SAR flight after a naval officer and one sub-officer disappeared near the South Orkney Islands in March 1998. One P-3B took part in the search for the small boat in which the men were sailing, together with the Argentine icebreaker ARA *Almirante Irizar*, two army AS 332 Super Pumas and one air force C-130 Hercules, plus one RAF Vickers VC10 and one C-130 operating from RAF Mount Pleasant in the Malvinas/Falklands. Although the boat was discovered on the islands, the men were never found.

Tornado F.Mk 3s assigned to No. 1435 Flight wear the Falkland Islands crest on the nose. The Tornado's 'Falklands fit' comprises two Sky Flash in the forward missile bays, four AIM-9L Sidewinders, armed cannon and no tanks. A VC10 tanker from No. 1312 Flight would provide tanker support.

Between 1 and 4 November 1999, one Orion, one air force C-130, one Meko 140-type 'Espora'-class frigate and one patrol ship took part in the first joint exercise between Argentina and the UK after the 1982 war. On the British side were an RAF Hercules and a Royal Navy frigate. During the exercise, SAR and maritime traffic control missions were performed. Since then, the Orions have partici-

P-3B Orion

Having been denied P-3s earlier, Argentina finally received US approval to acquire the type in the mid-1990s, following a change in US policy in the wake of Argentina's participation in Desert Storm. Under the terms of a 20 May 1997 agreement, six P-3B TACNAVMOD ('Super Bee') aircraft were chosen from US stocks at Davis-Monthan for a purchase price of US$7.3 million. Before delivery the aircraft were overhauled by Logistic Services International at Tucson, which also trained Argentine ground crew. A seventh airframe was purchased for spares, but was broken up by LSI and remained at Tucson. Argentine air crew began training with the US Navy's Orion FRS (VP-30) at Jacksonville in October 1997. The Electra Exploradors were retired in the same year.

Orions were delivered to Argentina wearing US Navy schemes – either the original grey/white (above) or the later Tactical Paint Scheme (below). Argentine navy markings were initially applied over these colours before the definitive scheme (left) was adopted.

Left: An important task for the Orion fleet is maritime traffic control, in particular monitoring fishing activities in the rich South Atlantic seas. This task is also performed by British aircraft flying from the Falkland islands. Aircraft from both countries now undertake joint exercises in fishery control and SAR tasks.

Escuadrilla Aeronaval de Exploración

pated in all the important exercises of the Argentine Navy, and undertaken SAR missions and flights to control illegal fishing.

4 July 2001 saw the first use of the real-time image transmission digital system in maritime traffic control missions over the economic exclusion zone. The system has a Canon EOS D2000 digital photo camera and an RCA CC4393 recorder mounted inside the cabin, which sends images to a computer that improves the quality, light and contrast and adds date and position of the photo. The image is then transmitted via HF communication equipment and arrives only seconds later at the Comando de Operaciones Navales (Naval Operations Command) at Base Naval Puerto Belgrano, allowing the command to see the situation at sea. In the near future, the plan is to transmit the images through satellite telephone, leaving the HF system for longer-range missions.

In November 2002 6-P-55 performed trial and demonstration flights with the FLIR Systems Star Fire II infra-red detection system. The flights were made over the sea and the Río de la Plata area from Base Aeronaval Almirante Zar and from Estación Aeronaval Ezeiza, in order to demonstrate the system's operation and its effective use for sea control tasks and SAR. Another consideration for updating the P-3B is installation of a modern radar.

Santiago Rivas and Juan Carlos Cicalesi

Argentina's Orions are scheduled to enjoy a long service life. Despite the limited funds available, upgrades are being studied to improve their effectiveness in the long-range maritime patrol role.

Aircraft of the Escuadrilla Aeronaval de Exploración

Aircraft type	c/n	Serial	Callsign	Delivered	Wfu/written off	Notes
P2V-5/P-2E	426-5095	0408	2-P-30/2-P-101	19/5/59	1964	ex-USAF 51-15920, ex-RAF WX502 and ex-LOT-CP. Sold for scrap in 11/65
P2V-5/P-2E	426-5132	0409	2-P-31/2-P-102	27/1/59	20/9/73	ex-USAF 51-12936, ex-RAF WX512 and ex-LOT-DW. To storage on 23/12/71. Sold for scrap
P2V-5/P-2E	426-5134	0410	2-P-32/2-P-103	25/10/58	12/69	ex-USAF 51-15938, ex-RAF WX516 and ex-LOT-GV. Sold for scrap
P2V-5/P-2E	426-5135	0411	2-P-33/2-P-104	10/5/59	20/9/73	ex-USAF 51-15939, ex-RAF WX513 and ex-LOT-KF. To storage on 1/2/70. Sold for scrap
P2V-5/P-2E	426-5146	0412	2-P-34/2-P-105	24/11/58	24/11/69	ex-USAF 51-15941, ex-RAF WX522 and ex-LOT-JY. Sold for scrap
P2V-5/P-2E	426-5155	0413	2-P-35/2-P-106	3/3/59	10/65	ex-USAF 51-15950, ex-RAF WX524 and ex-LOT-IJ. Sold for scrap
P2V-5/P-2E	426-5158	0414	2-P-36/2-P-107	19/9/58	7/11/65	ex-USAF 51-15953, ex-RAF WX527 and ex-LOT-AC. Crashed on 7/11/65 at Josafat mount (Brazil) during a UNITAS exercise, crew killed
P2V-5/P-2E	426-5166	0415	2-P-37/2-P-108	19/12/58	10/65	ex-USAF 51-15961, ex-RAF WX549 and ex-LOT-BA. Sold for scrap
P2V-5FE/EP-2E	426-5137	0541	2-P-101	1965	1977	ex-USN BuNo. 127773. Accident on 12/10/69 and repaired. Sold for scrap
P2V-5FS/SP-2E	426-5416	0644	2-P-103	23/5/70	15/9/76	ex-USN BuNo. 131535. Accident on 15/9/76 at South Shetland. Crew killed
P2V-5FS/SP-2E	426-5413	0682	2-P-104	7/9/72	1978	ex-USN BuNo. 131532. Sold for scrap
P2V-5FS/SP-2E	426-5404	0683	2-P-105	8/7/72	1977	ex-USN BuNo. 131523. Sold for scrap
P2V-7S/SP-2H	726-7234	0706	2-P-110	9/77	1980	ex-USN BuNo. 148349. Cannibalised for spares. Sold for scrap
P2V-7S/SP-2H	726-7257	0707	2-P-111	9/77	28/7/82	ex-USN BuNo. 148361. Sold for scrap
P2V-7S/SP-2H	726-7283	0708	2-P-112	11/77	30/8/82	ex-BuNo. 150280. At the Museo de Aviación Naval at Comandante Espora
P2V-7S/SP-2H	726-7233	0709	2-P-114	9/77	14/5/81	ex-BuNo. 148348. Sold for scrap
B200 Super King Air	BB-546	0748	4-G-47/2-G-47/6-P-47	21/8/79	(current)	1979 to 1982 with the escuadrilla
B200 Super King Air	BB-549	0749	4-G-48/2-G-48/6-P-48	21/8/79	(current)	1979 to 1982 with the escuadrilla
PC-6 Turbo Porter	2047	0686	4-G-2/6-G-2/4-G-4	1971	(current)	1979 to 1982 with the escuadrilla. Now serialled 0684/4-G-2
B80 Queen Air	LD-449	0687	G-81/2-G-81/6-G-82	26/12/71	1992	Sold as surplus
B80 Queen Air	LD-452	0689	G-83/1-G-83/6-G-83	13/4/72	1992	Sold as surplus
B80 Queen Air	LD-453	0690	1-G-84/6-G-84	4/4/72	7/3/90	Accident at Base Aeronaval Almirante Zar. Written off
L-188PF Electra	1102	0691	5-T-1/6-P-105	12/11/74	1998	Used for engine spares
L-188PF Electra	1120	0692	5-T-2/6-P-106	1974	12/1997	Explorador and tanker. At Museo de Aviación Naval
L-188PF Electra	1122	0693	5-T-3	1974	1995	Now at a bar in Buenos Aires suburb
L-188A Electra	–	–	–	–	–	Arrived in 1974. Used for spares
L-188A Electra	1123	0789	6-P-101	1/1/83	20/9/89	Explorador. Crashed at BAAZ on 20/9/89
L-188A Electra	1070	0790	6-P-103	26/8/83	1997	In storage at BAAZ
L-188A Electra	1067	0791	6-P-102	12/82	1997	Explorador. In storage at BAAZ
L-188A Electra	1071	0792	2-P-101/6-P-66	12/82	1984	Used for spares. Fuselage at Museo de Aviación Naval
L-188A Electra	1072	0793	6-P-104	25/9/83	2001	Wave. Assigned to Museo de Aviación Naval
P-3B Orion	5158	0867	6-P-51	30/10/98	(current)	ex-BuNo. 152718
P-3B Orion	5172	0868	6-P-52	11/7/99	(current)	ex-BuNo. 152732. Arrived with Argentine crew
P-3B Orion	5186	0869	6-P-53	8/12/97	(current)	ex-BuNo. 152746
P-3B Orion	5205	0870	6-P-54	16/9/98	(current)	ex-BuNo. 152761
P-3B Orion	5207	0871	6-P-55	16/2/97	(current)	ex-BuNo. 152763
P-3B Orion	5216	0872	6-P-56	27/6/99	(current)	ex-BuNo. 153419. Arrived with Argentine crew
P-3B Orion	5219	–	–	–	–	ex-BuNo. 153422. Used for spares

AIR COMBAT

His Corsair loaded with rockets and napalm, Lieutenant Ed Godfrey from VMF-212 heads north on an interdiction strike against Chinese supply lines and troop movements. The Corsair excelled at this work, and was also a stalwart of close support missions.

Marine Corsairs in Korea

Towards the end of World War II the Corsair was used extensively for ground attack, and by the time of the Korean War it was the F4U's primary task. Marine Corsairs were among the first UN aircraft pitched into battle after the Communist invasion, and although they were subsequently augmented by later, more capable, types, they remained in the thick of the fighting until the last day of the conflict.

Marine Corsairs in Korea

By early November 1950 it seemed that the war would be over by Christmas, and the propeller-driven types were mostly operating deep in North Korean territory. This scene at Yonpo depicts the F4Us of VMA-212, with F7F Tigercats and a landing C-47 behind.

The highly-publicised jet age engulfed the American military as early as 1948. The Cold War had not officially begun and there was a certain amount of international prestige at stake. It was very important to the fledgling United States Air Force to be able to claim to have an 'all-jet' air force, especially in the Far East. Its front-line fighter groups (8th, 18th, 35th and 49th) quickly traded their Mustangs for the new F-80 Shooting Stars. No one could have guessed that, in two years, the first major Cold War conflict would rely heavily on two outdated World War II prop fighters that were still available in large numbers: the North American F-51 Mustang and the Vought F4U Corsair. The latter was used in large numbers by both the US Navy and Marine Corps, and their magnificent contributions to that war have become legend.

When the Korean War began, Marine aircraft were operating from their stateside bases. In early 1950, there were very few indicators that any unusual trouble was brewing in the Far East. Once it began, Marine ground and air forces quickly accelerated their activity. Marine Aircraft Group 33 was in position to react first. It controlled three F4U squadrons that were ready for a fight: VMF-214 'Black Sheep', VMF-323 'Death Rattlers', and a night-fighter squadron flying the F4U-5N, VMF(N)-513 'Flying Nightmares'. Once they arrived in Japan, it was agreed that both day squadrons would operate from carriers to give them maximum

Left: Captain J.J. Geuss rests on the wing of his fully armed Corsair in 1951. The checkerboard nose markings belonged to VMA-312, which was operating from Kangnung (K-18) at the time.

Main photo: Marine Corsairs operated from both ship and shore. This pair of heavily armed VMF-323 aircraft is about to launch from a forward base. The aircraft were F4U-4Bs with four 20-mm cannon.

The 'Black Sheep' of VMF-214 were the first Marine Corsair pilots to go into action, flying from the deck of Sicily, where these scenes were recorded in the late summer of 1950.

striking power and mobility. After only one day of refresher flights, VMF-214 landed on USS *Sicily* (CVE 118) on 3 August 1950, for immediate combat operations.

First action

At 16.30 that afternoon, eight 'Black Sheep' Corsairs were armed with incendiary bombs and rockets. They launched on the initial Marine air offensive action of the war. Led by VMF-214's squadron commander, Lieutenant Colonel Walter E. Lischeid, they struck the town of Chinju and the nearby village of Sinban-ni. These two locations were full of enemy troops and the Marines aggressively wreaked havoc in both locations, until all of their ordnance was spent. Right after the main force launched, two more Corsairs from VMF-214 took off for Taegu to be briefed on the broad tactical situation, and this information was brought back to *Sicily* for squadron briefings.

The following day, 21 combat sorties were logged against North Korean troop build-ups on the 8th Army's southern flank. These strikes carried several types of ordnance and were low level in nature. Some of the pilots still did not have the number of flying hours in the aircraft that they would have liked, but that did not slow them down. The enemy paid a heavy price for their threat to US troops. This was still early enough in the war that the North Korean troops had no idea how deadly losing control of the sky could be. In the time leading up to the Inchon Landing, thousands of North Koreans were killed by close air support (mostly from F4U and F-51 attacks).

Three days after VMF-214 began its combat missions, the 'Death Rattlers' of VMF-323 were in place on the USS *Badoeng Strait* (CVE 116). Beginning on 6 August, they did not slow down until the final days of the war. Their first day in combat sent them against the strength of the North Korean forces that were clashing head-on with US 8th Army troops. Squadron records verify 30 sorties flown with devastating results. Until this date, F-51 Mustangs had been constantly hitting the same area, with a limited number of aircraft. The Corsair actions had to be a psychological blow to the North Koreans, because the use of napalm against them had more than doubled in just one day.

Marines at sea

The Marines' early use of the carriers gave them an advantage over their Air Force counterparts. With the UN forces wedged into the Pusan perimeter, the number of operational air bases was limited. A significant number of missions were being launched from bases in Japan, which made the missions long and tedious. The opposite was true for VMF-214 and VMF-323: they cruised off the coast of the Korean Peninsula and hit enemy troop concentrations and supply routes repeatedly during the daylight hours.

On one August mission recorded by the 'Black Sheep', their first wave of Corsairs caught a large column of trucks moving south on a main highway. By the time the second wave left the scene, over 25 trucks had been destroyed and many were off the road in ditches. The third wave came in and finished off anything that was left. If these two squadrons had been forced to launch their missions from Pohang or Japan, the truck column probably would have been able to deliver its supplies to North Korean troops involved in pressing the shrinking perimeter around Pusan.

This was only one of many outstanding examples of the brigade air-ground team in action. MAG-33 aircraft (three squadrons) constantly flew over friendly troops in need of help. This was worked in precision shifts that gave the enemy ground forces very little time to get organised. As reflected in Marine training, the aviation units operated in support of the brigade as their top priority. MAG-33 records show that between 3 August and 14 September

The 'Black Sheep' had just one day's training prior to embarking in Sicily for combat duty. This view, from November 1950, shows the destroyer escort which was on hand to pluck aviators from the water if they ditched.

VMF-214 hounded the North Korean army as it retreated north, but the Chinese intervention in November reversed the situation. Here a plane captain poses in his F4U-4B, which already bears an impressive tally.

Marine Corsairs in Korea

MAG-33's second day-fighter Corsair squadron, VMF-323, joined the fray three days after VMF-214 – on 6 August 1950. The squadron flew from USS Badoeng Strait before moving ashore to continue operations, initially from Pusan and then Kimpo. These lasted until June 1951, when the squadron withdrew to Itami in Japan to prepare for embarkation aboard USS Sicily on 5 June. During the cruise the squadron's F4U-4Bs were involved in enforcing the naval blockade off the northeast coast of Korea.

(the day before the Inchon Landing), VMF-214 flew a total of 670 sorties, VMF-323 498 and VMF(N)-513 343. These numbers were very impressive considering the short time span involved.

On 8 August, a Corsair from 'Death Rattlers', returning from a mission, landed safely back on the *Badoeng Strait*, but as the pilot taxied back to the parking area, the Corsair landing behind him hit the deck hard and bounced over the landing barrier, ramming into and destroying the first aircraft. The pilot was unhurt. This became the first recorded Marine Corsair loss of the war.

First combat loss

Worse was to come for VMF-323 and they were only days into the war. Two days later, Captain Vivian M. Moses was in a division strafing targets near the village of Chin dong-ni. His Corsair was hit hard by anti-aircraft artillery and, as a result, was losing oil and had to ditch in enemy territory. Fortunately, helicopters from VMO-6 were in the area and he was picked up, unhurt, and returned to the carrier. This became the first Marine Corsair loss of the war over enemy territory. However, it was not over for Captain Moses. The following day (11 August), he launched again for a mission in the area west of Kosong. This time his luck ran out as his F4U-4B (tail number 97492) was riddled by ground fire; in an attempt to crash land, Captain Moses died. Within one 24-hour period, VMF-323 had sustained three aircraft losses, bringing its total to four. At the time, it was less than one week into combat operations.

The Marine day squadrons usually recovered on the carriers at last light and were loaded and ready to launch by first light the next morning. This initially gave the North Koreans the impression that the night belonged to them, but they were in for a rude awakening. The Marines filled this gap with their 'Flying Nightmares' squadron equipped with the all-weather F4U-5N Corsair. Three night attack aircraft were able to inflict extensive damage on any road or rail traffic movement after dark: the F4U-5N Corsair, F7F-3N Tigercat, and the Air Force's B-26 Invaders. Their crews were responsible for denying the North Koreans a chance to build up supplies for any type of offensive, particularly after the Chinese entered the war in November 1950.

In addition to its three regular F4U squadrons, MAG-33 had a small detachment of F4U-5P reconnaissance aircraft assigned. T/Sgt George Glauser was one of the Corsair recce pilots to fly low-level photo-runs over Inchon harbour prior to the amphibious landings.

Dive-bombing

Captain John S. Perrin
VMF-214 'Black Sheep'

The US Marine landing at Inchon turned around the war almost instantly. The ground forces moved rapidly and, with extensive air support, extracted a heavy toll on defending troops. Many outstanding feats were performed by Marine aviation during those few days, and one of them was by Captain Perrin.

"Our troops had met some heavy resistance at Yongdungpo, which was an industrial suburb of Seoul. The advance had bogged down due to heavy machine-gun fire coming from a metal warehouse. I was flying wing with Lieutenant Colonel Walter Lischeid and we had just launched from the USS Sicily [CVE 118], so we were only minutes from the target. I was carrying eight 250-lb [113-kg] frag bombs and one 500-lb [227-kg] V-T bomb fused to go off above ground. Colonel Lischeid was carrying only napalm, so I was the one that would have to go after the warehouse. Working with a forward air controller, my biggest worry was the fact that our troops were only about 50 yd [46 m] from the target, and the possibility of collateral damage was a factor. The building was easy to spot because it was all metal, as opposed to the other structures in the area.

"I would make my bomb run parallel to the road that separated the target from the friendly troops. This reduced the chances of hitting our guys. I was at 8,000 ft [2438 m] when I rolled over and headed down. This was critical, because you had to have time to line the target up and with the fast air speed that would be increasing, you didn't have much time. As my aircraft was in the vertical, headed down, my speed was approaching 400 kt [740 km/h]. It was easy to line up the warehouse and I waited until my altitude was down to about 2,000 ft [610 m] before releasing the 500-pounder. You have to stay in the dive long enough to make sure your bomb is not affected by any *g* that would throw it off course. When released, it should only be under the influence of gravity.

"The most dangerous part for the pilot is figuring when to start pulling out. Your dive speed in a Corsair is very fast and when you start to pull out, you could grey out, which would make it difficult to recover before hitting the ground. In this case, I was able to get the nose up at about 1,000 ft [305 m] and it took a lot of grunting and concentration to stay alert. As the nose moves above the horizon, you pour the power to it and try to exit the area as fast as possible.

"My bomb exploded just as it reached the roof and the explosion continued as it penetrated. It took out everything in that warehouse, including the machine-guns. I felt very good about the perfect hit and I'm sure our troops on the ground felt the same way."

Captain John Perrin flies aircraft '16' during a mission from USS Sicily in 1950 in support of United Nation troops defending the Pusan perimeter.

VMF-212 was in the second tranche of F4U squadrons to be committed to the war in Korea. Here one of its Corsairs is off-loaded at Yokosuka, Japan, in September 1950.

Days after Marine troops had liberated Kimpo airfield following the Inchon landing, Captain Al Grasselli of VMF-212 is suited up and ready to fly a combat mission.

Marine records mention the 'Kosong Turkey Shoot', which occurred on 11 August 1950 and primarily involved Corsairs from VMF-323. It began early in the morning with an artillery bombardment of the town of Kosong, an attack that was standard procedure during any ground assault against entrenched enemy forces. Marine artillery '105s' blasted the target area long enough to soften it up. Overhead was a fully loaded division of VMF-323 Corsairs ready to be called in to assist when needed. For some reason, an enemy column of 100-plus vehicles was leaving Kosong and heading north – perhaps they were Soviet or Chinese advisors. The vehicles consisted of Soviet-made jeeps, motorcycles and trucks loaded with troops. When the artillery barrage ended, the Corsairs pounced on the column and lit it up quickly. When the first division's ordnance had been expended, more Corsairs were launched from *Badoeng Strait* and the destruction continued. These relentless attacks used mostly 5-in (12.7-cm) high-velocity aircraft rockets and 20-mm cannon. By the time the attack was halted, all of the vehicles were either burned out or in a ditch. It was one of the most effective attacks of the war, because daytime targets rapidly diminished as the war progressed.

The entry of just three Marine Corsair squadrons had a quick impact on the US effort, especially when working with ground troops. A new concept was being developed and it improved week after week until the war ended: during the day, the fighter-bombers would work over anything that moved, whether they were under the direction of a forward air controller or not; when the day attackers had to head for home at last light, the heavily-armed night-attack types would take over. Night operations for VMF(N)-513 went well and they usually worked one sector, in shifts, which meant that the North Koreans had no chance to catch their breath, except when the weather was so bad that all aircraft were grounded.

Reinforcements arrive

On 16 August 1950, MAG-12 received orders to move its two Corsair squadrons (VMF-212 and VMF-312) overseas. This move would also include a night-fighter squadron, VMF(N)-542, which was flying the F7F-3N Tigercat. Once these squadrons entered the war, a major change was made in the command structure.

About one week after the Inchon landing, the Marine Corps announced that the two MAGs in-theatre would swap squadrons, but their original squadrons remained intact. The three squadrons already in place (VMF-214, VMF-323 and VMF(N)-513) would continue to work from the carriers or land bases from which they had been operating.

For the Marines, the biggest operation since World War II was the Inchon landing on 15 September. This brilliant tactic was a life-saver for the UN's presence on the Korean Peninsula. The first major planning hurdle was

USS Badoeng Strait (CVE 116) made its second of three war cruises between 15 September 1951 and 1 March 1952. Aboard were the Corsairs of VMF-212.

Combat air patrol
Lieutenant Ed Godfrey
VMF-212 'Devilcats'

"I can remember several of the missions I flew and all of them had a certain amount of danger involved. One of these was a combat air patrol (CAP) over the carrier. We usually put two F4Us up overhead while the strike aircraft were launching to attack a wide array of targets, and once the deck was clear we would remain in position just in case of any intruders.

"On one of these occasions, our combat information centre picked up a fast-moving bogey flying out of China due east over the Yellow Sea. We were immediately vectored to the west on an intercept course with the lone, hostile aircraft. Before we had visual contact, it suddenly made a fast 180° turn back to the west. We don't know what the pilot had in mind, but with the speed it was travelling, it definitely was a MiG-15. Intelligence seemed to think it was just testing the reaction time of our task force. We would have taken it on and done our best to have prevented it from getting through. However, this was better than having to tangle with the MiG and perhaps end up in a one-man life raft between China and North Korea."

Some days were better for flying than others: this scene was recorded on the ice-covered deck of USS Rendova (CVE 114) in late November 1951. Ed Godfrey and the rest of VMF-212 were aboard for combat operations from 22 September and 6 December.

Above: A VMF-212 F4U is seen at an airfield close to the front line, having diverted in with a problem. It is parked next to a T-6 Mosquito forward air control platform, with which the the Corsairs regularly worked.

Left: Corsairs from VMF-212 share an airfield with F7F Tigercats and AD Skyraiders. All three types were involved in close support and continual harassment of Communist supply and troop movements.

Below: Although Communist forces hoped the cover of bad weather would allow them to move men and supplies, the UN forces continued to fly in very marginal conditions. Here a division of VMF-212 F4Us feels its way through the soup in the autumn of 1950.

to obtain recent pictures of the harbour and the low and high points of its famous tides. The only pictures in existence were those taken in 1948 by the USAF's F-15 (RF-61C) Reporters based at Johnson AB, but they did not provide the details needed to plan an amphibious assault. It was left to the Marine detachment flying its photo versions of the Corsair and Tigercat to do the low-level work in this area.

The images showed that there was a potential for disaster on the eastern side of Wolmi-do, where numerous dug-in artillery positions would exact a toll from any amphibious craft trying to move on Inchon. These emplacements were cleverly constructed among heavy tree cover, which made them very difficult to spot from the air. All this defensive preparation had been finished by the North in about 10 weeks.

Tree-burning mission

The first order of business was to burn out the foliage, and that is where the first strike by Marine Corsairs was concentrated. A total of 28 F4Us participated in the 'softening up', during which eight from the 'Black Sheep' and six from the 'Death Rattlers' scorched the entire side of the island. The effect was immediate. When the second wave of 14 Corsairs arrived, the smoke was so thick that they had to circle until it had cleared enough to allow them to see their targets.

'Black Sheep' F4Us follow USAF F-51s on to the runway as they launch for a close support mission in 1950. Between its first mission on 3 August and 14 September 1950, VMF-214 flew 670 combat sorties.

The second day of attacks was just as intense as the first, except that the tenacity of the enemy guns had been lessened as a result of the previous day's heavy air strikes. US Navy destroyers moved in close in an attempt to get the gun positions to open up. When they did, orbiting Marine Corsairs armed with rockets and napalm dived down and silenced them. When their ordnance was used up, the aircraft dropped down low and acted as spotters for the ships' guns.

The Inchon planners allowed five days to make the area safe for the landing. These closely co-ordinated attacks by Marine and Navy aircraft paved the way for a safe and very successful landing at Inchon by Marine ground forces. The critical amphibious assault took place after dawn on 15 September 1950.

Even though intelligence sources had determined that the five-day barrage had all but neutralised North Korean defences, nothing was left to chance. Before dawn on the day of

Air Combat

VMF-323 continued to fly from Badoeng Strait *in the crucial days of December 1950, when the UN was trying to stabilise the Chinese advance. In the spring the squadron operated from shore bases, beginning with Pusan (K-1), where these aircraft are seen.*

the landing, the escort carriers *Badoeng Strait* and *Sicily* launched their Corsairs in support of the landing, while three of the Navy's Task Force 77's big 'Essex'-class carriers – *Valley Forge*, *Philippine Sea* and *Boxer* – were offshore preparing to launch their air wings of heavier attack aircraft. MAG reports state that Marine Corsairs were overhead at Inchon before 05.00.

During the first two days of the assault (after the landing), the weapons of choice for the Marine Corsairs were napalm and the four 20-mm cannon. A good example of their use was when ground troops met fierce resistance along the lengthy causeway. A platoon of North Korean regulars had decided to fight to the death rather than retreat, so the F4Us dived down and showered them with napalm. Marine command decided the aircraft would carry a heavy napalm load because it was the most effective way of dispatching enemy resistance, and it ensured that friendly forces would suffer fewer casualties.

Any efforts by the North Koreans to reinforce their positions in the 'hot' zones were futile. Marine and Navy aircraft jumped on them so quickly that nothing was able to get through.

On the first night after the landing, many enemy troops and equipment pulled back. At dawn the next day, VMF-214 aircraft were up with two divisions (eight Corsairs) loaded with rockets and 20-mm cannon. As predicted, they encountered Soviet-built T-34 tanks in the open on the road to Kansong-ni. Like a mad swarm of hornets, the heavily armed 'Black Sheep' pilots attacked the column and quickly knocked out the enemy armour. Tank crew members were seen abandoning their tanks and running into nearby thatched huts.

A few of the F4Us did not expend all their ordnance, so they followed the main road in search of trucks and smaller vehicles. While the Marines were observing and cheering the attack from their positions on a nearby hill, one of the Corsairs (F4U-4B tail number 97479) began a strafing run against a T-34 and never pulled out. The cause of the crash will probably never be known, but it is likely that the pilot, Captain William F. Simpson, was fatally hit by ground fire during his dive. His body was recovered as the Marines advanced through that area a short time later. This was the third Corsair loss for VMF-214 in the war.

The fight for Seoul

The vicious door-to-door battle for Seoul began two days after the Marines secured Inchon. It was assumed that most of the North Korean troops were involved far to the south, with very few reinforcements available. On 17 September, intelligence confirmed that approximately 3,000 enemy personnel were headed south from positions around the 38th Parallel and another small group was coming up from the Suwon area. The biggest problem encountered as the Seoul campaign got under way was the large amount of enemy armour stationed in the city and surrounding suburbs. VMF-214, armed with rockets and napalm, circled a specific area and, one by one, took out the tanks. Working on VMF-214's flanks were the 'Death Rattlers', carrying the same ordnance loads. As a hot spot would crop up, one of the squadrons would send a couple of Corsairs to deal with it. The North Korean casualty rate was very high.

The diamondback rattlesnake applied to the 'Death Rattlers' Corsairs was one of the most distinctive markings applied to any UN aircraft, but it was relatively short-lived. Replacement aircraft were not marked, and the snake disappeared due to fears that it increased conspicuity in dense AAA environments.

Hit over enemy lines

1st Lieutenant Vance L. 'Bud' Yount
VMF-323 'Death Rattlers'

"We were assigned to fly a mission against Chinese troops that were holding ground up around the village of Uijongbu. We were carrying a maximum load of ordnance: bombs, rockets and 20-mm ammunition for our guns. Our division pressed the attack and scored excellent hits all over the area that contained the hostile forces. Starting my final bomb run, I made a stupid statement that I thought I had seen something on the next ridge and was going to stay low after releasing my ordnance. This would give me a chance to check it out. Unfortunately for me, there was one lucky Chinese rifleman that got a critical hit on my Corsair and it was in one of the most vital places – my oil line.

"I felt the thud from the round as it hit. Suddenly, there was a puff of smoke in the cockpit and I watched my oil pressure drop down to 15 psi (normal was approximately 80 psi). I yelled out that I had been hit and was turning to head south. One of the pilots in my division, Lieutenant Jim Gleaves, pulled up alongside me and stated he was going to stick with me and for me to try to fly the aircraft back to base. [Gleaves had been in the movies before coming to Korea and was the stand-in for John Wayne in *Flying Leathernecks*. He was killed about a month after this

'Bud' Yount poses by a 'Death Rattlers' F4U on the deck of USS Sicily *in 1951. On 21 September 1950 MAG-33's three F4U squadrons were reassigned to MAG-12. The two daytime units were based on the carriers, while the night-time unit was ashore at Itazuke AB, Japan.*

mission.] I dropped a few degrees of flaps as he flew under my tail searching for the oil leak.

"Naturally, with the sharp drop in pressure, I was sure I had been hit in the oil cooler, which was a very vulnerable spot. All the other gauges were normal and the big fan in front was still going around. If the engine quit, I had two choices: bail out or belly in. Bailing out of a Corsair was 'iffy' at best and many pilots had been hurt doing it, so I figured my best option was to set it down in a flat place. In a situation like this, the first thing to do is to clean the aircraft up by jettisoning any unused ordnance and the external fuel tanks. This reduced the risk of the aircraft flipping over or veering sharply in an undesired direction once contact with the ground was made. My external tank went flipping through the air behind me, and I laughed when Lieutenant Gleaves radioed me and asked if that was me bailing out.

"Everything seemed to be going according to plan as I flew closer and closer to friendly territory. As my good luck would have it, I was able to nurse it back to Suwon AB and land safely. Once the damage had been assessed, we found that a lucky shot had hit my oil pressure gauge line, causing it to give an alarmingly low reading. Fortunately, both I and my Corsair were back in the war immediately, but it could have been a different story if I had gone ahead and bailed out before reaching friendly territory."

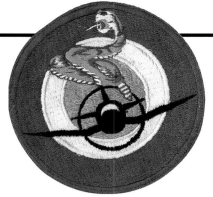

VMF-323 was activated in 1943, and remains today as one of the backbones of Marine airpower. It flew Corsairs throughout the Korean campaign. These views show the squadron's flightline in 1952 (left), a group portrait from Kimpo in 1951 (above left), and some of the squadron's maintenance personnel working on the reliable Pratt & Whitney R-2800-42W Double Wasp radial which powered the F4U-4B (above).

North Korea did possess one effective weapon in quantity: 120-mm (4.72-in) mortars. Its potential was respected by all the UN ground forces. Utilised from high ground, it could be costly to the Marines, not only in casualties but in lost equipment. The persistent attacks of VMF-214 and VMF-323 were credited with preventing the enemy from securing any high ground from which they could effectively use these mortars.

The Marine contribution to the war effort intensified on 19 September when two fresh Corsair squadrons became officially involved. VMF-212 'Devilcats' and VMF-312 'Checkerboards' flew into the recently-liberated Kimpo air base from Itami AB in Japan and launched their first combat sorties the following day (20 September).

Break-out from Pusan

With the success of the Inchon landing, the UN troops who had been squeezed into the Pusan perimeter launched a massive counter-attack, timed so that the North Koreans were already aware that they were going to be cut off. The lines broke quickly. As they began their long retreat to the north, some took up positions on the high ground in order to hit the advancing UN troops with mortar fire. This made them easy targets for the Marine F4Us that stayed close to the forward advancing troops during every hour of the day. One of the Marines on the ground described it this way: "The enemy troops that were firing at us were literally scraped off the hilltops by napalm and 20-mm fire by our Corsair cover."

The only chance that the retreating forces had to slow the UN advance was to try to hold to high ground along the routes and cause as many delays as they could. A good example of this happened on 20 September on Hills 80 and 85, which offered excellent views of the countryside and main roads. The Marine ground forces were attempting to get to the top of both hills when darkness came, so they backed off for the night. The North Korean troops a short distance to the north assumed that the Marines had obtained the pinnacles and were in the process of digging in. During the early morning hours, they assaulted both hills with automatic weapon fire, which the Marines could hear in the distance. By daybreak, the enemy realised that there was no one on either hill. This move proved to be a terrible mistake on their part.

The 'Death Rattlers' were overhead at first light, and when the Marine forces began their long trek up the hills, the Corsairs descended and covered the vantage points with napalm. By the time the ground troops reached the top, there was not enough left of the enemy to form stiff resistance. There had not been enough time for the North Koreans to dig in, so they were killed in the open.

As was expected from the continuous tenacious attacks, the 1st MAW set a personal record on 24 September when VMF-212 'Devilcats' recorded the outstanding feat of flying 46 combat sorties. They were evenly distributed between close air support and search and destroy missions. While the emphasis still centred on liberating Seoul, the F4Us ranged far enough to hunt down any vehicles exiting or entering the area. There was a lot of activity (moving targets) 20 miles (32 km) northeast of the city. It was not clear from existing records whether this was retreating forces moving north out of the Pusan area or reinforcements coming down to help in Seoul.

The northwest corner of Seoul was a flak trap. Any low-level flying was extremely hazardous but this was the only way the Corsair pilots could be effective, so the number of aircraft hit by ground fire was very high.

Right: The satisfaction of a good job well done – VMF-214 plane captain Sergeant Clarence Chick relaxes on the wing of his F4U-4B after it has been serviced, armed and made ready to go for the afternoon's close support 'push'.

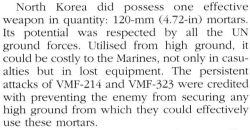

Unsung heroes – the plane captains kept their charges flying every day through a combination of sheer hard work and ingenuity. These VMF-214 personnel are at Taegu early in the war. Note the 12th Fighter-Bomber Squadron F-51Ds in the background.

Air Combat

F4Us in close support
Lieutenant Colonel Walter E. Gregory
Executive Officer, VMA-212 'Devilcats'

"We provided close air support as required. Our [USMC] troops trusted us to work extremely close to their positions, more so than other types. This was probably due to their peacetime training back in the states. We would hover over the area and the FAC would brief us in detail as to what we were going after. Then one of our Corsairs would make a dry run with the FAC talking him through it. This cut down the risk of hitting friendly troops. On many of these runs, we drew horrendous small arms fire. Sometimes, we figured it was a trap and radioed the FAC to call off the attack on that particular target. His only comment was that the ground fire was normal and to press on.

"On the bomb runs close to friendly forces, we had a rule that if there was no communication with the FAC then there was no ordnance dropped. When using napalm, 20-mm ammo or rockets, the runs would be fast and not so steep; with bombs, the angle of dive was steeper. When things were a bit rough on the ground, it was amazing how close we were permitted to come.

"On one occasion, 12 aircraft from each of three squadrons were involved. Some were AD Skyraiders.

While waiting at altitude for his time to dive in for a bomb run, a 'Devilcats' pilot took this photograph of squadron-mates working over a supply dump next to a bridge. As well as supporting troops in contact, Marine F4Us joined the major campaign to prevent supplies reaching the Communist front lines.

The target was a round dome-type hill approximately 300 yd [275 m] in diameter. The enemy had honeycombed it and it was jammed with troops (probably a reinforced battalion in strength). The first element was ADs and they put two 1,000-lb [454-kg] GP bombs (delayed fuses) right into the dome. Right behind them were VMA-323 AU-1s loaded with 24 napalm tanks. Then we came in with our AU-1s, dropping clusters of small phosphorous bombs. These were deadly against ground troops and the burning substance stuck to their clothing and skin. We rendered that entire force ineffective and saved many Marine lives on the ground."

VMF-214 kept several of its aircraft overhead at all times during the day, depending on forward air controllers to guide them to the most urgent targets. These aircraft received heavy damage, two being shot down in the *mêlée* on 24 September. Some of the risk faced by all of the Marine F4U pilots was due to an order that had gone into effect after World War II, which stated that all armour around the air-cooler system be removed. It was a peacetime economy move and the order had never been given to put it back. This made the aircraft vulnerable to small arms fire that could damage the oil lines, causing it to be lost over hostile territory.

The following day (25 September) proved to be one of the darkest for Marine air operations in the war. Lieutenant Colonel Walter E. Lischeid, CO of VMF-214, was killed when he was hit by ground fire over the west end of Seoul and went down in flames. Within a couple of hours, the CO of VMF-212, Lieutenant Colonel Wyczawski, and VMF(N)-542's CO, Lieutenant Colonel Volcansek, in F7F Tigercats, were shot down in the same general area, but they survived and were rescued. Several other Corsairs survived the day with heavy battle damage, giving support personnel back at the bases an idea of the dangers the close air support pilots faced at such low altitudes.

Aggressive pursuit

The northward pursuit of fleeing North Korean forces kicked into high gear once Seoul had been secured. VMF-212 and -312 were relentless in their attacks against these forces. Between 24 September and 3 October, both squadrons kept all their aircraft in the air, managing very quick turnaround times because they were operating from land bases and were close to the enemy forces. The rationale behind Marine air operations on such an aggressive scale was that if there had been any let-up, the enemy would have tried to regroup in the hilly country and make a stand. This would have led to unnecessary delaying action and, more importantly, would have caused casualties among the Marine troops. With this policy, the North Koreans retreated rapidly and in complete disarray.

During this period, at least one incident stands out among the few missions that have been recorded in detail. On one clear and sunny day, VMF-312 pilots were running the roads leading north when they received word that retreating enemy troops had established a roadblock that slowed the movement of Marine troops. They responded with an immediate attack in their napalm-laden F4Us, transforming the roadblock into a charred mass. The remaining aircraft in the other 'Checkerboard' divisions still had their 500-lb (227-kg) bombs intact, and immediately proceeded to follow the road north. A few minutes later, they came upon a column of eight loaded North Korean trucks trying to exit the area. The first Corsair hit the last truck in the column and, before the second pilot could even initiate his pass, the secondary explosions had jumped up to the lead truck. In one pass with one bomb, seven trucks had been destroyed. Needless to say, the lead truck never had a chance.

As the war moved steadily northward, it looked as if the Marines were going to be called on to make another amphibious landing, this time at Wonsan. Before final plans could be executed, however, Republic of Korea troops secured the North Korean port and the assault was cancelled. Two Marine Corsair squadrons (VMF-214 and VMF-323) had been kept on carriers in order to support this action. Instead, VMF-312 flew immediately from its base at Kimpo on 14 October in order to take up operations at Wonsan. Its primary assignment was to support troops moving farther north.

In the meantime, VMF-214 and VMF-323 sailed on their respective carriers from the naval base at Sasebo, Japan. Their tasks were to keep the general area around Wonsan harbour

Captains Geuss and Bailey from VMF-312 are pictured by one of the squadron's F4U-4s in 1951. The aircraft is loaded with 5-in (12.7-cm) HVARs – which were often used against Chinese armour, and two napalm tanks – the weapon of choice against troop concentrations.

Below: This division of VMF-312 F4Us has completed its mission, expended all ordnance, and is heading back to Kangnung (K-18) in 1951. The mountainous terrain would suggest that the aircraft were still in North Korean airspace.

Like many Marine squadrons, VMF-312 had a sizeable Reservist contingent. This group portrait at Kangnung in 1951 shows the 'Glenview Boys', these pilots having all served before mobilisation with the USMCR squadron VMF-121 at NAS Glenview, Illinois.

VMF-312 'Checkerboards' pilot Captain J.J. Geuss poses by a pile of box-finned 100-lb (45-kg) bombs at Kangnung in October 1951. Corsairs dropped a large number of GP bombs, notably against the trenches which stretched along mountain ridges.

The Corsair force, exemplified by this F4U-4 of VMF-312 from Kangnung in 1951 (above), operated primarily in the area to the south of Pyongyang. Working alongside USAF F-51s, the main tasks were to attack troop concentrations and to ensure nothing moved on the roads by day. They also provided close support for UN troops along the MLR (Main Line of Resistance – front line). Losses were heavy in the latter role, particularly in the winter of 1951/52 when the scene below was recorded, owing to the aggressive tactics that were employed.

secure and to fly top cover for minesweeping operations. This continued until around 27 October. There was very little interference from North Korean forces, whose main objective seemed to be crossing into Manchuria to fight another day.

Wonsan remained a very dangerous area for the ROK and Marine troops. There were numerous pockets of resistance and the North Korean forces who were willing to fight made every effort to challenge the UN when they could. This is why the Corsairs flew such a wide radius around the area.

On 24 October, VMF-312 caught a column of approximately 800 North Korean troops near Kojo, some 40 miles (65 km) southeast of Wonsan. The aggressive F4U pilots inflicted heavy losses with a variety of ordnance.

Marine troops had to be protected regardless of the cost. The area that required round-the-clock surveillance was too large for one squadron to handle, so VMF(N)-513 flew some daylight sorties to fill the gap. As a result of the air cover, the North Koreans were never able to present effective delaying actions, at least on the fronts secured by ROK troops and Marines.

Enter the Chinese

At this time, the UN forces were approaching the Yalu River and talk was that the war would be over by Christmas. However, the first week of November sounded the death knell for any dreams of the war ending quickly. The Chinese poured across the Yalu River by the thousand and suddenly the Marines were in a situation where they were fighting for their lives under the worst of weather conditions. In the first major encounters between the Corps and the Chinese, the latter had the advantage of higher ground. Two squadrons (VMF-312 and VMF(N)-513) were called in to make the first strikes against the new aggressors. Using GP bombs, 5-in (12.7-cm) rockets and 20-mm rounds, their attacks kept the enemy off balance and rendered them relatively ineffective – but things were not over yet.

Of all the heroic efforts by Marine ground and air forces during November, one particular story should be told. It was not spectacular, but proved just how vital the F4Us were in providing assistance to the troops. The weapon that worried the troops most was the Russian-built T-34 tank, especially if they did not have heavy firepower with them. One of these tanks could easily jam a road and, in the mountainous terrain, there was no way to skirt around it: it had to be knocked out. On one such occasion, a T-34 blocked the Marines from advancing on the village of Chinhung-ni. A FAC radioed a division of VMF-312 Corsairs in the area and, within a minute, one peeled off and came straight down, holding his dive until almost point blank range. According to a witness, two puffs of smoke suddenly appeared under the wings as the pilot fired his remaining rockets. Both hit the tank, which erupted in a fireball.

December 1950 proved to be a trying time for all UN forces in Korea, and especially the Marines. The Chosin Reservoir campaign has

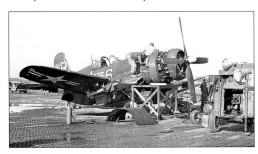
Above: Maintenance crew often worked miracles to repair battle damage in time for the F4Us to see action again the next day. When aircraft were too badly damaged to repair they became useful spares sources.

Right: The Corsair's compact dimensions suited it to serving aboard the smaller escort carriers of the 'Commencement Bay' class (Badoeng Strait, Bairoko, Point Cruz, Rendova and Sicily) and the single 'Independence'-class vessel Bataan. These bomb-laden VMA-312 F4U-4s are aboard the latter in March 1953.

Air Combat

F4Us from VMF/VMA-312 crowd the deck of Bataan *in 1952. Deploying Marine Corsairs to the carriers allowed them to hit targets of interest in North Korea after a much shorter transit than was possible from most of the available land base locations. The carriers sailed off both east and west coasts of Korea.*

Corsair pilot, leading a division of four aircraft, spotted a large number of troops trying to hide in a small village. His division immediately initiated a napalm attack and killed most of the troops. The survivors scattered from the burning huts and were gunned down as the Corsairs made several strafing passes with their 20-mm cannon.

Marine air operations

7 September through 9 October 1950 – Corsair day squadrons only

Squadron	days in action	combat sorties
VMF-214	16	484
VMF-323	22	784
VMF-212	19	607
VMF-312	10	288
Total		**2,163**

■ VMF-214 was pulled from combat operations on 25 September for replenishment in Japan.
■ During this period of approximately 30 days, 11 Marine Corsairs were lost. From this total, six pilots and one crewman were killed, and two pilots were wounded. The rest were rescued and resumed combat operations.

remained vivid and well-remembered through numerous books and articles. Not only was this rolling battle conducted against overwhelming odds but it took place in some of the worst weather conditions imaginable. The breakout at Yudam-ni on 3 December stands as one of the Marine Corsair pilots' finest days. With two observation aircraft from VMO-6 in the air from dawn to dusk, they pounded the Chinese forces unmercifully. They covered all sides of the move and constantly reported enemy troop movements and positions to the Marine columns. Records indicate that all five Corsair squadrons were involved, logging a total of 132 sorties that day.

F4U stats for 3 December 1950 (Yudam-ni)

Squadron	Sorties flown
VMF-212	27
VMF-214	36
VMF-312	34
VMF-323	28
VMF(N)-513	7

During this brief campaign, the F4Us operating from Yonpo were able to turn around quickly, so there were very few, if any, gaps in the coverage. They contributed about 100 sorties per day, while another 30 sorties were logged by the 'Death Rattlers' flying from the deck of *Badoeng Strait*. On 4 December, VMF-212 left Yonpo and flew back to Itami AB in Japan to prepare for redeployment aboard the carrier USS *Bataan*. About this time, USS *Sicily* was steaming into the area to take VMF-214 on board (7 December).

The month of December gave flight operations on the carriers a taste of a typical Korean winter. They faced winds stronger than 65 kt (120 km/h), heavy seas and freezing temperatures. Up to 3 in (7.6 cm) of ice had to be removed from the deck, and those Corsairs that spent the night on deck had to be moved below to thaw out. Flight operations were cancelled several times.

UN in retreat

The weather and the Chinese managed to hem Marine air operations into a corner. As UN forces began an orderly retreat to the south, the number of bases from which the F4Us could operate steadily dwindled. Most of the bases south of the 38th Parallel were overcrowded with USAF units, so the only places for the Marines to go was the carriers or a base in Japan. When Yonpo closed down, VMF-212 and VMF-214 went to the carriers and -312 began carrier qualifications so it could get back into action.

As the Chinese onslaught continued to roll south, Marine air support gave them little peace. With so many enemy troops in the open, the ordnance of choice was napalm. One

By the spring of 1951, the Korean War had completely changed in nature and appearance. The Communists had been stopped and pushed back to approximately the same line that had separated the north and south before the war started. However, the war still had another 27 months of fighting ahead before it ended. The high-profile, well-publicised portion of the war had ended after just 10 months, replaced by a new type of war similar to a cat-and-mouse game. The Chinese logistics personnel were assigned the job of moving supplies and equipment to the front lines,

VMA-312 spent much of its war afloat on carriers. Here the squadron's F4U-4Bs are seen on Bataan *in mid-1952.*

Above: Lt Tom Archer, one of the 'Checkerboards' pilots, poses on the deck of USS Bataan *(CVL 29) in February 1953. During this cruise VMF-312 was aboard the carrier from 9 February to 8 May 1953.*

Left: Ordnancemen, accompanied by canine helper, load 5-in (12.7-cm) HVARs (High-Velocity Aircraft Rockets) on to VMF-312 F4Us on the deck of USS Bairoko *(CVE 115) in 1952.*

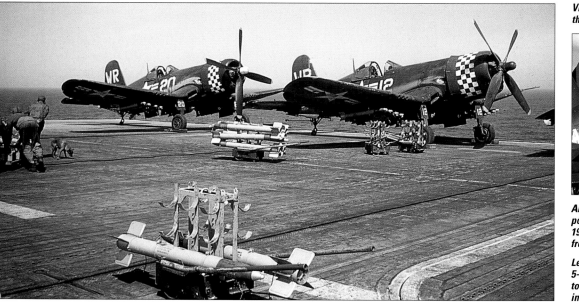

Marine Corsairs in Korea

The attack-dedicated AU-1 Corsair was introduced to the theatre by VMA-212 in June 1952. The squadron designation changed from VMF around the same time, and the Marine night-fighting units were also receiving the jet-powered F3D Skyknight. This AU-1 is parked at Taegu (K-2) in 1952 while it awaits minor repairs.

enough to not only maintain daily requirements but also to build up reserve to sustain a major offensive. This was the only way they could gain significant real estate. On the other side, the UN ground forces were ordered to hold what they had, and the air arm was assigned the job of stopping all convoys and trains coming south, thus preventing the Chinese from gaining any momentum on the ground.

The Marine Corsair squadrons were just like their counterparts in the Air Force fighter-bomber squadrons in that they did not receive much coverage from the media. They did their job extremely well on a daily basis and they received very little attention outside their ranks. The American public had grown tired of the war and the only subject that had enough action to make the front page of newspapers was the high-profile aerial duels being fought between F-86 Sabres and swept-wing MiG-15s. This trend continued until the war ended on 27 July 1953.

Tangling with MiGs

However, one incident involving the Corsair did stir a lot of attention and it involved a vicious fight between the World War II-era Corsair and the new MiG-15. Captain Jesse Folmar, a pilot in VMF-312, launched from USS *Sicily* for an interdiction mission against Chinese troop concentrations around Chinnampo. His wingman was Lieutenant Walter Daniels. They could not find any enemy activity, so they moved into another area looking for targets. As they began a turn to move up the coast, two MiG-15s appeared and began to set up a firing pass on both Corsairs. The Marines immediately jettisoned their ordnance and turned into the MiG formation.

One of the MiGs made a pass, and as it turned up and to the left, Captain Folmar turned inside of it and fired a burst from his 20-mm cannon. His rounds impacted all along the side of the MiG and it began to leave a trail of fuel vapour that quickly turned into black smoke. At this instant, the stricken jet began decelerating rapidly and moments later the pilot ejected. Folmar stated that the pilot appeared to be a ball of smoke and, when his chute opened, he could tell that the pilot's

Like VMF-212, VMF-323 received the AU-1 Corsair towards the end of the war, and also changed designation to VMA-323. The AU-1 was the production version of the XF4U-6, and was dedicated to ground attack duties, with a more powerful R-2800-83W engine developing 2,300 hp (1716 kW) with water injection. It had an extra pair of rocket rails under the outer wing panels for a total of 10.

Reserves and FACs

Captain Jay W. Hubbard
VMF-214 'Black Sheep'

"There were two special groups of Marine pilots that deserve far more credit than they have been given. First, the recalled Marine Air Reserve pilots, who filled the cockpits that had been manned by the regulars who had made the initial deployment and, having finished their required missions, were now back stateside teaching what they had learned to the new pilots in training. The Reservists contributed heavily to our success. Next, the night-fighter attack squadrons of MAG-12, which also included some Reservists, deserve great credit. They worked under the most dangerous conditions deep in the valleys of mountainous North Korea with little or no light. We will always owe a great debt of gratitude to those pilots.

"The basic schedule at K-18 for the day squadrons was to alternate between close air support, both pre-briefed and 'on-call', on one day, and deep interdiction missions as far north as the Yalu River on the next. The deep mission targeted all transport components and routes, marshalling sites and armour – if we could uncover it. All of this was interspersed with armed reconnaissance.

"We worked closely with USAF airborne controllers (callsign HAMMER). The ones that were working far north of the MLR were in F-51 Mustangs and they did a great job of calling out targets to us. We responded immediately, and this went on from dawn to last light. Most of us that covered certain sectors knew when anything had been changed and this usually led us to some well-camouflaged targets. Our work called for low-level flight and that was extremely dangerous because of the ground fire. We lost quite a few pilots while carrying out these missions."

A VMF-214 Corsair launches with a full load of rockets, frag bombs and napalm. Although the USMC supported all UN troops, their priorities remained with their own forces.

Air Combat

Corsairs attack by night
First Lieutenant Harold E. Roland
VMF(N)-513 'Flying Nightmares'

Flying down low and attacking enemy supply lines, vehicles, and troop positions proved to be extremely dangerous for fighter-bomber pilots. Doing this during the daytime was hazardous enough, but to carry out these attacks at night over mountainous terrain sometimes went beyond the call of duty. Captain Roland flew numerous missions in the Corsair during the hours of darkness. His input sheds a little light on how VMF(N)-513 worked its missions during the war.

"The roads, well north of the bomb line in the Sibyon-ni and Chorwon areas, were thick with Chinese trucks, as they were constantly trying to build up enough supplies for their troops to stage an offensive. They would run with headlights on in convoys of 10 to 15, with short separations between convoys. From a distance, their headlights would glow in the sky not unlike a busy road system in the US. They were all free game, but the small, abrupt mountains in the area, and of course never knowing exactly where even the flat ground was, caused us to be cautious about going right down after them without light.

"I learned in the first few flights that pickling off bombs and rockets or strafing from 1,000 ft [305 m] produced very poor results. The truck drivers were very bold, perhaps learning from the other night intruders that ordnance delivered from that altitude was really not a serious threat. They would usually keep driving when you tried to attack from a 'safe' altitude. Even if they stopped and turned their lights off, we could not claim anything. Our standard was that you must see a truck burning in order to claim it. Sometimes, we would strafe up and down a convoy and they would be immersed in total darkness, with no indication of what effects our guns had. Thus was born the joke that 'they were carrying a load of wet cabbages'.

"We -5N Corsair pilots discussed the situation and agreed that we had to hit an individual truck to do the needed damage. We then abandoned the tactic of bombing or strafing up and down a road that contained trucks and instead singled out individual vehicles or tightly bunched groups. The F7F Tigercat pilots were at a disadvantage here due to their larger, less manoeuvrable aircraft.

"Fortunately, we had C-47 flare ships assigned to our sectors along the main supply routes. These aircraft were carrying hundreds of 1,000,000-candlepower magnesium flares. If their aircraft had happened to get hit by ground fire, it would have become an instant, giant flare lighting the countryside like a small sun. Without these heroes, we could not have done our job, as their continuous dropping of flares at lower altitudes allowed us to stay down and work over the trucks. A lot of our outstanding successes in this mission had to be credited to them."

The single night-fighter Corsair unit worked deep into the mountainous areas of North Korea, attacking truck convoys and locomotives. This VMF(N)-513 F4U-5N is loaded with bombs for the forthcoming night's mission.

G-suit was on fire. The MiG-15 hit the water in a vertical position. The fight was not over, however.

More MiGs had come into the area, and a minute later, according to Captain Folmar, "orange golf balls" went streaming by the left side of his cockpit. Before he could react, a jolt hit his Corsair and an explosion followed. Looking out the cockpit, he could see that his left wing was gutted all the way to his inboard gun. His luck had run out and, with the aircraft uncontrollable, he bailed out. Fortunately, the MiGs departed the area quickly and Folmar was rescued less than eight minutes after he hit the water. It had been an incredible jet kill for a World War II prop fighter, even though it did not end the way it should have.

When moving off the coast of North Korea, the carriers had to be careful to keep ample sea room between the ship and the numerous islands and mud banks. The standard was to

VMA-212's AU-1s continued to pound the forces north of the MLR until the last days of the war, the unit having joined the fray on 19 September 1950. The pilot braving the cold in front of a heavily laden AU-1 (left) is Lieutenant Russ Janson.

Marine Corsairs in Korea

After the Chinese ground forces entered the war in early November 1950, the night flying F4U-5Ns of VMF(N)-513 began operating from Pusan, in the far south of the country. They were crammed in with several UN squadrons, and faced a long flight up to the Communist supply lines that were their principal targets.

stay outside the 100-fathom (600-ft/183-m) mark. During the warm months this did not create a problem for pilots, but during the freezing winter months it meant they had to fly over 75 miles (120 km) of open sea before they reached land. This translated into a potentially dangerous situation, because if they developed engine trouble after launch or if they were returning to the carrier with battle damage, they had to contend with the possibility of bailing out or ditching in freezing temperatures.

During these periods, the pilots wore protective 'poopy suits', but if one had even a small leak a pilot could freeze to death in minutes – and this happened many times. Also, a rescue helicopter that had to travel 65-70 miles (105-112 km) to pull out a pilot would not arrive in time to save him. These dangers, though, were the least of a Corsair pilot's worries, because they still had to face intense and accurate ground fire, especially around the enemy's high-value assets such as rolling stock and ammunition dumps.

Manpower build-up

During the last half of 1951 and the first few months of 1952, the Chinese were constantly testing UN reactions with an endless stream of convoys moving south from Manchuria. It did not take a master tactician to realise that with no air or naval support, the only way the Chinese were going to drive the UN forces off the Korean Peninsula was to build up enough supplies and ammunition to sustain an effective offensive. They did have one major advantage, though: unlimited manpower.

A very good example of what the fighter-bombers faced is found in records for May 1951. A conservative estimate of the number of Communist trucks moving on the MSRs (main supply routes) at night was 54,000+, seven times the number recorded for January 1951. This jammed the roads and taxed the night interdiction aircraft (B-26, F4U-5N and F7F-3N) beyond what they were capable of effectively handling. In one 60-day period during this elevated activity, 20 Marine aircraft were lost and only six pilots rescued.

One of VMF(N)-513's enlisted personnel crouches by with a fire extinguisher as a pilot cranks up this F4U-5N's engine for a post-maintenance test hop. The location was Kangnung (K-18), from where the 'Flying Nightmares' (badge, right) operated in 1951.

To counter this, Fifth AF initiated well co-ordinated air attacks against these supply routes and troop build-ups, always including Navy and Marine air assets. Marine records show that on a typical day when one of these planned attacks was being carried out, the entire strike force logged 800 sorties. Of this total, Marine Corsairs flew 101 sorties, slightly more than their norm. VMF-312 records stated that on these surges, they would fly as many as 16

Air Combat

Above: Quick reactions by an F-86 pilot sitting on alert at Suwon (K-13) captured on film this crash of a Marine F4U.

US Marine Corps Corsair losses in the Korean War (by month)				
	1950	1951	1952	1953
January		7	10	0
February		5	2	4
March		9	7	1
April		15	6	2
May		12	13	6
June		9	6	1
July		23	7	3
August	8	10	5	
September	6	10	9	
October	7	15	6	
November	4	6	5	
December	6	11	4	
Annual totals	31	132	80	17
Total	**260**			

Lieutenant Ernie Banks stands by a VMF-212 Corsair which made it back to Suwon after being badly shot up over Chinese lines in April 1951.

Landing aboard a carrier with battle damage could be very hazardous. This VMA-323 aircraft has suffered a relatively benign nose-over on Sicily in 1951.

Vought's F4U was renowned as a tough bird which brought many of its pilots back safely. The VMF-312 aircraft at Kunsan (above) shows evidence of its lucky escape from Chinese AAA, while the VMF-323 aircraft below made it as far as Chunchon (K-47 – main base of the Mosquito T-6 FACs) before landing, during the course of which the propeller was bent.

sorties in the morning and another eight in the afternoon. Marine participation in these attacks was not limited to supporting their own troops, and on several missions they were involved solely with Army units. On one such mission, VMF-312 lost its Executive Officer, Major Daniel Davis: on his eighth firing pass against ground targets, the aircraft took multiply hits, and in pulling up, Davis lost a wing which sent his Corsair spiralling into the ground.

All the VMF/VMA Corsair squadrons that served in Korea had to be versatile because they not only worked from various carrier decks but also from several land bases. Their maintenance personnel have to be given the credit, for they were constantly shifting equipment, packing and unpacking. Like their pilots, they served the cause with great dedication and effort. The only squadron that seemed to have some stability was VMF(N)-513, which eventually settled down at Kunsan during the final months of its Corsair and Tigercat operations. It held one of the most unusual distinctions in military aviation history, in that at one time it operated full complements of three different aircraft types in combat: the F4U-5N, F7F-3N and the new F3D Skyknight.

Overdue replacements

As the war wound down, the Marines had two attack squadrons ready to relieve two of their long-term Corsair units, VMA-312 and VMA-323. The 'Polka Dots' from VMA-332 relieved the 'Checkerboards' from -312 about six weeks before the war ended, arriving in time to fly some combat missions from the carrier. According to Volume Five of *US Marine Operations in Korea* (published in 1972), VMA-323 'Death Rattlers' bowed out of the Korean War on 3 July 1953, only about three weeks before the war ended. It had achieved the distinction of being the longest-serving Marine tactical (VMA/VMF) air squadron during the Korean War. The squadron's final combat mission on 2 July brought its total Korean operations to 20,827 sorties and 48,677.2 hours. VMF-214 had flown the first combat mission of the war for the Marines but was reassigned to CONUS in November 1951, thereby setting up VMA-323 to achieve this distinction.

The amount of ordnance expended by Corsair pilots was unbelievable. This performance was not only reflected in every Marine and Navy pilot, it was also the norm for the Air Force. These statistics reflect the calibre of training and the standards upheld by the American military. One of the 'Checkerboard' pilots from VMF-312 kept accurate records of

Marine Corsairs in Korea

Left: The rotation and replacement of Marine squadrons in Korea did not begin until early June 1953, when VMA-332 replaced VMA-312. The squadron saw about six weeks of combat, this 'Polka Dots' machine seen returning from a mission in July.

Below: VMA-332 pilots line up In early June 1953 on the deck of USS Point Cruz (CVE 119) as they prepare to go into combat off the coast of North Korea. The 'Polka Dots' remained on station for several weeks after the war ended.

each and every mission. Captain J.J. Geuss states that in his lengthy combat tour (78 missions) in the Korean War, he delivered 54 napalm canisters, 257 rockets, 21 250-lb (113-kg) GP bombs, 215 100-lb (45-kg) bombs, 16 500-lb (227-kg) bombs, and 49 1,000-lb (454-kg) bombs, and expended most of his 20-mm rounds against a wide array of targets. This extraordinary record could easily describe the performance of every Corsair pilot who was lucky enough to make it through a complete tour.

The early exploits of this select group of pilots in the Korean War have become legend, but from the spring of 1951 through the end of the war two years later, it was as if they had disappeared from the radar screen. Nothing could be farther from the truth. Their dedication to protecting the ground forces, and the loss of 260 Corsairs in combat and related incidents, is tribute enough to the Marine regulars and Reservists who answered the call and put their heart and soul into their performance.

Warren Thompson

Right: One of VMA-332's F4U-4Bs lines up on the catapult in 1953. The squadron remained with USS Point Cruz until mid-December.

Below: VMA-332 took over VMA-312's aircraft, which soon gained the replacement squadron's polka dot markings and 'MR' tailcodes. This aircraft was seen at Kunsan AB (K-8) around the time the war ended.

WARPLANE CLASSIC

North American F-100 Super Sabre

First of the 'Century Series', the F-100 ushered in the supersonic era. Rapidly superseded by more capable types in its original fighter role, the 'Hun' found its true calling in the less glamorous world of fighter-bombers. As well as providing the backbone of the USAF's Cold War tactical nuclear forces in the late 1950s, it dominated the 'in-country' war in South Vietnam. The F-100 was also the platform which pioneered the defence suppression role.

A bomb-armed 'alert bird' coughs into life at Tuy Hoa in 1967, the J57 engine being started by cartridge. As well as pre-planned sorties, the F-100 force maintained several aircraft on alert so that they could provide rapid close air support if troops came into contact and required extra firepower. Tuy Hoa was the busiest of the four main F-100 bases in Vietnam (the others being Bien Hoa, Phan Rang and Phu Cat), and among the squadrons assigned to the resident 31st TFW was the 308th Tactical Fighter Squadron 'Emerald Knights', whose 'SM' tailcode is worn by this F-100D.

Top: Born in the Cold War, the F-100 achieved fame in the very 'hot' war in Vietnam. Here a 306th TFS F-100D with a full bomb load heads to its target. The 306th was assigned to the 31st TFW at Tuy Hoa, which had five F-100 squadrons assigned during the period 1967-70, making it the most important of the Super Sabre bases in southeast Asia.

Above: Of all the Cold War projects, few were as visually impressive as the ZELL programme, in which F-100s (and later F-104s) were to be blasted into the air from a shelter or, as here, a towed trailer. In the event of a nuclear exchange, recovering the aircraft was left to the discretion of the pilot to find whatever landing strip he could – but once the Mk 7 bomb had been delivered to its target, the continued usefulness of the aircraft had all but expired.

North American's F-100 Super Sabre was designed as the replacement for the uncontested 'top dog' of the fighter world of the early 1950s, the F-86 Sabre, and even echoed the name of its illustrious forebear. The original Sabre had claimed a 12:1 kill:loss ratio over the MiG-15 in Korea and became the standard fighter of leading Western air forces. The Super Sabre took on this mantle, and at first seemed to be repeating the earlier aircraft's incredible success. The F-100 was the USAF's first operational aircraft capable of exceeding the speed of sound in level flight. It was the first in a glamorous, glittering and famous 'Century series' of fighters which used F-10x designations. The aircraft enjoyed a relatively long service life (by the standards of the day) and gained an enviable combat record in Vietnam. The Super Sabre was the aircraft of the USAF's elite 'Thunderbirds' aerobatic team for several years. The F-86 had looked merely 'pretty', whereas the F-100 was good-looking in an altogether sleeker, more business-like and shark-like manner.

Following the success of the P-51 Mustang and F-86 Sabre, North American Aviation hoped that its first supersonic fighter would repeat the success of its progenitors. Insofar as the aircraft was built in bigger numbers than any of the other 'Century Series' fighters, the company succeeded, though its production total of 2,294 aircraft was barely one-quarter of that of the F-86 Sabre/FJ Fury family.

Although the Mustang and Sabre together brought NAA the status of America's premier fighter-builder, the F-100 lost the company its lead and, indeed, NAA never built another production fighter after the Super Sabre. Moreover, while the Sabre had dominated Tactical Air Command in the fighter and fighter-bomber roles, and had been pressed into service as a dedicated all-weather fighter interceptor, the F-100 aircraft failed to become the single, dominant fighter in the USAF inventory, being augmented by a number of other types, several of which demonstrated greater suitability or effectiveness in particular roles. In fact, the Super Sabre became something of a low-cost makeweight in TAC's inventory: a good, versatile foot soldier which could usefully augment the more powerful F-105 Thunderchief in the air-to-ground role, but whose fundamental weaknesses in even this role led the Air Force to the almost unprecedented step of acquiring the US Navy's A-7 Corsair to replace it.

Nor was the F-100 a great export success for NAA in the way that the Mustang and Sabre had been (let alone the F-84), and saw service only with the Republic of China Air Force, and in Denmark, France and Turkey.

Genesis of the F-100

While other fighter manufacturers worked on more ambitious designs, some of which went far beyond what was then realistically achievable, NAA's plans for a replacement for its highly successful F-86 Sabre were rather more modest and more feasible, and the company hoped that its new fighter would be rapidly developed and would therefore be in service long before its bolder competitors. In the event, development of the fairly conservative F-100 was not appreciably less protracted than that of its rivals, and a host of faster, bigger and more useful fighters was soon in service alongside it.

NAA's initial hope was that it could produce a supersonic Sabre by doing little more than sweeping back the wing more sharply (to 45°) and adding a new turbojet engine. North American Aviation's Raymond Rice and Edgar Schmued began work on a series of company-financed design studies in February 1949. They first proposed an area-ruled, 45°-winged, Mach 1.03 'Advanced F-86D', powered by a General Electric 'Advanced J47', which was rated at 9,400 lb st (41.80 kN) dry and 13,000 lb st (57.81 kN) with afterburning. The USAF rejected it and asked instead for a new day fighter. NAA then submitted a similar 'Advanced F-86E', with a new, slimmer fuselage and a new nose intake. It soon became clear that any gains in performance would be modest, thanks to the steep rise in aerodynamic drag that takes place as near-sonic speeds are reached.

NAA's next proposal was known simply as the NA-180 Sabre 45, (the '45' indicating the wing sweep), which combined features of the 'Advanced F-86D' and the 'Advanced F-86E'. It bore little resemblance to the F-86 Sabre except in basic configuration and structural design, and was significantly larger, faster and more powerful than the original aircraft, with different controls and a revised tail surface layout.

The Sabre 45 was to be powered by the new Pratt & Whitney J57-P-1 turbojet. This new axial-flow engine was slimmer than the centrifugal-flow engines then in common use, and thus allowed a narrower, lower-drag fuselage. It was also considerably more powerful, rated at 15,000 lb st (66.71 kN) with afterburner. By comparison with contem-

North American tested the 45° sweepback wing and low-set tailplane of the 'Sabre 45' on the first prototype YF-86D.

porary British axial-flow turbojets such as the Rolls-Royce Avon and Bristol Siddeley Sapphire, the J57 was heavy, crude and under-powered; however, it was reliable and cheap, and was being procured in huge numbers for a wide range of programmes, including the lumbering B-52 Stratofortress. Its afterburner was more advanced, efficient and reliable than those fitted to British engines, restoring the balance somewhat. When engaged, the afterburner delivered about 50 per cent extra thrust in an instant, giving the pilot a real 'kick in the back' and producing an impressive torch of gold and lilac flame and 'shock diamonds'.

These features gave the Sabre 45 an estimated maximum speed of Mach 1.3 (860 mph; 1384 km/h) at 35,000 ft (10668 m) and a respectable 580-nm (667-mile/1074-km) combat radius at an estimated combat gross weight of 23,750 lb (10773 kg). The aircraft was originally perceived as a radar-equipped fighter interceptor to replace the F-86D, but this role was given to the new F-102, and NAA was directed to draw up the Sabre 45 as a simple day fighter, without radar and with an armament of four T-130 (later redesignated as M39) 20-mm (0.787-in) cannon. This suited NAA down to the ground, since dispensing with radar allowed it to use a large, unobstructed nose intake to feed the new J57 engine, which, like all early axial-flow turbojets, was more vulnerable to disturbed airflow.

Priority for a Sabre follow-on

With the Korean War raging, the Sabre 45 was the right aircraft at the right time. The F-86 was achieving an astonishing degree of air superiority over the MiG-15, but it was clear that this was due more to pilot training and quality and to tactics than it was to any superiority of the aircraft. Indeed, the technical parity of the MiG-15 was a real worry, making a Sabre replacement an urgent priority for the USAF.

In October 1951, the USAF Council decided to press for the development of the Sabre 45 proposal, although some key figures believed that the design would be too costly and complex to fulfil the basic day fighter role. NAA had hoped that the USAF would order two Sabre 45 prototypes, one for aerodynamic testing and one for armament trials. However, the Air Force Council wanted the Sabre 45 in service much sooner than a conventional procurement process would have allowed, believing that two prototypes would not be enough to achieve early operational status. Accordingly, on 1 November 1951, the USAF issued a Letter Contract for two Sabre 45 prototypes, plus 110 NA-192 production aircraft. Ten of the latter were to be 'production test' aircraft. Full production was to be initiated even before initial flight testing was completed. This approach subsequently became known as the Cook-Craigie plan, after the two USAF generals who spearheaded it. The NA-180 prototypes were therefore built on definitive production jigs and

52-5754 was the first Super Sabre, one of two YF-100A prototypes. In initial configuration it had a pitot probe mounted on top of the intake, with a short probe underneath. The positions were subsequently reversed, as featured in the photograph below.

This fine portrait of the first YF-100A over Edwards AFB captures the simple lines and (for its day) dramatic sweepback of the wing. NAA revised the F-100's fin to the unfortunate short design, but it was too late to test it on either of the prototypes, which both had a fin shape more akin to that of later aircraft.

Above: To the untrained eye the vertical fin originally fitted to the F-100A never appeared big enough for the task, and this was to prove the case. When flying with supersonic tanks, like this AFFTC aircraft, the directional stability problems were exacerbated.

Right, top: 55-5756 was the first production F-100A. Series aircraft appeared soon after the second of the two prototypes, and the first were pitched into the flight test campaign while others were being built. Seventy aircraft had been completed with the inadequate short fin before the necessary alterations could be introduced to the line.

Above right: The arrival of the sixth production F-100A at Nellis AFB generated considerable interest among Air Force personnel. The Nevada base was later to play host to the F-100 weapons and tactical evaluation unit, and the 'Thunderbirds' display team.

With examples of its illustrious forebear – the F-86 Sabre – in the background, an F-100A sits on the ramp at Nellis AFB, Nevada. The Research and Development Unit conducted operational evaluation of new TAC aircraft, later evolving into the Air Warfare Center.

tooling, and large stocks of components began to be stockpiled for mass production. One month later, the new aircraft was designated the F-100.

The Sabre 45 mock-up had been inspected on 9 November, when more than 100 configuration change requests were received. In response, North American Aviation reshaped the fuselage with an even higher fineness ratio and provided an extended clamshell-type cockpit canopy, while the horizontal tail was moved to a position below the extended chord of the wing. It was hoped that this would improve controllability at high angles of attack and would help to prevent pitch-up following the stall. The latter was a common and often deadly phenomenon in many early swept-wing aircraft, and had even been dubbed the 'Sabre dance' after numerous accidents in the F-86. One YF-86D (50-577) was modified with a similar low-set horizontal tail arrangement, and proved its efficacy.

Subsequently, on 23 June 1952, NAA was directed to make provision for external weapons racks and to replace existing bladder tanks with non-self-sealing tanks in order to achieve a 400-lb (181-kg) weight saving. Additional changes were approved on 26 August 1952, at the conclusion of the final design stage. The air intake lip was given a sharp edge in order to improve airflow to the engine at supersonic speeds, and the nose was lengthened by 9 in (23 cm). The decision was also made to reduce the thickness/chord ratio of the horizontal and vertical tails to 0.035, by shortening them and increasing their chord. This final change came too late to be incorporated on the two YF-100 prototypes, then nearing completion, but was incorporated on production aircraft, 250 more of which were ordered at the same time.

Super Sabre takes to the air

The extent of these design changes was sufficient to warrant the prototypes being redesignated YF-100As. The first of these (52-5754) was completed at NAA's Los Angeles factory on 24 April 1953, and was then moved in great secrecy to Edwards AFB. Company test pilot George S. 'Wheaties' Welch made the type's maiden flight there on 25 May 1953, exceeding the speed of sound on this first 55-minute flight and then again on the second (20-minute) flight later that day. Subsonic at low level, the F-100 could exceed Mach 1 above 30,000 ft (9144 m) even in level flight, and on 6 July 1953 the YF-100A achieved a speed of Mach 1.44 in a long dive from 51,000 ft (15545 m). The two prototypes were powered by Pratt & Whitney J57-P-7 engines, then nominally rated at 9,220 lb st (41.00 kN) dry and 14,800 lb st (65.82 kN) with afterburner, although they were de-rated in the prototype aircraft.

The second prototype (52-5755) followed the first into the air on 14 October 1953, and five days later the aircraft was formally shown to the press. George Welch flew the aircraft past the press grandstand just feet from the ground and around Mach 1, creating a sonic bang which left broken windows in the airport administration building at Palmdale.

The first production aircraft (F-100A, 52-5756) flew on 29 October 1953. On the same day, Colonel F.K. ('Speedy Pete') Everest flew the first YF-100A to set a world speed record of 755.149 mph (1215.26 km/h) over a 15-km (9.3-mile) course, whose entry and exit gates were marked by columns of smoke from piles of burning vehicle tyres. This was the last world speed record set at less than 100 ft (30 m). Everest had previously flown the F-86D chase aircraft for the YF-100's first flight, and had won a beer from Welch by correctly predicting that the aircraft would achieve Mach 1 during its maiden flight. Everest had previously averaged 757.75 mph (1219.45 km/h) on the required four runs over a 3-km (1.86-mile) course – beating the record set by the Douglas F4D Skyray, but not by the required 1 per cent margin.

NAA immediately claimed to have produced the world's first aircraft capable of breaking the sound barrier in level flight, as previous swept-wing contemporaries had managed supersonic flight only in a dive. This was probably true, since MiG's SM-9 (forerunner of the MiG-19) did not break the sound barrier in level flight until 5 January 1954. NAA's claim to have produced the world's first in-service supersonic fighter are open to doubt, however, since the F-100's introduction to service was disrupted by technical problems.

Service test pilots who flew the YF-100As found that the aircraft outperformed the Air Force's existing types by a handsome margin, though they noted a number of short-

F-100 Super Sabre

comings that might affect the introduction into service. Visibility over the nose during take-off and landing was rated as poor, the touchdown speed was too high, low-speed handling was 'rather poor' (with a marked tendency of the YF-100A to yaw and pitch near the stall, and an uncontrollable wing drop), and longitudinal stability in high-speed flight was considered inadequate. Without afterburner the rate of climb was considered too slow, taking 16 minutes to reach 40,000 ft (12192 m). Everest himself considered that the aircraft should not be released for service until some of the deficiencies had been rectified, but he was overruled.

Catastrophic problems

During early test flights, some rudder flutter problems had been encountered, but the installation of hydraulic rudder dampers from the 24th aircraft solved this. (A pitch damper was installed in the 154th and subsequent aircraft.) Worse was to come. Unfortunately, the aircraft's testing had not been fully completed before these early deliveries and the directional stability of the aircraft had been miscalculated: four aircraft were lost during the first year of testing due to inertia roll-yaw coupling – which was directly attributable to the short tail fin. The problem proved especially serious when underwing drop tanks were carried. On 12 October 1954, NAA test pilot George Welch was killed when his aircraft disintegrated while carrying out a maximum performance test dive followed by a high-g pullout in the ninth production F-100A (52-5764). The loss of Welch, who had shot down four enemy aircraft at Pearl Harbor and 14.5 more during the remainder of World War II, was felt keenly by North American.

A senior RAF evaluation officer, Air Commodore Geoffrey D. Stephenson, Commandant of the Central Fighter Establishment, was killed when his F-100A crashed at Eglin under similar circumstances. Another aircraft was lost on 9 November, although pilot Major Frank N. Emory was able to eject safely. The entire fleet was grounded on 12 October 1954, with 68 aircraft accepted or in service and 112 more completed and awaiting delivery.

Despite this, and even while the F-100 fleet was still grounded, NAA's 'Dutch' Kindleberger received the prestigious Collier Trophy from President Eisenhower. The

Fixing the fin

trophy was awarded by the National Aeronautic Association to recognise 'the greatest achievement in aviation in America' during the preceding year.

The decision was quickly made to reshape the fin and rudder, lengthening it to a form that resembled the original YF-100A surfaces, and adding about 27 per cent more vertical tail area. The F-100's large length:span ratio led to huge yaw excursions during rolling pull-out manoeuvres, producing aerodynamic loads sufficient to break the tailfin. A larger, stronger tailfin ensured adequate stability to Mach 1.4 and prevented these yaw excursions. The artificial feel systems for the aileron and stabiliser powered controls were also modified.

Service deliveries

At the end of November 1953, Tactical Air Command's 479th Fighter Day Wing at George AFB started equipping with short-finned F-100As, and initially all went well. The unit became operational on 29 September 1954. Deliveries of the modified F-100As began from the Los Angeles factory in the spring of 1954. From the 184th F-100A (e.g., the last 19 aircraft) these modifications were incorporated on the line, before the aircraft even flew, but all surviving F-100As (starting with the 34th aircraft and a random trial batch of 11 jets) were quickly brought up to the same standard. The grounding order was rescinded in February 1955 (by which time the MiG-19 was entering service) and the final F-100As were delivered in July 1955. The 479th Fighter Day Wing at George AFB finally achieved Initial Operational Capability with the modified F-100A in

Above: The short fin of the initial production F-100A provided insufficient directional stability, resulting in a series of roll-yaw coupling crashes. In December 1954 NACA modified an F-100A from the High-Speed Flight Station with a lengthened fin, the aircraft being seen at Edwards with a standard F-100A from the Air Force Flight Test Center.

Left: North American also schemed a larger fin – adding 27 per cent more area – to cure the F-100A's problems. This became the standard design, and was retrofitted to short-finned aircraft as well as being adopted on the production line. This aircraft, the 12th F-100A and second Block 5 aircraft, served with the Wright Air Development Center.

Seen in September 1954, this short-finned aircraft was designated EF-100A, and was used for tests by the Wright Air Development Center. The 'E' prefix, not to be confused with the present-day 'electronics', stood for 'exempt' and was changed to 'J' in 1955 to signify temporary assignment for test duties.

One of the more unusual F-100 operators was the 1708th Ferrying Wing, parented by the Military Air Transport Service. The 1708th was responsible for ferrying Super Sabres to the operational units, and a few 'Huns' – including this F-100A – were on strength to train pilots.

The first F-100 unit was the 479th Fighter Day Wing at George AFB. Its first aircraft were short-finned, but modified aircraft were later supplied.

Enough F-100As were made available to equip three ANG squadrons, which operated Sidewinder-equipped 'Huns' in the air defence role. Arizona's 152nd FIS (right) at Tucson converted from F-84Fs in May 1958 as an ADC-gained air defence squadron. Other ANG F-100A recipients were Connecticut's 118th TFS at Windsor Locks (below, summer 1960 to January 1966) and New Mexico's 188th FIS at Kirtland (April 1958 to the spring of 1964).

September 1955, although Project Hot Rod conducted by the USAF Air Proving Ground Command at Eglin AFB to evaluate the suitability of the F-100A for operational service revealed that the type still suffered from major operational deficiencies.

Most F-100As were phased out of the active inventory in 1958, some being transferred to the Air National Guard, and the rest being placed in storage at Nellis AFB.

The first Air National Guard unit to receive the F-100A was New Mexico's 188th FIS, which converted in April

1958, followed by the 152nd FIS (Arizona ANG) and the 118th TFS (Connecticut ANG). By 1960 the ANG had reached a peak inventory of 70 F-100As, and many of these returned to active service during the Berlin crisis of 1961, when Air National Guard and Air Force Reserve units were mobilised and called to active duty. Some of these F-100As were retained by the USAF even after the ANG personnel had been released from active duty in early 1962, and were used for aircrew training.

Some 118 F-100As (more than 58 per cent of total F-100A production) were transferred to the Chinese Nationalist Air Force, which used them until 1984. The aircraft had been withdrawn from USAF service by 1970.

The F-100A had a poor safety record, even after the tail modifications. About 50 F-100As were lost in USAF service, and Taiwan lost 49 more in accidents. F-100 landings were often described as 'controlled crashes' and one senior USAF F-100 pilot opined that the aircraft's optimum role was static display.

Early aircraft were powered by the 9,700-lb st (43.14-kN) Pratt & Whitney J57-P-7 engine (15,000 lb st/66.71 kN with afterburning), and the final 36 F-100A aircraft (from the 167th aircraft on) were built with the J57-P-39 engine, which produced 1,000 lb (4.45 kN) more thrust. From the 101st aircraft, the F-100As were also delivered with the extended wings originally intended for the F-100C fighter-bomber. Production of the F-100A eventually totalled 203 aircraft, although the USAF had increased its order to 273 aircraft (plus one static test example) on 26 August 1952. The last 70 aircraft were actually delivered as F-100Cs.

Design features

The final F-100As were broadly representative of the definitive F-100 design, since all further revisions and modifications were relatively minor, and all further production versions shared the same basic structure and configuration.

The Super Sabre's low-mounted 45° wing was dramatically swept, by the standards of the day, but, less obviously, was also extremely thin. The wing had a thickness:chord ratio of 0.082, compared to that of 0.10 for the F-86, and the wing leading-edge glove reduced the overall ratio to 0.07. This made it exceptionally thin and helped reduce drag at transonic speeds. With no room in the blade-like wing, the mainwheels retracted inwards into wells in the centre fuselage.

Like the F-86's wing, the Super Sabre's wing leading edge featured five-segment automatic slats. These increased the lift at take-off, delayed wing buffet, improved lateral control at low speeds, and permitted tighter turns. The ailerons were mounted inboard on the wing in order to reduce the tendency of the wing to twist during aileron deflection at high speeds, thereby preventing aileron reversal.

No fuel was carried in the wing of the F-100A, being accommodated instead in five non-self-sealing bladder tanks inside the fuselage. They had a total capacity of 750 US gal (2839 litres). Two 275-US gal (1041-litre) underwing drop tanks could be carried.

Behind and below the wing was a one-piece, powered, geared slab tailplane, activated by a pair of independent, irreversible hydraulic systems, with artificial 'feel' provided via springs and bungees. The rudder was unusually short, its height constrained by the position of the fuel dump pipe (which had been near the top of the short-finned F-100A's

Pratt & Whitney J57

The J57 (JT3C civil designation) was a landmark engine which ushered in a new era of fuel economy, reliability and pressure ratio. Pratt & Whitney built over 15,000, while Ford Motor Company production topped 6,200. The engine was adopted for numerous fighter types, as well as the B-52, KC-135, 707 and DC-8 large aircraft. It also directly spawned the JT3D/TF33 turbofan and the larger JT4A/J75 turbojet. The J57 was a two-spool axial-flow engine, with the inner low-pressure spool and outer high-pressure spool combining to provide a pressure ratio of 12.5.

Versions of the J57 installed in the F-100 were the J57-P-7 and J57-P-39 (both of 14,800 lb/65.86 kN thrust with afterburner), and the J57-P-21/21A (16,000 lb/71.2 kN). Many aircraft were later modified with the afterburner section from the F-102, while some received the afterburner section of the F-106's J75.

Today it is standard practice to design aircraft so that the engines slide out, but in the 1950s it was the vogue to remove the entire rear fuselage on a trolley, as demonstrated by a QF-100D drone on Tyndall's 'Death Row'.

tailfin, but was now about halfway down the tail). Despite its small size the rudder was extremely powerful. The final control surface was a retractable 'barn door' speed brake, which was mounted below the rear fuselage.

Unlike many of its contemporaries, the F-100 was not area-ruled and the humped, shark-like fuselage was relatively simple. The F-100 retained a plain pitot intake in the nose, like that of the F-86. This allowed a straight-through airflow to the compressor, without the complication of bifurcated intakes on the fuselage sides or in the wing roots. Unlike the F-86's intake, that fitted to the F-100 was broad and flat, somewhere between a rectangle and an oval in cross-section, allowing a broad, flat intake duct to flow easily under the pilot's cockpit and over the nose-wheel bay and cannon bays.

Cannon armament

The Super Sabre's armament consisted of four 20-mm (0.787-in) Pontiac M39 cannon, mounted in pairs on each side of the fuselage below the cockpit. The M39 had been tested in Korea on modified F-86Fs as the T-160, and had proved extremely effective. The weapon, like the British ADEN cannon and the French DEFA 552, was a belt-fed revolver-type cannon derived from the German Mauser MG 213C. While the British and French opted for a 30-mm (1.18-in) cannon, the US designers (from the Illinois Institute of Technology, working under contract for General Electric's Pontiac division) chose 20-mm to give a faster rate of fire. The M39 could fire up to 1,500 rounds per minute at a muzzle velocity of 3,300 ft (1006 m) per second.

In the F-100, ammunition capacity was usually limited to 200 rounds per gun in order to reduce the risk of rounds 'cooking off' in their ammunition boxes, though some F-100Ds could carry 257 rpg. The F-100F two-seater carried just 175 rounds for each of its two cannon. These powerful weapons were aimed using a lead-computing A-4 ranging gunsight. Lead was computed automatically, using ranging information from the ranging radar whose antenna was mounted inside the upper part of the engine inlet lip, covered by a flush-mounted dielectric panel.

The F-100's J57 engine was closely cowled in the rear fuselage – with so little room around it that the rear part of the F-100 proved almost impossible to paint. Early USAF F-100s had their 'Buzz' numbers on the rear fuselage, but they kept burning off and eventually had to be moved. Camouflaged aircraft soon became heavily weathered in this area. The F-100's rear fuselage culminated in a crude but effective variable-area afterburner nozzle, with petals

Superficially similar to the late F-100A, the C model was far more capable, having a 'wet' wing and underwing hardpoints for the carriage of tanks and weapons. It also had 'universal' (i.e. nuclear) wiring.

RF-100A – Slick Chick

The most mysterious of the 2,294 Super Sabres were the six RF-100A Slick Chick aircraft converted for the reconnaissance mission. Few details have emerged regarding their operations, from either the USAF (in Germany and Japan) or the RoCAF (from Taiwan). The camera installation entailed the removal of the guns and the installation of a bulged bay for five cameras. The advent of the faster and more capable RF-101 Voodoo put an end to the prospect of any further RF-100 conversions.

An F-100C from the 336th TFS/4th TFW leads two 333rd TFS F-100Fs. The aircraft formed the '9th AF Firepower Team'.

Below: F-100Cs went to Europe in 1956, equipping the 36th Fighter-Day Wing headquartered at Bitburg, West Germany. The role was tactical nuclear strike, the aircraft carrying a single Mk 7 'Blue Boy' on the left intermediate pylon and three drop tanks in the '1-E' configuration. The wing maintained a daytime nuclear-armed Victor alert status, whereas the later F-100D units had round-the-clock capability. The 36th had five squadrons scattered around several bases: this snow-dusted aircraft is from the 53rd Fighter-Day Squadron, seen at its Landstuhl base in January 1958.

opening and closing to vary the jetpipe diameter. This nozzle was only the visible portion of a 20-ft (6.1-m) long, 5,000-lb (2268-kg) structure which was fastened to the back of the 14-ft (4.28-m) long J57 engine.

The F-100 was even more innovative under the skin. The aircraft made extensive use of new manufacturing and construction techniques, and of new materials, particularly heat-resistant titanium. This was the first time that this metal had been used in large quantities in an aircraft, and North American actually used 80 per cent of all the titanium produced in the United States until 1954 in the manufacture of the Super Sabre. Titanium is an extremely strong and light metal and is more resistant to heat than aluminium, but is more brittle and more difficult to machine, and therefore costs more to work with. The aircraft made greater use of automatically machine-milled components, increasing machine tool costs but allowing the use of simpler sub-assemblies with smaller parts counts. Thus, while the F-86 wing box had used 462 parts and 16,000 fasteners, the integrally-stiffened F-100 wing box used 36 components and only 264 fasteners. Wing skins were tapered from root to tip and were machined into integrally-stiffened components. Massive single-alloy sheets were transformed into fuselage sides in a huge stretch press.

This use of advanced materials and manufacturing techniques was made possible by massive industry-wide investment which fed into a whole range of aircraft programmes. In Britain or France, where there might be only one or two simultaneous projects, such investment was virtually impossible. Nor could the Europeans draw upon the same range of research programmes, having had no direct counterparts to the succession of X-planes, high-speed wind tunnels and the like.

Strategic reconnaissance

Small numbers of F-86s had been successfully converted to high-speed tactical reconnaissance aircraft, and it was always likely that the F-100's sheer speed performance would make it a candidate for a similar conversion. Sure enough, in late 1953, NAA received a request to modify six F-100As as RF-100A reconnaissance aircraft, each carrying five Chicago Aerial Survey cameras. The F-100's supersonic capability promised to give it a degree of immunity from interception, and the new cameras, although not as advanced as the Hycon being developed for the U-2, could resolve golf balls on grass from 53,000 ft (16154 m).

A seven-man team under Don Rader began detailed design work immediately, and in September 1954 six F-100As (53-1545 to -1548, 55-1551 and 55-1554) were taken from the production line to be modified as unarmed RF-100A photographic reconnaissance aircraft under what became (on 7 December 1954) Project Slick Chick. Even with the removal of the cannon armament, the cameras and reconnaissance systems could not fit inside the existing fuselage, and a distinctive bulge had to be added under the fuselage belly, extending back from Station 80 below the cockpit to Station 267, almost at the wing trailing edge. The aircraft was also bulged on the forward fuselage sides, below the cockpit.

Together these fairings covered the camera stations which accommodated split vertical K-38 cameras (with their 36-in/91.4-cm lenses laid horizontally along the belly, on each side of the nose gear bay, each having a periscopic device to look down), and with a triple arrangement of oblique K-17s and a K-17C vertical camera roughly level with the wing leading edge. The camera ports were covered by electrically-powered doors to protect the expensive optics. Avionics systems were revised and relo-

Right: An F-100C from the 4th TFW takes off from Seymour-Johnson in 1959. The standard underwing store was the 275-US gal (1041-litre) supersonic 'banana' tank on the intermediate pylons, but a 200-US gal (757-litre) tank was available for carriage on inboard or outboard pylons. The F-100D introduced even larger tanks: a 335-US gal (1268-litre) 'banana' tank and a 450-US gal (1703-litre) subsonic tank.

Left: Seen in 1963, these F-100Cs are from the 166th TFS, Ohio ANG. During the Pueblo Crisis the squadron was deployed to Kunsan.

Below: Armed with napalm and bombs, a 152nd TFTS F-100C makes a dive attack. The Arizona unit trained F-100 pilots for the ANG, having earlier been an ADC fighter unit. It had the distinction of operating all four F-100 versions: A, C, D and F.

F-100Cs in the Air National Guard

At its peak the F-100C served with 11 ANG squadrons: 110th/MO, 119th/NJ, 120th/CO, 121st/DC, 124th/IA, 136th/NY, 152nd/AZ, 166th/OH, 174th/IA, 184th/KS (below) and 188th/NM. The first units to receive the type were the DC and New York squadrons, which began flying F-100Cs in mid-1960, and by the end of 1962 a further five had been equipped. The rundown of the C fleet began in June 1970 when the New Jersey Guard gave up its aircraft for F-105Bs. The last ANG F-100C squadron was Iowa's 124th TFS at Des Moines, which flew the type until mid-1975 before upgrading to F-100Ds.

cated – the gun ranging radar gave way to a new navigation system – and there was provision for the pilot to wear a pressure suit for very high altitude flying. The former aft electronics bay accommodated a new 830-US gal (3142-litre) fuel tank, giving the Slick Chick an endurance of up to 5½ hours.

The mission profile called for a lot of high-speed flight using afterburner, so the RF-100A carried four drop tanks rather than the usual two. New inboard pylons sway-braced to the fuselage carried a pair of 200-US gal (757-litre) tanks, and there were 275-US gal (1041-litre) tanks outboard.

Secret ops

A thick shroud of secrecy still covers the RF-100As that participated in politically-sensitive reconnaissance missions over Soviet and Soviet-occupied territory during the Cold War. It is known that the aircraft sometimes carried spurious serial numbers (one was photographed masquerading as 53-2600, actually an F-89 Scorpion).

Three of the Slick Chicks were shipped to Europe aboard the USS *Tripoli*, being flown from Burtonwood to Bitburg on 16 May 1955 to form Detachment 1 of the RB-57-equipped 7407th Support Squadron. The remaining trio joined the 6021st Reconnaissance Squadron at Yokota, Japan on 2 June 1955.

The aircraft's engines were replaced frequently in an effort to obtain the best possible high-altitude performance. The original J57-P-39 engines were initially tweaked to give an afterburning thrust of 16,000 lb st (71.16 kN), before being replaced by F-100D-type J57-P-21s. They then reverted to the J57-P-39, which proved better above 40,000 ft (12192 m), before being fitted with J57-P-21As in 1957.

The European Slick Chicks flew from Bitburg, Rhein Main, Hahn, Fürstenfeldbruck and Incirlik, accelerating to supersonic speed over West Germany before crossing the border with the East. They then flew in a straight-line reconnaissance run before starting a climbing turn, gaining altitude before starting to dive (still in the turn) and then heading into a return run. The aircraft thus flew keyhole-shaped flight plans, accounting for the 'Keyhole' name applied to the missions. 53-1551 was abandoned by its pilot near Bitburg in October 1956, and the unit then used an unmodified F-100C for training before the Detachment disbanded on 1 July 1958.

In the Far East, the Yokota-based aircraft flew similar missions. One aircraft (53-1548) was lost on 23 June 1955 and the surviving one returned to Inglewood in June 1958.

The four surviving RF-100As were then transferred to the Nationalist Republic of China Air Force on Taiwan, arriving in December 1958 and January 1959 and equipping the 4th Squadron at Taoyuan. It has generally been believed that these aircraft flew operational overflight missions over the People's Republic of China, but recent reports suggest that this may not have been the case and that the aircraft suffered problems which prevented their operational use. They were supposedly retired in December 1960.

Two aerobatic display teams flew the F-100C – the 'Thunderbirds' between 1956 and 1963, and USAFE's 'Skyblazers' (illustrated) from late 1958 to early 1962. In the F-100 era the latter team was provided by the 36th TFW at Bitburg, operating seven aircraft which were additional to the wing's operational inventory.

Project Rough Rider

Under Project Rough Rider, a specially instrumented JF-100F, 56-3744, of the Wright Air Development Center was used to record the intensity of thunderstorms and lightning strikes in the vicinity of Oklahoma between 1960 and 1967. The aircraft was specially instrumented with gust vanes on the nose boom, temperature and hail probes on the underside, electrical field measuring equipment on the wing tips, hail erosion protection strips on the leading edges, and cameras on the left wing tank to photograph water droplets or ice crystals.

Two views show the Rough Rider JF-100F. The bucking bronco badge was picked as the project logo as an apt characterisation of the experience of flying through thunderstorms. Note the modified port wing tank which was used to house some of the recording equipment.

Below right: This Nellis-based aircraft was used to test the GAM-83A (later AGM-12) Bullpup missile, which led to the modification of 65 F-100Ds to carry the weapon.

Below: This gaudily striped F-100D was assigned to the commander of the 474th TFW at Cannon AFB, the markings representing the colours of the constituent squadrons. In May 1962 the 474th sent one of its squadrons on TDY to Takhli, initiating deployments to Thailand in the face of growing fighting in Laos. The wing's 428th TFS was the first to lose an aircraft in combat in southeast Asia, on 18 August 1964 while escorting an Air America ResCAP mission over Laos.

Fighter-bomber

Failing to set the world on fire in the day fighter role, and enjoying only modest success as a reconnaissance platform, the F-100 was still an aircraft in search of a role when the F-100A entered front-line service. During late 1953, slippages in the Republic F-84F Thunderstreak programme led Tactical Air Command (TAC) to recommend that a version of the Super Sabre should be developed with a secondary fighter-bomber capability. The use of the term fighter-bomber was perhaps slightly misleading, since what was actually required was a tactical nuclear strike aircraft; it would be used to blunt any Soviet armoured advance into western Europe and would employ small, low-yield nuclear weapons, providing the US and its allies with a measure of flexible response to Soviet aggression. The USAF hoped to gain an interim fighter-bomber capability pending full availability of the F-84F and of aircraft like the F-105. This fighter-bomber-capable Super Sabre version became the F-100C.

It was not then apparent that providing such a capability would save the Super Sabre from an ignominious early retirement, since the type's poor performance in its designed role had not then been exposed.

The groundwork for the F-100's successful adaptation as a fighter-bomber had begun in October 1952, even before the YF-100A had made its maiden flight. Mindful of the F-86's poor radius of action and of its successful use in the fighter-bomber role in Korea, the USAF requested that NAA should examine the possibility of developing 'wet' wings for the Super Sabre. It followed this up in July 1953 with a request that the new wing should also be strong enough to carry ordnance.

On 31 December 1953 the USAF directed that the last 70 F-100As on the order be modified as NA-214 (F-100C) fighter-bombers. The importance of the fighter-bomber programme was underlined on 24 February 1954, when the Air Force ordered an additional 230 F-100Cs. The F-100C would become the first version of the Super Sabre to serve with the USAF in significant numbers, and manufacture eventually totalled 476 production aircraft. On 27 May, orders had been increased to 564 F-100Cs, though this was subsequently amended on 27 September so that many of the aircraft were completed as F-100Ds.

The fourth production F-100A (52-5759) was taken out of the flight test programme and modified to serve as the YF-100C prototype. This aircraft flew for the first time on 26 July 1954. It was not possible to incorporate integral fuel tanks in this fully-finished airframe, which also initially featured the short stubby fin common to early F-100s. The aircraft did, however, receive other modifications planned for the production F-100C.

The F-100C's wing was extensively modified with four additional hardpoints for either fuel tanks or weapons. The six underwing stations could together accommodate up to 5,000 lb (2268 kg) of stores including bombs, up to 12 5-in (12.7-cm) high-velocity air rockets, fuel tanks, napalm, and even 'special stores' in the form of the Mk 7 nuclear bomb.

Both wingtips were extended by 12 in (30.5 cm); this increased the wing area from 376 to 385.21 sq ft (34.93 to 35.79 m²), improving the roll characteristics and reducing the stalling speed. The wingtip extensions proved so useful that they were also incorporated into the F-100A production line, beginning with the 101st example.

The first production F-100C (53-1709) rolled off the line on 19 October 1954 and was conditionally accepted by the USAF on 29 October, though all Super Sabres were then still officially grounded. The aircraft (still fitted with an F-100A-type short tailfin) made its maiden flight on 17 January 1955 in the hands of Al White. This aircraft had the wet wing with integral fuel tanks. In this new leak-proofing system, all bolts that fastened the wing skin to the spars were specially sealed with injected material; this turned the wing into a 451-US gal (1707-litre) tank and brought total internal fuel capacity to 1,602 US gal (6064 litres).

The F-100C also had provision for a removable, non-retractable inflight-refuelling probe below the starboard wing and for single-point pressure refuelling. The F-100A had relied on individual gravity refuelling of its five fuselage tanks.

On 20 August 1955 USAF Colonel Harold Hanes flew the first production F-100C in two runs over a 15- to 25-km (9.3- to 15.5-mile) course laid out on the Mojave Desert. In doing so, he established a new world speed record of 822.135 mph (1323.062 km/h). This was the first record set at high altitude (the aircraft flew the course at 40,000 ft/12192 m) and also marked the first 'official' supersonic speed record. Chuck Yeager's flight in the Bell X-1 had been excluded because the aircraft had been dropped from a mother ship rather than taking off under its own power.

The first few F-100Cs were powered by the J57-P-7 engine, rated at 9,700 lb st (43.14 kN) dry and 14,800 lb st (65.82 kN) with afterburner. The next aircraft used the similarly-rated J57-P-39. From the 101st F-100C, the chosen powerplant was the J57-P-21, rated at 10,200 lb st

(45.36 kN) dry and 16,000 lb st (71.16 kN) with afterburner. The -21 engine also produced more thrust at high altitude, increasing the aircraft's speed at altitude by about 40 mph (64 km/h) and reducing the time taken to climb to 35,000 ft (10668 m) by about 10 per cent.

Deliveries of the F-100C to TAC's 450th Fighter Day Squadron at Foster AFB in Texas began in April 1955, and this unit became fully operational on 14 July 1955. On 4 September 1955, Colonel Carlos Talbott flew his F-100C the 2,325 miles (3742 km) across the USA from coast to coast, achieving an average speed of 610.726 mph (982.841 km/h). Colonel Talbott was awarded the Bendix Trophy for this flight.

The F-100C served only relatively briefly with the USAF's front-line fighter-bomber wings (including units in Japan and at Bitburg, Fürstenfeldbruck, Hahn and Landstuhl in West Germany, Camp New Amsterdam in the Netherlands, and Sidi Slimane in Morocco) before being superseded by the F-100D. The ex-USAF F-100Cs were then passed to the Air National Guard, whose first squadrons began receiving the F-100C during mid-1959.

ANG F-100C units were among those mobilised on 1 October 1961 in response to the Berlin Crisis, some deploying to Europe to reinforce NATO, others remaining within ConUS. All were demobilised in August 1962. Four ANG F-100C squadrons (from Colorado, New York, Iowa and New Mexico) were subsequently called up for service in Vietnam.

Although its front-line service was relatively brief, the F-100C was at least a success in its new role, unlike the F-100A. The Super Sabre was, at last, a good advertisement for NAA and the USAF and, accordingly, the type was chosen to equip the USAF's 'Thunderbirds' flight demonstration team. The team operated F-100Cs from 19 May 1956 until 1964, when they were replaced by F-105s; this, however, lasted only six shows until a major accident grounded the team and forced a conversion back to the F-100D. Successively designated the 3600th, 3595th and 4520th Air Demonstration Flight, the 'Thunderbirds' flew a seven-aircraft team and used 14 F-100Cs. The team's Super Sabres were modified with simple smoke-generating equipment that injected diesel oil into the jet efflux. This oil was carried from the modified rear fuselage tank in thin pipes which ran along the rear fuselage, beside the fin root. They also had their cannon, gun camera, gunsight, ranging radar and autopilot removed, and the 'slot' aircraft had a stainless steel fin leading edge and a relocated VHF antenna under

F-100Ds equipped numerous Fighter-Bomber Wings at CONUS bases, all of which were redesignated as Tactical Fighter Wings on 1 July 1958 (as were the Fighter-Day Wings). This aircraft was assigned to the 356th TFS of the 354th TFW, and is seen at Aviano, Italy, in 1959 during a regular reinforcement exercise deployment. In 1965 the wing deployed its aircraft to Ramey AFB, Puerto Rico, and San Isidro AB, Dominican Republic, during the crisis in the latter country.

ZELL – Zero Length Launch

By the mid-1950s the vulnerability of allied airfields to enemy attack was becoming a cause for some concern. Few USAF tactical aircraft were more reliant on having a long stretch of concrete available from which to take off than the F-100. Under the Zero Length Launch (ZELL) programme, the F-100 would be provided with a means of getting airborne from dispersed sites or from an airfield with damaged runways; it consisted of a simple mobile wheeled launch platform, with cradles which supported the aircraft's main undercarriage units and held the F-100 in the selected launch attitude. The aircraft was fitted with a massive 130,000-lb st (578.15-kN) Rocketdyne M-34 solid fuel rocket booster that burned for four seconds, propelling the aircraft to its climb-out speed of 275 kt (509 km/h) at an acceleration of up to 4 g. The booster was then jettisoned. It was found that the F-100D could be attached to its ramp and prepared for take-off by a five-man team well within two and a half hours (90 minutes, according to some sources). This capability was demonstrated during the 1958 Air Force Fighter Weapons Meet, when the test F-100D was fitted to and launched from its trailer in front of the press at Nellis AFB's Range 1 complex at Indian Springs.

NAA actually modified two F-100Ds for ZELL testing, 56-2904 and 56-2947. The first (-2904) made the first launch on 26 March 1958 in the hands of NAA's test pilot, Al Blackburn. The second launch went less well: the M-34 refused to jettison throughout Blackburn's one-hour flight, which he eventually had to end by ejecting. Blackburn flew 14 more launches, at different weights and with different combinations of stores, including the final launch on 26 August 1959, which was made at night from a simulated hardened shelter at Holloman AFB, New Mexico. Many of these launches were made with a Blue Boy Mk 7 'shape' under the port wing and a fuel tank to starboard, reflecting the system's deadly purpose, which was to ensure that the USAF could get its retaliatory nuclear strike airborne after an enemy attack. Four more launches were made by the

Several ZELL launches were made with a dummy Mk 7 nuclear bomb on the port wing, reflecting the 'real-world' scenario for which ZELL was proposed.

USAF project pilot, Captain Robert F. Titus (who went on to become a MiG-killer in Vietnam, though not in the F-100). It was concluded that any combat-ready pilot would be competent to make a ZELL launch, and the last 48 F-100Ds (148 according to some sources) were delivered with ZELL provision.

The second ZELL-equipped aircraft, 56-2947, launches from a mock-up of a nuclear hardened shelter, built at Holloman AFB, New Mexico, as part of the ZELL studies.

If the ZELL idea had been adopted, the trailer would have allowed the F-100 force to be dispersed widely. Here a complete ZELL rig is on display at Nellis AFB in 1960.

Little John *was an F-100D assigned to the 531st TFS, which was parented by the 39th Air Division at Misawa AB, Japan. The primary role of this unit was to stand nuclear alert, including a forward deployment at Kunsan in Korea.*

PACAF had three nuclear-capable F-100 wings in Japan in the late 1950s: the 8th TFW at Itazuke (above), 18th TFW at Kadena (right) and the 21st TFW at Misawa. The 18th TFW aircraft, carrying a practice bomb dispenser, is from the 67th Tactical Fighter Squadron.

Two wings operated F-100Ds from bases in eastern England. The 20th Fighter-Bomber Wing (TFW from 1958) received F-100Ds at Wethersfield in 1957 (below). It was joined in January 1960 by the 48th TFW (below right) at Lakenheath, a refugee from the withdrawal of US nuclear forces from France. The principal role was tactical nuclear strike, and the aircraft would have deployed forward to the continent in time of tension.

the nose. 'Thunderbird' Super Sabres were painted primarily in natural metal, but had a huge 'Thunderbird' motif on their bellies plus red, white and blue markings on the nose, wingtips and tail units.

F-100Cs were also used by another aerobatic display team, this one being the 36th FDW's 'Skyblazers', a four-aircraft team that flew displays throughout Europe. Its aircraft were modified in just the same way as the 'Thunderbirds' aircraft and wore a similar scheme, with jagged nose, tailplane and wingtip stripes in red, white and blue, a blue tail bedecked in white stars, and a red-and-white candy-striped trailing edge.

With Super Sabre orders at last being placed in quantity, the decision was taken to find a second production source and North American's Columbus, Ohio, plant was chosen on 11 October 1954. Although they were no different to the Inglewood-built F-100s, Super Sabres built at Columbus were known as NA-222s to the company and used NH rather than NA block designations. Twenty-five F-100Cs were built at Columbus, the first (55-2709) making its initial flight on 8 September 1955. They were followed by 221 F-100Ds.

Rewinged heavyweight

The final single-seat F-100 variant was by far the best, featuring a host of improvements that addressed most of the shortcomings of the earlier variants. This was the F-100D, which went under no fewer than four separate company design numbers (NA-223, -224, -235 and -245). It was built in the largest quantities, with 1,274 aircraft following the 203 F-100As and 476 F-100Cs off the Inglewood and Columbus production lines.

The F-100D was an improved fighter-bomber variant with internal electronic countermeasures (ECM) equipment, an AN/APS-54 tail warning radar on the trailing edge of an enlarged fin, an AN/AJB-1 low-altitude bombing system (LABS) to allow nuclear toss-bombing, and a 1,200-lb (544-kg) increase in maximum take-off weight. The landing speed of an even heavier Super Sabre would have been nothing short of dangerous and, as a result, the F-100D also featured a redesigned wing.

The new wing had increased chord at the root, expanding the total wing area to 400.18 sq ft (37.18 m^2). This resulted from the installation of flaps on the trailing edge, which were slightly less swept than the rest of the trailing edge and gave a slight kink to the aircraft's planform. To

The 'Hun' in Europe

Nowhere was the F-100 more important than in Europe, where it equipped five fighter-bomber wings during the peak period of 1957 to 1961. They were initially based in France (48th FBW at Chaumont, 388/49th FBW at Etain-Rouvres and 50th FBW at Toul-Rosières), West Germany (36th FDW at Bitburg) and England (20th FBW at Wethersfield). All US nuclear-capable forces were removed from French soil in 1959/60, the F-100 wings redeploying to the UK and West Germany under Operation Red Richard.

Right: Two fighter-bomber wings moved from France in 1959 to West Germany: the 49th relocated to Spangdahlem in August and the 50th TFW (illustrated) moved to Hahn in December. Although the 49th was moved to Holloman AFB in 1968, it remained committed to NATO.

Below: In West Germany the Super Sabre was initially assigned to the 36th Fighter-Day Wing at Bitburg, which operated F-100Cs in the nuclear strike role. The wing transitioned to the F-105D from May 1961.

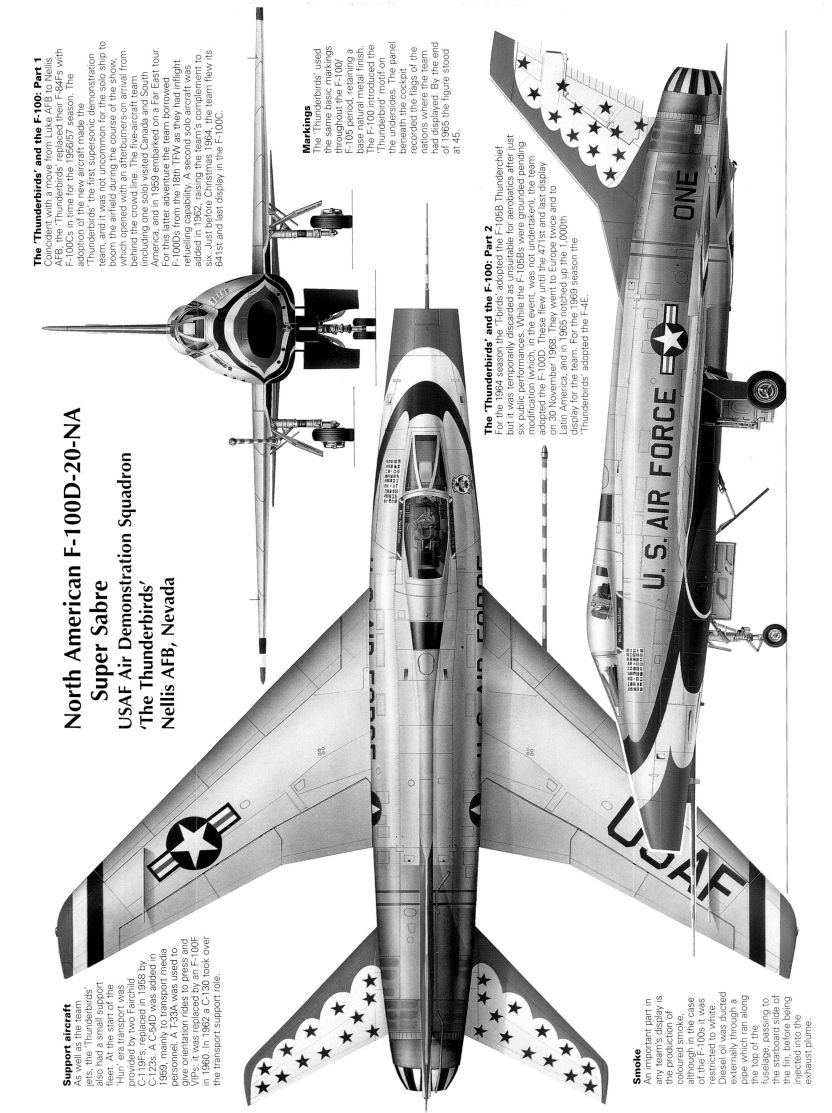

North American F-100D-20-NA Super Sabre
USAF Air Demonstration Squadron 'The Thunderbirds'
Nellis AFB, Nevada

The 'Thunderbirds' and the F-100: Part 1

Coincident with a move from Luke AFB to Nellis AFB, the 'Thunderbirds' replaced their F-84Fs with F-100Cs in time for the 1956/57 season. The adoption of the new aircraft made the 'Thunderbirds' the first supersonic demonstration team, and it was not uncommon for the solo ship to boom the airfield during the course of the show, which opened with an afterburners-on arrival from behind the crowd line. The five-aircraft team (including one solo) visited Canada and South America, and in 1959 embarked on a Far East tour. For this latter adventure the team borrowed F-100Ds from the 18th TFW as they had inflight refuelling capability. A second solo aircraft was added in 1962, raising the team's complement to six. Just before Christmas 1964, the team flew its 641st and last display in the F-100C.

Markings

The 'Thunderbirds' used the same basic markings throughout the F-100/F-105 period, retaining a base natural metal finish. The F-100 introduced the 'Thunderbird' motif on the undersides. The panel beneath the cockpit recorded the flags of the nations where the team had displayed. By the end of 1965 the figure stood at 45.

The 'Thunderbirds' and the F-100: Part 2

For the 1964 season the 'T-birds' adopted the F-105B Thunderchief but it was temporarily discarded as unsuitable for aerobatics after just six public performances. While the F-105Bs were grounded pending modification (which, in the event, was not undertaken), the team adopted the F-100D. These flew until the 471st and last display on 30 November 1968. They went to Europe twice and to Latin America, and in 1965 notched up the 1,000th display for the team. For the 1969 season the 'Thunderbirds' adopted the F-4E.

Support aircraft

As well as the team jets, the 'Thunderbirds' also had a small support fleet. At the start of the 'Hun' era transport was provided by two Fairchild C-119Fs, replaced in 1958 by C-123s. A C-54D was added in 1959, mainly to transport media personnel. A T-33A was used to give orientation rides to press and VIPs; it was replaced by an F-100F in 1960. In 1962 a C-130 took over the transport support role.

Smoke

An important part in any team's display is the production of coloured smoke, although in the case of the F-100s it was restricted to white. Diesel oil was ducted externally through a pipe which ran along the top of the fuselage, passing to the starboard side of the fin, before being injected into the exhaust plume.

Warplane Classic

54-1966 began life as a standard F-100C, but was loaned to North American for conversion into the TF-100C two-seater. It had an F-100D-style fin, and was lengthened by 36 in (0.91 m). It lacked any combat equipment as the original intention was for an austere conversion trainer.

Prior to flying with the thrust-reverser fitted, the NF-100F was tested in the 40 x 80-ft (12.2 x 24.4-m) wind tunnel at NASA Ames (Moffett Field, California). The aircraft later became a DF-100F drone director, and was one of three Super Sabres to use the NF-100F designation.

make room for the flaps, the ailerons were moved outboard. The flaps reduced landing speeds considerably, but increased pilot workload, since they had to be raised immediately upon touchdown to increase the 'weight on wheels' and maximise the F-100's braking efficiency. The flap normally extended to a full-down 45° position, though some aircraft had an intermediate 20° setting for take-off.

Other minor changes included provision of explosive jettison for the underwing pylons, which had previously relied on gravity for emergency separation. The aircraft could carry six M117 750-lb (340-kg) bombs, four 1,000-lb (454-kg) bombs, or a single Mk 7, Mk 28EX, Mk 28RE or Mk 43 nuclear weapon. Air-to-air missile armament, previously tested and demonstrated on a batch of six modified F-100Cs, was added from the 184th F-100D with accommodation for four GAR-8 (later AIM-9B) Sidewinder infrared-homing AAMs. This aircraft also introduced a new centreline hardpoint for nuclear weapons carriage, avoiding the asymmetry problems associated with the port intermediate pylon used previously. The same F-100D was originally planned to introduce an improved autopilot, but this was not ready in time, and did not make its debut until the 384th F-100D, although earlier aircraft were subsequently retrofitted with the new equipment.

The first F-100D made its maiden flight (in the hands of Dan Darnell) on 24 January 1956, and the first Columbus-built example (NA-224 55-2734) followed on 12 June. Deliveries to front-line units began in September 1956, initially to the 405th Fighter-Bomber Wing at Langley AFB, Virginia. The new variant rapidly replaced the F-100C in most USAF Super Sabre wings.

Like the F-100A and the F-100C before it, the F-100D suffered teething problems, particularly regarding an inaccurate fire control system, an unreliable electrical system, and poor integration of LABS with the autopilot. There were also engine bearing and afterburner fuel system problems, inadvertent and uncommanded bomb releases, and inflight-refuelling probes that sometimes fell off the wing during high-g manoeuvres.

The response was a series of *ad hoc* in-service modifications that led to almost immediate loss of fleet-wide configuration control, so that no two F-100Ds on a typical flightline were quite identical. From 1962, 700 F-100Ds and F-100Fs were modified and standardised under Project High Wire, also gaining a spring-loaded airfield arrester hook. Modified aircraft were given new block numbers, so that, for instance, an F-100D-25-NA would become an F-100D-26-NA after High Wire modifications.

In service, 65 F-100Ds were modified to carry the Martin GAM-83A Bullpup optically-guided air-to-surface missile. The first GAM-83A-equipped F-100D squadron made its operational debut in December 1960, but the weapon proved problematic and was withdrawn after combat experience in Vietnam revealed its shortcomings. NAA went to great expense to develop a buddy refuelling pod for the F-100D, and although such a device was built, tested and cleared, it was never used in service.

Most F-100Ds did feature a redesigned refuelling probe, which kinked upwards closer to the pilot's natural eye-line in order to make refuelling slightly easier. This was perhaps just as well, since the F-100D made greater use of the technique than any previous fighter during its support of a number of long-range deployments. In November 1957, 16 F-100Cs and 16 F-100Ds undertook Operation Mobile Zebra, flying from George AFB, California, to Yokota in a gruelling 16 hours. Subsequent F-100Ds were delivered to the Fifth Air Force by less direct means. They were first flown to McClellan AFB, California, where they were partially dismantled and cocooned, and then towed 15 miles (24 km) on public roads to the Sacramento river, from where they were barged to NAS Alameda. There they were transferred to a US Navy carrier which conveyed them to Yokosuka, where they were moved to lighters which ferried them across Tokyo Bay to Kisarazu AB, where they were de-cocooned, re-assembled and then flown to Itazuke!

When the Lebanon crisis blew up in May 1958, F-100Ds from Myrtle Beach AFB, South Carolina, were immediately deployed to the region, arriving there the same day on which the order to deploy was received.

The F-100D saw extensive service in Vietnam, after which some 335 were transferred to the ANG. After a false start in which the team's original F-100Cs were briefly replaced by wholly unsuitable F-105s, the 'Thunderbirds' used the F-100D from July 1964 until November 1968, when they began to transition to the F-4E Phantom. On 21 October 1967 an air show accident at Laughlin AFB in Texas resulted in a 'Thunderbirds' F-100D (flown by future Chief of Staff Captain Merrill A. McPeak) disintegrating after a fatigue failure of the wing. The 'Thunderbirds' were temporarily grounded and a 4-g manoeuvre limit was imposed in Vietnam, where there had been several unexplained losses. A modification to the wing box solved the problem.

Two-seat trainer

The previous generation of jet fighters had been relatively benign, allowing a young pilot to transition from a P-51 Mustang, an F-80 Shooting Star or a T-33 jet trainer to an F-86 without undue difficulty. Thus, no two-seat trainer

Reverse-thrust

56-3725 was the first F-100F two-seater to be built, and after its initial test duties were completed, it was modified for a research programme and assigned the designation NF-100F. Wright Air Development Center removed the afterburner section and replaced it with a thrust reverser. Blown flaps were installed, as was an enlarged airbrake. The aircraft was used for research into steep approaches and high-speed landings in support of the North American X-15 and Boeing X-20 Dyna-Soar programmes.

Below: Wearing AFSC and WADC badges, this is the NF-100F thrust-reverser testbed. Note the massive airbrake which is deployed for display, and the fairing under the intake.

F-100B and F-107A

The USAF's F-100 fighter-bombers performed an invaluable deterrent role during the Cold War, but the first time the aircraft looked as though it would be needed for a real 'shooting war', it was as a fighter interceptor.

The F-86 had spawned a number of highly successful dedicated all-weather fighter interceptor derivatives, and NAA hoped to produce a similar version of the Super Sabre. Work on a faster follow-on to the F-100A began in 1952, under the company designation NA-212. The new variant was expected to retain the F-100A's original swept wing planform but to have a much thinner wing cross-section and a 5 per cent thickness/chord ratio, while the fuselage was to have an increased fineness ratio and was to be area-ruled to reduce drag. Despite its thin section, the wing contained integral tanks, and there was to be no provision for the carriage of external fuel tanks. The aircraft was to be powered by an upgraded 16,000-lb st (71.15-kN) J57 engine with a variable-area inlet duct and a convergent-divergent afterburner nozzle. This was expected to confer a maximum speed of about Mach 1.8 at high altitude.

Interestingly, the F-100B (as the NA-212 was already tentatively designated) was expected to operate from primitive and unprepared forward airfields, and dual-wheel main landing-gear units were to be provided. This basic F-100B day

North American built three YF-107A prototypes, this being the first. One of the primary roles envisaged for the type was tactical nuclear strike, for which its 'special weapon' was carried semi-recessed in a central fuselage bay.

fighter formed the basis of a further derivative, the F-100I or F-100BI all-weather interceptor. It had an all-rocket armament and an AI radar in a new nose radome, and a forward fuselage redesigned with a variable-area air intake under the nose.

In November 1953, North American began adapting this evolution of the NA-212 back to the fighter-bomber role, adding six underwing hardpoints, providing single-point pressure refuelling capability, and redesigning the windshield and canopy to improve the pilot's view over the nose. In the face of continuing indifference from the USAF, the NA-212 programme was scaled back on 15 January 1954, when NAA president Lee Atwood abandoned plans for full production, reducing the programme's scope to a simple comprehensive engineering study. When interest in the project was revived, it was clear that the USAF was interested in the fighter-bomber configuration of the NA-212, and all work on the F-100B interceptor project was terminated, though the nose radome and chin intake of the interceptor version were retained.

The resulting F-100B gained an order on 11 June 1954, when the USAF authorised a contract for 33 F-100B fighter-bombers. On 8 July 1954, the designation of the aircraft was changed to F-107A. Further development of the F-107A, which became a dual-role interceptor and fighter-bomber, resulted in an aircraft considerably removed from the F-100 (with novel features such as over-fuselage variable-flow intake and one-piece tailfin), though it is worth noting that the aircraft flew successfully in prototype form before losing out to the Republic F-105 Thunderchief.

This view of the second YF-107A shows the type's derivation from the F-100, although the road from the initial NA-212 design to the eventual F-100B/F-107A hardware saw a great many changes being introduced. By the time the aircraft first flew on 10 September 1956, only the general wing structure design remained from the F-100.

versions of the F-84 and F-86 entered service, and the accident rate remained acceptable, if high.

From the beginning, however, the Super Sabre suffered a worrying accident rate. With its highly swept wing, tricky handling characteristics, and alarmingly high landing speed, it soon became clear that a two-seat trainer version would be useful in helping pilots convert to the aircraft. North American completed a private-venture design study for a supersonic trainer version of the F-100 day fighter on 10 May 1954, and on 2 September 1954 the USAF offered to loan the company a standard F-100C (54-1966) for conversion to TF-100C trainer configuration. Before this aircraft could fly in its new guise, the USAF's requirement for the trainer had firmed up, and in December 1955 a contract was issued for 259 TF-100Cs. The TF-100C was seen as a conversion trainer, pure and simple, and was expected to lack all operational equipment.

NAA test pilot Alvin S. White made the TF-100C (NA-230) prototype's first flight in its new configuration on 3 August 1956. The aircraft had the enlarged F-100D-type tailfin, but retained the C model's flapless wing. It was subsequently lost on 9 April 1957, when it failed to recover during a spinning test. 54-1966 was destined to be the only NA-230, since the USAF changed its mind about the TF-100C and decided that it needed a rather different trainer, one that would retain full operational capability in order to teach pilots weapons aiming and delivery and that would be useful for standardisation and continuation training, not just type conversion. In the event, the resulting trainer's operational capability even extended to the carriage of nuclear weapons, and the only concession made was to reduce the cannon armament from four to two and ammunition capacity from 200 to 175 rounds per gun in order to save internal volume and weight.

Production two-seater

North American accordingly designed the F-100F (NA-243), closely basing the new trainer on the latest F-100D configuration. The short-lived TF-100C was felt to have proved the aerodynamic changes to the two-seat Super Sabre, and the first F-100F to fly was the first production aircraft (56-3725), which made its maiden flight in the hands of Gage Mace on 7 March 1957. An early F-100F was used by Major Robinson Risner (a Korean War ace) to retrace Lindbergh's journey from New York to Paris to cele-

An F-100F from the 356th TFS refuels in 1960 during a deployment to Europe. The squadron was part of the 354th TFW, home-based at Myrtle Beach, South Carolina. Two-seaters had full combat capability – including nuclear wiring – although they had two instead of four guns. The early, straight refuelling probe, as fitted to this aircraft, was later replaced in the F-100F and most Cs by the kinked probe introduced during F-100D production.

F-100 type conversion was concentrated at Luke and Williams, within the Phoenix suburbs. The two units operated a large number of F-100Fs between them.

Left: 'Thunderbird Nine' was part of the team's support organisation. Three F-100Fs were assigned to the team between 1960 and 1968 and were mainly used to give lucky VIPs and reporters a chance to experience the thrill of flying in the Super Sabre.

F-100As and F-100Fs line up on a Taiwanese base early in the type's RoCAF career. The aircraft in the foreground still wears its USAF 'buzz' number and has yet to receive Nationalist Chinese insignia. The As had all been hastily modified prior to delivery, including the fitment of F-100D fins.

From 1961 to 1966 the French Super Sabre force was based in West Germany, armed with US-owned and controlled nuclear weapons. The nuclear commitment was lost when the aircraft returned to France. The disbandment of EC 3 allowed a third squadron to form at Toul: EC 3/11 'Corse'.

brate the 30th anniversary of the historic flight of 'Lucky Lindy'. Deliveries began in January 1958.

At the specific request of the Pacific Air Forces, the final 29 two-seaters (F-100F-20-NA company designation NA-255) were built to a revised standard. They were fitted with an enhanced navigation system including an AN/ASN-7 dead-reckoning computer, and a PC-212 Doppler. The aircraft also featured modified flaps with a span-wise duct built into the leading edges, which directed air from the lower surface of the wing over the upper surface of the flap. This primitive 'flap-blowing' increased efficiency and reduced buffeting during landing. The modified flaps were limited to a full deflection angle of 40°, instead of the usual 45°.

F-100 production ended in October 1959 with the delivery of the last of 339 F-100Fs. This brought total Super Sabre production to 2,294 aircraft, including the 359 F-100Cs and F-100Ds built at Columbus.

Against Communist China

Tension between the Communist mainland People's Republic of China and the Nationalist Republic of China on the island of Formosa (Taiwan) erupted into conflict in August 1958. Communist Chinese forces bombarded the

Chinese Super Sabres

Nationalist China became the first foreign recipient of the Super Sabre as a result of the 1958 Qemoy crisis, when Communist China threatened to invade Taiwan. Four squadrons' worth of F-100As, plus F-100Fs, were handed to the RoCAF in some haste to bolster the island's defences. The last of these survived in service until 1984, and a number were preserved. The immaculate F-100F above wears the markings of the 2nd TFW at Hsinchu, and is spuriously marked with the serial number of a single-seat F-100A.

Nationalist enclave of Qemoy (a small island just off the mainland) and air battles raged above the Formosa straits. The AIM-9B-equipped RoCAF F-86s fared well against the Communist MiG-15s and MiG-17s, but were out-numbered. They were also too slow to cope with the MiG-19s then entering service with the PLA or to operate in the reactive interceptor role. When it became clear that the mainland was planning to invade Qemoy as a preliminary to the 'liberation' of Taiwan itself, the USA intervened.

USAF units in the area (including the F-100D-equipped 511th TFS at Ching Chuan, Taiwan, and the 354th TFW at Kadena, on Okinawa) were placed on higher states of alert, and six F-100Fs and 80 F-100As were hastily supplied to the RoCAF from August 1958, allowing the immediate conversion of four squadrons to the new type. In that year, the RoCAF also received the four surviving Slick Chick RF-100A reconnaissance aircraft. Thirty-eight more F-100As and eight more two-seaters were supplied from 1970-71.

All but four of the Taiwanese F-100As (53-1569, 53-1581, 53-1651 and 53-1662) were heavily modified before delivery, receiving a new F-100D-type tailfin with AN/APS-54 tail warning radar, a radio compass, an arrester hook, and new inboard pylons wired for the carriage of GAM-83A (AGM-12) Bullpup ASMs or twin AIM-9 Sidewinder AAMs. In this form, the aircraft were broadly equivalent to the F-100D, albeit without flaps and integral wing tanks, and were sometimes known as 'F-100A Rehabs'.

Details of the F-100's combat use by the RoCAF remain secret, though it is understood that the RF-100As suffered such poor availability that they flew no operational overflights, and were scrapped in 1960.

Other foreign users

The available stock of surplus F-100As went to Nationalist China, so other foreign recipients of the F-100 received later versions of the aircraft. All three remaining foreign F-100 users received their aircraft under the terms of the US Military Assistance Program (MAP), under which aircraft were effectively donated free of charge to selected NATO allies.

The first of these MAP Super Sabres went to France, where the first of 12 F-100Fs was received on 1 May 1958, and the first of an initial batch of 68 F-100Ds followed on 18 May. These 80 aircraft equipped Escadre de Chasse 11 at Luxeuil and Escadre de Chasse 3 at Reims. Many were

MAP F-100s for France

The initial batch of 80 F-100D/Fs for France was divided between EC 1/3 'Navarre', EC 2/3 'Champagne' at Reims, and EC 1/11 'Roussillon' and EC 2/11 'Vosges' at Luxeuil. In early 1961 all four squadrons moved to West Germany as part of 4th ATAF, the EC 3 units going to Lahr while EC 11 moved into Bremgarten. EC 3 transitioned to the Mirage IIIE shortly before it returned to France in 1967, while EC 11 brought its F-100s back to Toul-Rosières, where it absorbed the aircraft from EC 3. This led to the formation of a third squadron, EC 3/11 'Corse'. Serving for some years in a natural metal finish (above), most French Super Sabres were later painted in tactical camouflage, as demonstrated by the EC 1/11 example below. A flight of seven F-100Ds and one F-100F was established in Djibouti on 1 January 1973 as EC 4/11 'Jura'. The unit's aircraft sported a desert-style camouflage, and later adopted a large sharkmouth. One of the aircraft is seen at its Djibouti base in January 1974 (below right).

Turkey's F-100 force went to war twice in Cyprus, in 1964 and again 10 years later. Having received its first aircraft in 1958, the Turkish air force was – amazingly – still adding to its F-100 fleet in the early 1980s, and the type remained in THK service until 1987.

drawn from France-based F-100 units, but all were overhauled, modernised and upgraded in Spain before delivery to the Armée de l'Air.

The French Super Sabres were soon in action. In Algeria, French forces made extensive use of close air support aircraft in their war against the Armée de la Liberation Nationale, but increasingly sophisticated rebel weapons soon made the piston-engined T-6s vulnerable. From 1959, therefore, Super Sabres of EC 1/3 flew missions against pre-planned targets, taking off from their base at Reims and recovering to Istres to refuel on their return journey.

The two French Super Sabre wings moved to Lahr and Bremgarten in West Germany in February and June 1961, respectively, where they undertook nuclear strike duties for the 4th Allied Tactical Air Force using US Mk 7 bombs. Despite the delivery of 20 more F-100Ds as attrition replacements, EC 3 re-equipped with Mirage IIIs in September 1966, allowing the addition of a third Escadron within EC 11.

When President Charles de Gaulle pulled France out of NATO's military command structure in 1967, USAF units in France were withdrawn. The remaining West Germany-based Armée de l'Air units moved back to the vacated air bases on French soil and lost their strike commitment.

Turkish 'Huns'

Turkey was the largest foreign operator of the Super Sabre, and used the type for longer than any other. From 1958, the Türk Hava Kuvvetleri (THK) received some 270 Super Sabres, including 111 F-100Cs, 106 F-100Ds and 53 F-100Fs. Turkey was thus the only overseas operator of the F-100C, but was also the most important operator of the two later versions, as well. The THK received 14 F-100Ds and nine F-100Fs in 1958, allowing 111 Filo at Eskisehir to convert. Thirty-two F-100Ds and two more F-100Fs were delivered in 1959, followed by three F-100Ds and two F-100Fs in 1960. 113 and 112 Filo received these aircraft, settling at Erhac and Eskisehir.

On 8 August 1964, Turkey's F-100s went to war, attacking Greek Cypriot National Guard units and EOKA terrorists after attacks against Turkish villages in the north of Cyprus. One 111 Filo F-100D crashed while attacking a landing craft, and the pilot was captured and killed after ejecting. Between October 1965 and 1969, 112 Filo had to revert to the F-84F due to a shortage of Super Sabres, but re-converted following the delivery of 16 more F-100Ds and two F-100Fs, which also allowed 182 Filo at Erhac to convert.

TF-100F – Danish two-seater

The TF-100F designation covered the second batch of 14 two-seat aircraft delivered to the Danske Flyvevåbnet, which differed in some respects from the original 10 F-100Fs received by Denmark. Like the single-seaters, they were fitted with Martin-Baker seats.

Some 20 more F-100Ds and two F-100Fs arrived in 1970. In 1972 Turkey received 36 F-100Cs, and 47 more followed in 1973, with 28 more being taken on charge in 1974. 113 and 182 Filos at Erhac were redesignated as 171 and 172 Filos, and three new F-100 units formed with the new aircraft: 181 Filo at Diyarbakir, and 131 and 132 Filos at Konya.

The final MAP F-100 recipient was Denmark, which received F-100Ds and Fs to replace its two three-squadron wings of ageing straight-winged Republic F-84G Thunderjet fighter-bombers in the close air support, air defence and maritime attack roles. Danish pilots had evaluated the F-100 in 1957, and in August 1958 a small group of pilots and ground crew was trained on the aircraft at Myrtle Beach, though F-100s were not formally offered until 1959.

Esk 727 at Skrydstrup took delivery of three F-100Fs and 17 F-100Ds from May 1959, and 31 F-100Ds and seven F-100Fs followed in 1960 to re-equip Esk 725 at Karup and Esk 730 at Skrydstrup. The Danish F-100s were extensively modified in service, gaining Martin-Baker Mk DE5A ejection seats, a Decca Type 1664 Roller Map, and a Saab BT-9J bombsight. Attrition was heavy, a full one-third of the fleet

A Turkish pilot runs to his F-100D during a practice scramble. In the fighter role the aircraft could be fitted with inverted 'Y' racks for the carriage of four AIM-9 Sidewinders.

Denmark's F-100s were initially flown in natural metal finish, but subsequently received an all-over dark green scheme. Even the aft fuselage was painted, although the paint rapidly burnt through to form a characteristic pattern. During the 1960s the Danes had three squadrons of F-100s, but in 1970 one converted to the more capable Saab Draken, leaving Esk 727 and 730 at Skrydstrup.

Warplane Classic

Seen in 1960, this F-100D served with the 510th TFS, part of the 405th TFW at Clark AB in the Philippines. A year later, the squadron sent the first operational F-100 detachment to southeast Asia, six aircraft arriving at Don Muang airport (Bangkok) on 16 April 1961. The detachment lasted until late in the year, and was officially for air defence purposes, but in reality was more of a sabre-rattling exercise against growing Communist actions in the region.

being lost in accidents. With insufficient aircraft to equip all three squadrons, Esk 725 converted to the Draken in 1970. Fourteen more ex-USAF F-100Fs were delivered from March 1974; they were fitted with Martin-Baker seats and redesignated as TF-100Fs in order to differentiate them from Denmark's original F-100Fs.

At war in Vietnam

US sources have often (wrongly) stated that the Super Sabre made its combat debut in Vietnam: F-100s had already flown combat missions in RoCAF service, and Armée de l'Air Super Sabres had flown live bombing missions over Algeria. Vietnam was not even the first combat use of the USAF's Super Sabres, since the Slick Chicks had come under hostile fire during their secretive reconnaissance missions. The war in southeast Asia was, however, probably the most important event in the aircraft's long career, and was the only large-scale conflict in which it was involved.

At its peak, the Super Sabre equipped a total of 16 USAF wings and four of those saw service in Vietnam, operating there between 1966 and 1971, and flying over 360,000 missions. This notable figure was all the more impressive when compared with the 259,702-mission total clocked up by 16,000 P-51 Mustangs during World War II. Those F-100s deployed to Vietnam averaged 1.2 sorties per day and the type demonstrated an 80 per cent readiness rate. The aircraft were hard-worked, and by 1969, the average flight time per surviving aircraft was 5,100 hours. Some 198 F-100s were lost in combat and 44 more in in-theatre accidents, with the loss of 87 pilots killed, five missing in action and five prisoners of war.

By the time the US involvement in Vietnam began, the F-100 was firmly established as a fighter-bomber, its weaknesses in the day fighter role having been widely recognised and acknowledged. Even in the fighter-bomber role, the aircraft's career was seen as coming towards an end, and five front-line wings had already converted to the F-104 and F-105 during 1959-62. The remainder seemed set to follow fairly rapidly.

It was therefore slightly ironic that the F-100D was one of the first US combat aircraft deployed to southeast Asia, and that (at least nominally) it did so as an air defence fighter. Following a string of events in disintegrating Laos, six F-100D/Fs from the 510th TFS at Clark AB, Philippines, were deployed to Don Muang airport in Thailand on 16 April 1961 under Operation Bell Tone. The aircraft were nominally there to provide air defence for the Thai capital, though the air threat was negligible and the F-100 was ill-

F-100 vs. MiG-17

The F-100 Super Sabre's history is a saga of superlatives. The 'Hun' flew faster, higher, and farther than its predecessors. It set speed records. It flew more individual sorties in the Vietnam war than any other fighter. It guarded against Soviet attack during tense moments in the Cold War. In fact, the F-100 Super Sabre did almost everything a modern fighter could do – except shoot down an enemy aircraft. Incredibly, despite its decades on the cutting edge of combat aviation, the F-100 was never credited with a single air-to-air victory. Since the high priests of the fighter profession regard an aerial 'kill' as sacred on the altar of their religion, the Super Sabre's other achievements can never compensate for the fact that it was never a MiG killer.

Or was it?

As far as official records are concerned, the facts are clear. No F-100 ever shot down an enemy aircraft. No enemy aircraft ever shot down an F-100, either. But veterans of the earliest days in Vietnam – a brief interval when the F-100 was employed as an escort fighter before being relegated to air-to-ground duty – say the official records are wrong. They say US Air Force Captain Donald L. Kilgus (pictured in his MiG-killing aircraft, below) shot down a North Vietnamese MiG-17 on 4 April 1965. They also say Kilgus was denied credit for an aerial victory that should have placed in his record, not because the MiG didn't fall out of the sky, but because errors were made.

Kilgus, a fighter pilot with the 416th Tactical Fighter Squadron at Da Nang, South Vietnam, flew the first mission on 2 March 1965 when the United States launched Operation Rolling Thunder – a campaign against North Vietnam that eventually lasted more than three years. "In those early days, we were just beginning to see heavy air fighting in the region around Hanoi," Kilgus said in an interview in 1990. "Big air battles would become familiar to us later, but in the beginning it was all new."

Just a month into the Rolling Thunder campaign, the first air-to-air engagement of the Vietnam war took place on 3 April 1965 when Soviet-built MiG-17 fighters of the North Vietnamese Air Force fired on a US Navy F-8 Crusader with no result. The next day marked a series of air-to-air battles, a tragic setback for the US, and a controversial dogfight for Kilgus.

On 4 April 1965, numerous air strikes went into North Vietnam. The setback occurred when North Vietnamese MiG-17s popped out of heavy clouds and shot down two Air Force F-105 Thunderchiefs piloted by Captain James A. Magnusson and Major Frank E. Bennett. Both F-105 pilots lost their lives. Both were members of the 354th Tactical Fighter Squadron, 355th Tactical Fighter Wing, flying from Korat Air Base, Thailand, operating that day as ZINC flight. Magnusson, at the controls of F-105D 59-1764, apparently was killed almost immediately, perhaps by cannon fire that struck his cockpit. Bennett, however, who was piloting F-105D 59-1754, should have survived. He nursed his crippled aircraft out to the Gulf of Tonkin and ejected safely. For a moment, he appeared to be safe on the surface of the Gulf, ready to be picked up, but somehow Bennett got tangled in his parachute and drowned before help could arrive.

North Vietnamese gunfire also downed an A-1H Skyraider (bureau number unknown), killing Captain Walter Draeger. Another F-105 pilot, Capt. Carlyle 'Smitty' Harris, at the controls of aircraft 62-4217 (from the 44th TFS 'Vampires', 18th TFW at Korat) was shot down, survived, and became one of the earliest American prisoners of war. He pioneered the 'tap code' later used by prisoners to communicate from one North Vietnamese cell to another.

Because there were certain things the outside world did not know that day, the air battle was reported as a stunning defeat for the United States. Americans simply were not accustomed to coming out second best in fighter-versus-fighter combat. Press reports focused on the dramatic loss of the two F-105s to MiGs.

Many years later, when North Vietnam's records became available, it became known that the North Vietnamese lost three MiG-17s that day. It appears that the North Vietnamese actually shot down two of their own MiGs with their own ground fire – possibly the same two MiGs that bagged the F-105s. Neither side has ever confirmed the circumstances of the loss of the third MiG listed as a casualty in Hanoi's records. The press did not immediately report a dogfight that day, between F-100 Super Sabre pilot Kilgus – assigned to escort the F-105s – and a MiG-17 Kilgus was certain he shot down.

"We saw something come up out of the haze," Kilgus said. "And one thousandth of a second later ... it's a MiG. I turned into him, jettisoned my auxiliary fuel tanks, and in that instant he turned 90 degrees to face me." Kilgus spotted a second MiG. The first overshot and missed him. Kilgus shook off the second, manoeuvred abruptly, and found himself behind the first.

"I said, 'I'll get in range.' I pulled my nose up. All four guns are in the belly of the airplane, so I pulled up the nose and just fired enough so he'd see those 20-millimetre cannons winking." The F-100 was armed with four 20-mm Pontiac M39E cannons with 1,200 rounds, although most aerial victories in Vietnam were achieved with air-to-air missiles.

"Knowing I was in an advantageous position because I was above him, I allowed him to get a little separation on me. I went on afterburner and saw 450 knots on my air speed indicator. He was now going straight down and I was thinking, 'He's playing chicken,' knowing that because his plane is lighter he can pull out of a dive faster than I can. I was preoccupied with my gunsight. This was while going straight down and turning the gun switch to hot. My mind was saying, 'When are we going to pull out?'

"I fired a burst. Now, training comes into play. I tried to remember everything I'd learned, and began shooting seriously at him at [an altitude of] 7,100 feet. I said to myself, I wouldn't worry about how

F-100 Super Sabre

suited to deal with it. It did, however, place a flight-strength unit of the USAF's principal jet fighter-bombers in the area. The deployment increased to squadron size from 18 May 1962, with 18 aircraft manned by rotational Saw Buck deployments from Cannon and England AFBs.

Following the shoot-down of an RF-8 over Laos on 22 May 1964, eight F-100Ds from the 615th TFS redeployed to Da Nang and mounted the first retaliatory strike on 9 June 1964, against targets in the Plaines des Jarres in Laos. Colonel George Laven led the mission in 54-2076, an aircraft he had hand-picked because its 'last four' matched those of the P-38 he had flown during World War II, when he had become an ace.

Eight F-100Ds from the 615th TFS deployed to Da Nang again after the Gulf of Tonkin incident, and began flying intensively, mainly in the escort role. The Super Sabre force suffered its first recorded combat loss soon afterwards, when an F-100D (56-3085) was shot down on 18 August 1964 over Laos. The aircraft were increasingly committed to close air support and AAA-suppression missions, and losses began to mount.

From 14 December 1964, the F-100s flew Barrel Roll sorties against NVA forces just across the border in Laos, often operating in squadron strength. The longer-ranged F-105 (which carried a heavier load than the Super Sabre) increasingly displaced the F-100 as the focus of attacks switched to North Vietnamese targets during Flaming Dart, though the F-100s usually accompanied the Thunderchiefs to provide fighter cover and flak suppression. Operation Rolling Thunder began in March 1965. The Vietnamese People's Air Force began flying combat missions on 3 April, and the following day, MiGs tangled with USAF F-100s for the first time. Captain Donald Kilgus fired on one MiG-17, observed strikes on the enemy aircraft and claimed a probable kill. Other aircraft engaged more MiGs with AIM-9s, without apparent success (though post-war research revealed that three MiGs failed to return), and the F-100 was withdrawn from the fighter cover role. 4 April marked the only occasion on which F-100s encountered enemy fighter opposition.

From November 1965, the F-100s were camouflaged, losing their gaudy unit colours and receiving two-letter squadron tailcodes in their place. Nose art remained, however, and increasingly began to flourish.

F-100Ds from the 416th TFS are seen at Da Nang (above) and Tan Son Nhut (above left). The 'Silver Knights led something of a nomadic existence in southeast Asia, deploying as part of Operation Sawbuck to Clark and on to Da Nang in March 1965, and then to Bien Hoa in mid-June. In November 1965 they went to Tan Son Nhut, then Phu Cat in April 1967, and finally to Tuy Hoa in May 1969. Having been in the warzone since the start, the unit finally returned to England AFB, Louisiana, at the end of September 1970.

much ammo I was using because this was my last chance to hit him. I saw puffs and sparks on the vertical tail of the MiG, and very shortly thereafter I didn't see anything. I could have been at 580 knots. I won't embroider the story by saying I got spray from the Gulf of Tonkin on my windshield, but I pulled out at the last minute." Kilgus said he saw "sparks" on the MiG and "major pieces coming off it." There was no other action, that day, which could explain the North's confirmation that a third MiG was shot down.

Don McCarthy, a Waterford, Conn., historian who has studied both American and North Vietnamese accounts of the battle, is "absolutely certain" Kilgus's cannon shells brought down the MiG-17. "The Air Force only credited Kilgus with a 'probable' kill," McCarthy said in a telephone interview. "The reason has never been clear." The Air Force has never released an official list of aerial victories intended for public consumption. Kilgus's claim was also supported by Capt. (later Lt. Col.) Ralph Havens, another member of the F-100D flight that day.

Kilgus died in a traffic mishap in the Washington, D. C. area in 1995. In all he flew three tours in Vietnam, in the O-1 Bird Dog, F-100D Super Sabre, and F-105 Thunderchief. He never made an issue of his claim to have downed a MiG-17, but he believed the MiG went down and all available evidence – including North Vietnamese records – seems to confirm it. Shortly after those early Rolling Thunder missions, the F-100 was shifted to air-to-ground duties.

Robert F. Dorr

After its brief flirtation with the escort fighter mission in the first few weeks of Rolling Thunder, the 416th TFS – and the F-100 community in general – became embroiled in 'mud-moving'. This 'Silver Knights' aircraft, seen on a mission in mid-1965, carries a typical mix of iron bombs and napalm tanks.

Right: The escalating seriousness of the war in southeast Asia was mirrored in the adoption of warpaint in 1965. Officially called T.O.1-1-4, the three-tone SEA camouflage scheme spread rapidly through the USAF's tactical forces. This is a 416th TFS aircraft, parked next to another which retained its natural metal finish.

Virtually all targets in South Vietnam were 'soft', which usually meant Viet Cong forces. Area weapons such as cluster bombs, 'daisy-cutters', napalm and 2.75-in rockets were used. This shot captures the dispersal of a full rocket salvo.

By October 1966, one F-100 unit served with the 27th TFW at Tan Son Nhut, three squadrons of F-100s formed the 35th TFW at Phan Rang, and three more squadrons formed the 3rd TFW at Bien Hoa. Additional F-100Ds were based with the 37th TFW at Phu Cat from 1 March 1967, and an eventual total of five squadrons served with the 31st TFW at Tuy Hoa from November 1966. Many of these units were manned by ANG squadrons on rotational deployment, and flew F-100Cs. By June 1967, only five regular USAF F-100 squadrons remained in the USA. Despite some maintenance issues, the F-100D briefly gained the best maintenance record of any aircraft in the Vietnam combat zone, and always enjoyed a better availability rate than most fast jets in-theatre.

Although the F-105 and F-4 could carry a heavier bomb load farther than the F-100, the Super Sabre's ability to drop from low altitudes brought it a reputation for accuracy, and the type was preferred by many forward air controllers for precision strikes in support of troops in contact. After 1966 the F-100 was the only USAF aircraft in-theatre using probe-and-drogue refuelling, and it became increasingly difficult to find tanker support, since most tankers were fitted with the boom system. These factors together led to the F-100D fighter-bombers generally operating in South Vietnam.

The F-100s were gradually withdrawn from combat as attrition took its toll and increasing numbers of newer aircraft (especially F-4E Phantoms) became available, and the last F-100Ds finally left Vietnam in July 1971. By then, the Super Sabre had proved its mettle in two more specialised roles, both of which required the use of the two-seat F-100F.

Misty FAC

The use of airborne forward air controllers (FACs) to direct and correct the efforts of fast jet close air support aircraft had proved invaluable in Vietnam, although their piston-engined O-1s and O-2s were slow and vulnerable. The obvious solution was to put a FAC in the back seat of a fast jet, and in 1967 the Commando Sabre programme was instituted to do just that.

Led by Major George 'Bud' Day, the Misty FAC F-100Fs were assigned to Det 1 of the 612th TFS, borrowing four F-100Fs from the 416th TFS. Trial missions from Phan Rang began on 15 May 1967, before operations proper commenced (flown from Phu Cat) on 28 June. The aircraft were crewed by a pair of volunteer pilots, both of whom had to have logged 25 combat sorties and 1,000 flying hours before joining the programme. The back-seater carried a comprehensive set of detailed maps, a handheld 35-mm 'strike camera' (actually an SLR with a telephoto lens), and handled communications with the fighter-bombers. The front-seater found the targets and marked them, using the Misty F-100F's armament of two seven-shot white phosphorus rockets.

Pueblo Crisis – Air National Guard F-100Cs in Vietnam

When the North Koreans seized the USS *Pueblo* on 24 January 1968, US active-duty forces were too stretched to react to this new crisis, so President Johnson mobilised a number of ANG squadrons two days later. Included were eight units with F-100Cs. Of these, two were deployed to Myrtle Beach to conduct F-100 training, two went to Kunsan in Korea, and the remaining four were sent to join the fray in Vietnam (120th to Phan Rang, 174th to Phu Cat, 136th and 188th to Tuy Hoa).

Right: Arriving at Phan Rang on 3 May 1968 to augment the 35th TFW, the 120th TFS of the Colorado ANG was the first Guard unit to fly combat missions, mounting its first sortie on 8 May. The squadron's 5,905th and last was flown on 8 April 1969.

Below: Iowa's 174th TFS ('HA' tailcode) deployed from Sioux City to Phu Cat in May 1968, flying as part of the 37th TFW. The other two F-100C units joined the 31st TFW at Tuy Hoa.

The DC ANG's 121st TFS was one of two units (along with the 119th TFS/New Jersey) which formed a wing at Myrtle Beach during the Pueblo mobilisation to augment the active-duty training effort.

North American F-100D-75-NA Super Sabre

416th Tactical Fighter Squadron
37th Tactical Fighter Wing
Phu Cat AB, South Vietnam

Phu Cat was established as the last of the major F-100 bases in the southeast Asia theatre in the spring of 1967, acquiring the 355th TFS, 416th TFS and Det 1/612th TFS – the last being the Misty FAC unit. In May 1968 an ANG F-100C squadron was added. In April/May 1969 the 37th TFW F-100 wing was broken up, the four squadrons either relocating to Tuy Hoa and Phan Rang, or returning home.

My Gal Sal III was typical of the hundreds of Super Sabres which fought in Vietnam. It survived its combat deployment, later serving with the Ohio ANG's 162nd TFS, before being assigned to the 113th TFS, Indiana ANG, the last 'Hun'-equipped squadron in the USAF.

Cameras
The F-100D carried a KA-71 or KB-18 strike camera in the lower port fuselage. For better recording of attacks a Combat Documentation Camera pod could be installed under the port wingroot. This fairing housed a forward-facing N-9 camera, with a downward/aft-facing DBM-4C in the back. An alternative was a centreline camera pod, occasionally seen on aircraft during the war in Vietnam.

Speedbrake
The F-100D was fitted with a ventral spee[d]
hydraulic rams. Early F-100Ds could not ca[rry]
capability was introduced at Block 20, the
which prevented the speedbrake deploying
more elegant solution was the adoption of
which allowed it to clear the three principa[l]
the MN-1A practice bomb dispenser and t[he]

Refuelling probe
The probe was attached underneath the leading edge of the starboard wing, fuel being piped into the integral wing tank. The probe was mounted low enough to provide adequate clearance for the leading-edge slats. Originally straight, the probe was modified to a kinked shape to raise the receiving nozzle closer to the pilot's eyeline, making the tricky task of refuelling a bit easier.

Weapons
Most F-100Ds had seven stores pylons available. The theoretical maximum bomb loads were four 1,000-lb (454-kg) stores, eight 750-lb (340-kg) bombs and 10 500-lb (227-kg) bombs, although the latter two loadouts required the use of triple ejector racks. In practice the F-100 rarely carried more than one store per pylon, and in most conventional loadouts the intermediate pylon was used for the carriage of fuel tanks. As here four stores was the usual load, comprising bombs, rockets or napalm. In the nuclear role the aircraft was cleared to carry the Mk 7, B28, B43, B57 and B61 weapons. One store would be carried, on either the port intermediate pylon (Mk 7, B28, B43) or centreline (B57 and B61). Some F-100Ds were modified to carry the AGM-12 Bullpup (inboard pylons only), while others were wired for the carriage of AIM-9 Sidewinders on twin-rail Y-racks, also on the inboard pylons. The AIM-4 Falcon was tested, but not adopted.

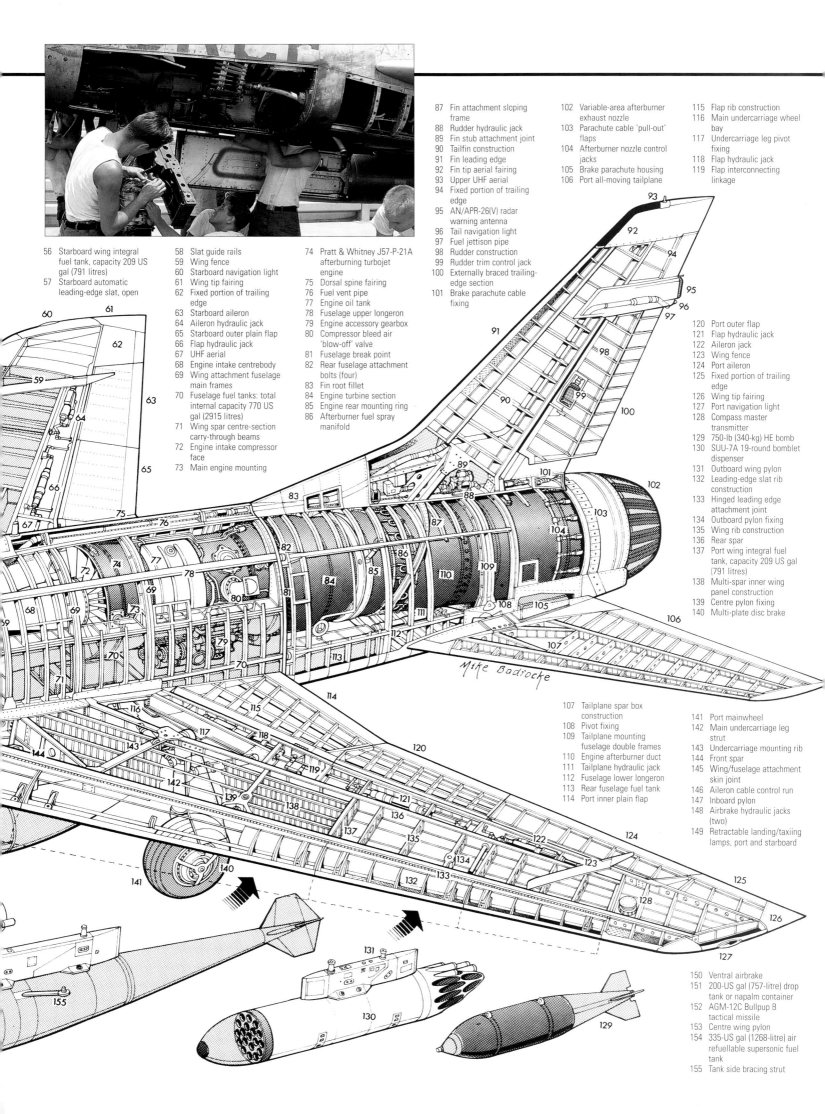

Inside the Super Sabre

F-100 specifications

Dimensions
Wingspan: YF-100A – 36 ft 7 in (11.15 m); F-100A/C/D – 38 ft 9 in (11.81 m)
Length: YF-100A – 46 ft 3 in (14.10 m); F-100A/C – 47 ft 1¼ in (14.36 m); F-100D – 50 ft 0 in (15.24 m); F-100F – 52 ft 3 in (15.93 m)
Height: YF-100A – 16 ft 3 in (4.95 m); F-100A – 13 ft 4 in (4.06 m); F-100A modified 15 ft 8 in (4.77 m); F-100C – 15 ft 6 in (4.72 m); F-100D/F – 16 ft 2¾ in (4.95 m)
Wing area: YF-100A – 376 sq ft (34.93 m²); F-100A/C 385 sq ft (35.77 m²); F-100D/F 400 sq ft (37.16 m²)

Weights
Empty: YF-100A – 18,135 lb (8226 kg); F-100A – 18,185 lb (8249 kg); F-100C – 19,270 lb (8741 kg); F-100D – 20,638 lb (9361 kg); F-100F – 21,712 lb (9848 kg)
Gross: YF-100A – 24,789 lb (11244 kg); F-100A – 24,996 lb (11338 kg); F-100C – 27,587 lb (12513 kg); F-100D – 28,847 lb (13085 kg); F-100F – 31,413 lb (14249 kg)
MTOW: F-100C – 32,615 lb (14794 kg); F-100D – 34,832 lb (15800 kg); F-100F – 39,122 lb (17746 kg)

Powerplant
YF-100A – one Pratt & Whitney XJ57-P-7 rated at 8,700 lb st (38.71 kN) dry, or 13,200 lb st (58.74 kN) with afterburning
F-100A/C – one Pratt & Whitney J57-P-7/39 rated at 9,700 lb st (43.16 kN) dry, or 14,800 lb st (65.86 kN) with afterburning
F-100C/D/F – one Pratt & Whitney J57-P-21 rated at 10,200 lb st (45.39 kN) dry, or 16,000 lb st (71.2 kN) with afterburning
F-100D/F – one Pratt & Whitney J57-P-21A rated at 10,200 lb st (45.39 kN) dry, or 16,920 lb st (75.29 kN) with afterburning

Fuel
Fuel capacity (internal): YF-100 – 1,307 US gal (4947 litres); F-100A/F – 1,294 US gal (4898 litres); F-100C – 1,702 US gal (6443 litres), F-100D – 1,739 US gal (6583 litres)
Fuel capacity (total): F-100C/D – 2,139 US gal (8097 litres)

Armament
Internal: F-100A/C/D – four 20-mm Pontiac M39 cannon with 200 rpg, or 257 rpg (some F-100D); F-100F – two 20-mm Pontiac M39 cannon with 175 rpg
External: F-100A – 2,000 lb (907 kg); F-100C/F – 5,000 lb (2268 kg); F-100D – 7,040 lb (3193 kg)

Performance
Maximum speed at 35,000 ft (10668 m): YF-100A – 634 mph (1020 km/h); F-100A – 740 mph (1191 km/h); F-100C – 803 mph (1292 km/h); F-100D – 765 mph (1231 km/h); F-100F – 760 mph (1223 km/h)
Initial climb rate: YF-100A – 12,500 ft (3810 m) per minute; F-100A – 23,800 ft (7254 m) per minute; F-100C – 21,600 ft (6584 m) per minute; F-100D – 19,000 ft (5791 m) per minute
Service ceiling: YF-100A – 52,600 ft (16032 m); F-100A/F – 44,900 ft (13685 m); F-100C – 38,700 ft (11796 m); F-100D – 36,100 ft (11003 m)
Combat ceiling: F-100A/F – 51,000 ft (15545 m); F-100C – 49,100 ft (14966 m); F-100D – 47,700 ft (14539 m)
Combat radius: YF-100A – 422 miles (679 km); F-100A/F – 358 miles (576 km); F-100C – 572 miles (920 km); F-100D – 534 miles (859 km)
Maximum range: YF-100A – 1,410 miles (2269 km); F-100A/F – 1,294 miles (2082 km); F-100C – 1,954 miles (3144 km); F-100D – 1,995 miles (3210 km)

The four 20-mm cannon were easily accessed through panels in the lower fuselage, and the breech assembly hinged out for maintenance (above right). Ammunition was held in tanks located either side of the cockpit, feeding rounds to the staggered guns via side-by-side belts (above). These scenes of ground crew servicing and arming the weapons were recorded during combat operations in Vietnam.

North American F-100D cutaway

1. Pitot tube, folded for ground handling
2. Engine air intake
3. Pitot tube hinge point
4. Radome
5. IFF aerial
6. Ranging radar
7. Intake bleed air electronics cooling duct
8. Intake duct framing
9. Cooling air exhaust duct
10. Cannon muzzle port
11. UHF aerial
12. Nose avionics compartment
13. Hinged nose compartment access door
14. Inflight refuelling probe
15. Windscreen panels
16. A-4 radar gunsight
17. Instrument panel shroud
18. Cockpit front pressure bulkhead
19. Rudder pedals
20. Gunsight power supply
21. Armament relay panel
22. Intake ducting
23. Cockpit canopy emergency operating controls
24. Nosewheel leg door
25. Torque scissors
26. Twin nosewheels
27. Nose undercarriage leg strut
28. Philco-Ford M39 20-mm cannon (four)
29. Kick-in boarding steps
30. Ejection seat footrests
31. Instrument panel
32. Engine throttle
33. Canopy external handle
34. Starboard side console panel
35. Pilot's ejection seat
36. Headrest
37. Cockpit canopy cover
38. Ejection seat guide rails
39. Cockpit rear pressure bulkhead
40. Port side console panel
41. Cockpit floor level
42. Control cable runs
43. Gun bay access panel
44. Ammunition feed chutes
45. Ammunition tanks, 200 rounds per gun
46. Power supply amplifier
47. Rear electrical and electronics bay
48. Cockpit pressurisation valve
49. Anti-collision light
50. Air conditioning plant
51. Radio compass aerial
52. Intake bleed air heat exchanger
53. Heat exchanger exhaust duct
54. Secondary air turbine
55. Air turbine exhaust duct, open

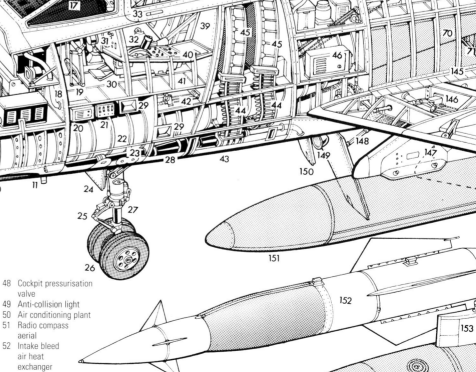

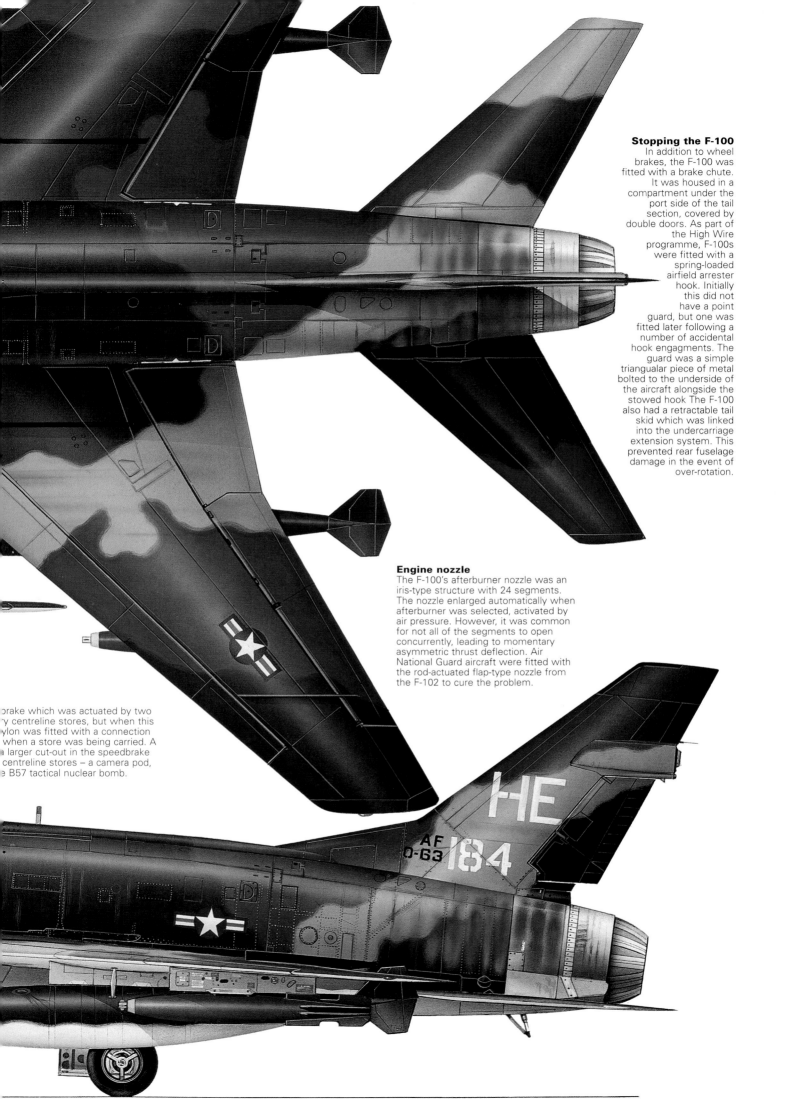

Stopping the F-100
In addition to wheel brakes, the F-100 was fitted with a brake chute. It was housed in a compartment under the port side of the tail section, covered by double doors. As part of the High Wire programme, F-100s were fitted with a spring-loaded airfield arrester hook. Initially this did not have a point guard, but one was fitted later following a number of accidental hook engagments. The guard was a simple triangualar piece of metal bolted to the underside of the aircraft alongside the stowed hook The F-100 also had a retractable tail skid which was linked into the undercarriage extension system. This prevented rear fuselage damage in the event of over-rotation.

Engine nozzle
The F-100's afterburner nozzle was an iris-type structure with 24 segments. The nozzle enlarged automatically when afterburner was selected, activated by air pressure. However, it was common for not all of the segments to open concurrently, leading to momentary asymmetric thrust deflection. Air National Guard aircraft were fitted with the rod-actuated flap-type nozzle from the F-102 to cure the problem.

brake which was actuated by two
y centreline stores, but when this
ylon was fitted with a connection
when a store was being carried. A
larger cut-out in the speedbrake
centreline stores – a camera pod,
B57 tactical nuclear bomb.

Misty pilots were an elite group, their number including two future USAF Chiefs of Staff (Ronald Fogleman and Merrill McPeak) and the round-the-world record breaker, Dick Rutan. The mission was hazardous, and many aircraft were hit by ground fire as they orbited the target area at low level. Seven Misty FAC pilots were killed in action and four more became POWs. Thirty-four aircraft were lost between August 1967 and May 1970, when they were replaced by Wolf FAC, Stormy FAC and Tiger FAC F-4s. The Misty FAC F-100Fs also flew reconnaissance and ResCAP missions, acting as on-scene controllers and co-ordinators during combat SAR missions. Missions often involved inflight refuelling and could last up to six hours, with four inflight refuelling contacts.

Wild Weasel

Arguably the most hazardous missions flown by the Super Sabre were those flown by the Wild Weasel F-100Fs. The first Vietnamese SA-2 SAM sites had been found and identified by USAF reconnaissance aircraft in April 1965, and by July, seven had been identified. An F-4C was shot down by an SA-2 on 24 July 1965, and only then were attacks against them authorised. Such attacks were hampered by sanctuary areas around Hanoi, Haiphong and Phuc Yen, and by the insistence that sites first be positively identified by low-level reconnaissance photos. If a missile was deployed outside the huge 'off limits' areas, it could be moved by the time any attack was mounted.

One response was to provide individual strike aircraft with ECM equipment that would blind the enemy radar, and podded ECM equipment was quickly developed and deployed. It soon became clear that the best way to defeat the surface-to-air missile threat would be to destroy the guidance radars on which they relied (or force them to shut down), not least because each such radar could control several missiles.

The best way of achieving this would be to send out pathfinder aircraft, equipped with both the radar-homing sensors necessary to detect and locate the enemy radars, and the armament to destroy them. This tentative conclusion had been reached in 1964, when a number of F-100Fs equipped with QRC-253-2 homing equipment were used against HAWK SAMs during Exercise Goldfire. A task force led by General K.C. Dempster, formed on 3 August, rapidly reached the same conclusion, and directed Bendix and Applied Technology Inc. (ATI) to come up with solutions.

Suitable self-protection had been available for larger aircraft for some years, but it was too large, too heavy and too expensive for tactical aircraft. ATI had already developed smaller, lighter equipment for the U-2, using the new-fangled transistor, and was given a contract (chalked on a blackboard and photographed, according to legend) to produce a suitable radar warning receiver, missile launch detector and tuned radar intercept receiver for installation on a test F-100F (58-1231).

North American quickly installed hand-built equipment in four F-100Fs (58-1221, -1226, -1227 and -1231). It included an AN/APR-25 radar homing and warning receiver, which was capable of detecting and locating the SA-2 fire control radar's standard S-band signal (as well as the C-band signal emitted by upgraded SA-2 systems and the X-band signals emitted by AI radars). Its output was

displayed on a threat panel and a 3-in (7.62-cm) diameter CRT display. An AN/APR-26 tuned crystal receiver could detect the power change in the L-band Fan Song guidance radar which would indicate an imminent launch, illuminating a red light in the F-100F's cockpit.

Finally, an IR-133 receiver was fitted, having greater sensitivity than the APR-25 and thus offering longer range

Phan Rang operations: an F-100D and an F-100F from the 612th TFS drop 'daisy-cutter' bombs (top), while a 352nd TFS F-100F lands back at the base (above). Phan Rang's F-100s were parented by the 35th TFW.

Commando Sabre – fast FAC over Vietnam

Under the project name Commando Sabre, and using the callsign MISTY, Det 1 of the 612th TFS flew F-100Fs on dangerous fast-FAC missions. The detachment started and ended its career at Phan Rang, but for the most part operated from Phu Cat. Although the unit had its own tailcode allocated ('HS'), it often flew 'HE'-coded aircraft from the co-located 416th TFS (above). The duration of the fast-FAC missions required much use of inflight refuelling (left). The aircraft carries the standard Misty load of two seven-round pods with white phosphorus 'Willie Pete' rockets for marking targets.

A pair of 309th TFS F-100Ds arrives at Tuy Hoa after a combat mission in April 1970, during the last few months of combat for the Super Sabre in Vietnam. After landing, the aircraft taxied to the disarming pits, where armament specialists made safe the aircraft before it continued to its dispersal. This aircraft, Thor's Hammer, carries the 31st TFW wing badge on the fin. It has been fitted with AN/APR-25 radar warning equipment, as denoted by the antenna under the intake lip.

Below: A Wild Weasel I F-100F takes on fuel from a KC-135A. The Wild Weasel aircraft were unmarked apart from national insignia. The operations of the unit were a closely-held secret at the time.

Below right: Two Wild Weasel aircraft are seen on the flightline at Korat in Thailand.

warning. It also provided a degree of threat classification through signal analysis. The modified aircraft were also fitted with a KA-60 panoramic strike camera and a dual-track tape recorder.

Combat deployment

The four modified Wild Weasel I F-100Fs were used to form the basis of a new unit, the 6234th TFW, staffed with five volunteer F-100 pilots, five back-seaters drawn from B-52 and B-66 EWOs, and 40 support staff. The unit deployed to Korat RTAFB on Thanksgiving Day, 1965, where it came under the control of the 388th TFW.

The first Wild Weasel F-100F combat mission was flown on 3 December 1965, with two F-100Fs accompanied by four F-105s. Using the codename Iron Hand, the F-100Fs identified and marked the radar site, which was then attacked by the accompanying F-105Ds. These early Wild Weasels were armed with a pair of 12-shot LAU-3 rocket pods, although they were soon replaced by bombs. Subsequent missions tended to use a single F-100F, and sometimes involved fighter top cover and an accompanying reconnaissance aircraft. The mission was extremely hazardous, and an F-100F was lost to AAA on 20 December. The first confirmed kill of an SA-2 radar was achieved on 22 December, using 304 high-explosive armour-piercing rockets and 2,900 rounds of 20-mm (0.787-in) ammunition. A number of SA-2 missiles were attacked at the same time, still under their camouflage netting. The attack was one of those which won Captain Alan Lamb, who later destroyed two more SAM radars, a well-deserved (but much-delayed) Silver Star; all six crews were awarded DFCs.

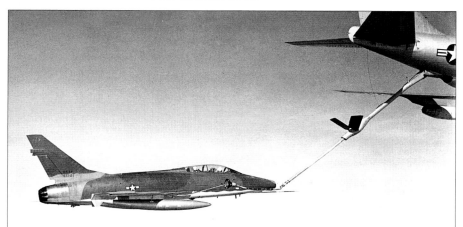

Three additional Wild Weasel I F-100Fs were deployed to southeast Asia on 27 February 1966, and the aircraft were soon leading every strike against targets in North Vietnam. From the spring of 1966, the F-100Fs carried AGM-45A Shrike anti-radiation missiles, which could passively detect enemy radar emissions and follow them all the way back to their source. The first combat use of the Shrike was made on 18 April 1966.

Many potential targets for the Wild Weasel F-100Fs were deep inside the areas which USAF aircraft were forbidden to attack without individual specific permission. The extent to which the war was then being micro-managed was soon clear to one F-100F pilot, Major Donald L. 'Buns' Frazier, who requested permission to attack one particularly troublesome SAM site near Hanoi. His request went up the chain of command and, to his astonishment, he received a personal phone call in response from the then-President, Lyndon Johnson, authorising the attack. This really was clearance from what the troops called the HMFIC ('Head Motherf***er in Charge'), and the mission was flown the next day.

A second Wild Weasel F-100F was lost to AAA on 23 March 1966. In return for the programme's two losses, F-100F Wild Weasel Is claimed nine confirmed SAM radar kills, and an unknown number of other enemy radars were forced off the air long enough for strike packages to get through unmolested. Having proven the Wild Weasel concept, the F-100F Wild Weasel I was replaced by Wild Weasel III conversions of the F-105F and later by the F-4-based Wild Weasel IV.

At war over Cyprus

Turkey's newly expanded F-100 fleet (strengthened between 1972 and 1974 by the addition of 111 F-100Ds and 20 F-100Fs) went to war again in July 1974. Following the overthrow of the Greek Cypriot leader Archbishop Makarios, Turkish residents of the island came under increasing pressure, and on 20 July Turkey acted to protect them. Six F-100 squadrons (111, 112, 131, 132, 171 and 172 Filos) supported the Turkish invasion of northern (Turkish) Cyprus, flying ground attack missions against Greek Cypriot National Guard positions on the first day of the operation.

F-100s from 111, 112 and 181 Filos (and a handful of F-104Gs) scored an unfortunate 'own goal' when they attacked and sank a Turkish Navy 'Gearing'-class destroyer, mistaking it for a similar Greek vessel. Eight F-100s were lost during the Cyprus operation, one in a take-off accident, two to engine failures, and at least two to hostile ground fire. The fate of the remaining three aircraft lost was not officially confirmed, but all eight pilots survived.

Prospective F-100 versions

NAA had hoped to export much larger numbers of Super Sabres, and there were a number of design studies for new versions of the aircraft. Although the F-100E, F-100G, F-100H, F-100M, F-100P, F-100Q and F-100R designations were never assigned, other versions were allocated new designation suffixes. The F-100J, for example, was a projected all-weather interceptor version offered to Japan

Wild Weasel – hunting radars

Although only seven aircraft were involved, Project Wild Weasel was one of the most important tasks undertaken by the F-100, as it pioneered the use of fighter aircraft in the anti-radar, defence suppression role – now known as SEAD and an integral part of combat operations. The seven F-100Fs were fitted with equipment to detect and locate hostile radars, which could then be attacked with rockets and bombs carried by the F-100 itself, or by accompanying attack aircraft (notably the F-105). Later, the F-100Fs acquired the capability to fire the AGM-45 Shrike missile, which homed on radar emissions. The Wild Weasel force operated from Korat, Thailand, as a separate and secretive detachment from the Tactical Air Warfare Center at Eglin AFB, Florida, although it reported to the 6234th TFW. Combat operations (officially an 'evaluation') lasted from 20 December 1965 to July 1966, at which time the F-105F Wild Weasel III took over the Iron Hand anti-radar mission.

'Huns' in the Guard

Right: 21 Air National Guard squadrons flew the F-100D model, beginning in the spring of 1970 with the equipment of the 162nd TFS/Ohio and 175th TFS/South Dakota. This aircraft wears the mountain lion badge of Colorado's 'Mile High Militia'.

Below: Proof that there was still life in the old dog, this Ohio ANG aircraft wears three 'kill' markings (T-38s or F-5s) from an air defence exercise. The year was 1976, as denoted by the Bicentennial badge.

Right: Large numbers of ex-active duty F-100Ds were 'cascaded' to ANG squadrons in 1970/71. This aircraft of the 118th TFS/CT ANG was previously the leader's aircraft with the 'Thunderbirds' team, traces of the 'ONE' being visible on the aft fuselage.

Below right: For most of the F-100's fighter-bomber career with the ANG, training was provided by the Arizona ANG.

Below: Each unit had F-100Fs assigned for continuation training. This Georgia ANG aircraft has recently been transferred from new Mexico's 188th TFS, and still wears that unit's smart paint scheme.

through the Foreign Military Sales programme, while the F-100K and F-100L designations were applied to J57-P-55-powered versions of the F-100F and F-100D, respectively. The F-100N was intended as a simplified export version of the F-100D, having no nuclear capability and less advanced electronics. The aircraft was aimed at NATO's smaller, poorer nations.

The F-100S designation was applied to an ambitious 1964 proposal to re-engine F-100Ds and F-100Fs with a Rolls-Royce RB.168-25R Spey turbofan, as used in the British Phantom versions. It was hoped that this would produce a low-cost fighter-bomber that would have 50 per cent more payload than the basic Super Sabre, 30 per cent longer range, a shorter take-off distance and an improved climb rate. North American had hoped to convert existing USAF aircraft during their regular Inspect and Repair As Necessary (IRAN) maintenance, and had planned to establish a production line in France for 200 new-build examples of the F-100S for the Armée de l'Air. As an alternative, NAA also proposed using an afterburning Allison TF41 engine, but this was no more successful in winning customers, and the advanced F-100S, which could have given the Super Sabre a new lease of life, remained on the drawing board.

The final rundown

Without major modernisation, the F-100D and F-100F were looking increasingly anachronistic by the 1970s, and the type entered a rapid decline. USAFE retired its last F-100s in April 1972, when the 48th TFW at Lakenheath, England, completed conversion to the F-4D. The last front-line F-100s in USAF service were those of the 524th TFS at Cannon AFB, New Mexico, the last of the 27th TFW's squadrons to convert to the F-111D.

The Air National Guard's four F-100 squadrons in south-east Asia had returned home from Vietnam in 1969. All 26 ANG F-100 squadrons quickly converted to other, newer types between 1969 and 1979. The F-100 finally flew its last operational mission with the ANG on 10 November 1979, when First Lieutenant Bill Layne of the 113th TFS, 181st Tactical Fighter Group flew a sortie in 56-2979 *City of Terre Haute* at Hulman Field, Indiana. This aircraft was subsequently flown to the MASDC 'boneyard' by Brigadier General Frank Hettlinger, CO of the 122nd TFW.

Long after their front-line days were over, many F-100Cs performed useful work as target tugs. This aircraft wears the badge of the 4758th Defense Systems Evaluation Squadron on the fin. This squadron operated out of Holloman AFB and provided target facilities over the White Sands range.

Right: In Europe the last USAF F-100 user was the 48th TFW at RAF Lakenheath, whose last F-100Fs were flown out in mid-April 1972 – three months after F-4D Phantoms arrived. In the latter part of their career the 48th 'Huns' wore camouflage and tailcodes ('LS' = 493rd TFS).

Above: The 27th TFW at Cannon AFB was the last active-duty wing to operate Super Sabres, withdrawing its last aircraft (from the 524th TFS, illustrated) in late 1972. By that time the wing had been flying F-111s for three years.

control, but plans to retain a handful of aircraft for target towing came to nothing. Four F-100Ds flew to Alconbury and Woodbridge for crash rescue training and fire practice, and to Lakenheath and Wethersfield (former USAF F-100 bases) for display, and 36 more flew to Sculthorpe for storage and disposal. Nine were saved for preservation at UK museums, but the rest were scrapped by the 7519th Combat Support Squadron.

At one time, Denmark had hoped to replace all of its F-100s with Drakens, but funding was not available and the Danish F-100s were retained until replaced by F-16s in the early 1980s. The aircraft by then were showing their age, and a succession of technical problems led to them being grounded several times. Some aircraft were fitted with F-102-type Project Pacer Transplant afterburners, and some even received wingtip-mounted ALR-45D/APR-37D RHAWS, with distinctive wingtip pods. The RDAF made its last F-100 flight on 11 August 1982. Three TF-100Fs were retained for display, and six TF-100Fs went to Flight Systems Inc. (FSI). Some 20 surviving F-100Ds went to Turkey, together with two F-100Fs. One more F-100D became unserviceable in Italy en route to Turkey, eventually becoming Aviano's gate guard.

The last F-100As retired from service in late 1984 when Taiwan finally grounded its last Super Sabres, which by then equipped only two squadrons. The last of these ageing aircraft made its final flight on 5 September 1984. Taiwanese plans to pack the surplus aircraft with explosive and launch them at China as crude cruise missiles were quietly discouraged by the USA.

More 'Huns' for Turkey

The withdrawal of the last ANG and Danish F-100s allowed Turkey's F-100 force to be reinforced and, while everyone else was busy retiring the F-100, Turkey actually strengthened its Super Sabre fleet. Fifteen ex-ANG F-100Ds arrived in 1977-78, and 20 ex-RDAF F-100Ds and two F-100Fs were delivered in 1981-82. Thus, in 1980 182 Filo actually converted to the F-100 – the last new Super Sabre squadron to form anywhere in the world.

By 1983, 65 F-100Ds and 30 F-100Fs equipped two front-line units (181 and 182 Filos at Diyarbakir) and a tactical weapons training unit at Konya, which consisted of 131 and 132 Filos, giving basic 'academic' weapons training to pilots destined for front-line units flying the F-100 and the F-104. With a glut of F-104s available as NATO nations re-equipped with the F-16, Turkey's remaining front-line units re-equipped during March-April 1985, but 35 F-100Ds and 20 F-100Fs remained in use at Konya until 1 November 1987, when 132 Filo flew its last Super Sabre sortie. This marked the end of the F-100 in front-line military service,

Surviving French Super Sabres were withdrawn from service in 1976-78 as EC 11 re-equipped with the Jaguar. The last in service were the seven F-100Ds and single F-100F of EC 4/11 'Jura' at Djibouti. These aircraft wore huge gaudy sharkmouths, and four were specially equipped for the reconnaissance role.

Having been supplied under MAP provisions, the surviving Armée de l'Air Super Sabres were returned to American

In Europe the F-100 lived on into the 1980s. France returned its aircraft to US ownership in 1978, and most ended up at Sculthorpe in England (above) where they were scrapped. Denmark's fleet (right) served until 1982, the single-seaters being passed on to Turkey, which flew the type until 1987. Most of the two-seaters went on to another career as target tugs.

QF-100 – drone conversion

With large numbers of aircraft available, the Super Sabre was a natural choice to satisfy the growing demands of the USAF's FSAT (Full-Scale Aerial Target) programme, resulting in a total of 310 aircraft being converted to QF-100 status for consumption in missile trials and live-firing exercises. As well as radio control equipment, the QF-100s were fitted with a scoring system, countermeasures as required, and – for IR-guided missile shots – wingtip burners which attracted the missile away from the aircraft's engine in an attempt to enhance its survivability and so prolong its useful life. The aircraft above has a burner fitted, and other modifications such as a 'sugar-scoop' IR shield over the exhaust and blacked-out canopy. The red star was an obvious addition for an aircraft destined to be shot down by a USAF fighter.

but the aircraft continued to fly in important military support roles.

Target drones

Between 1973 and 1981, Sperry Flight Systems had converted 215 redundant F-102s to QF-102A and PQM-102A/B unmanned target drone configuration under the Pave Deuce programme. These were used for missile trials, but also as realistic targets for front-line pilots during William Tell weapons competitions and in other exercises. It was always planned that the Delta Dagger drones would be followed by similar drone conversions of the F-106 Delta Dart from 1986, but the rapid expenditure of QF-102s led to a requirement for another interim target drone.

In March 1982, therefore, Sperry received a contract for an initial trial batch of nine F-100 drone conversions. Two 'manned' YQF-100D prototypes (56-3414 and 56-3610) were produced at Sperry's Litchfield Airport facility, followed by six QF-100Ds and a two-seat QF-100F. The trials batch was followed by 89 more QF-100Ds and two more QF-100Fs. Unmanned operations by the 82nd Tactical Aerial Targets Squadron (part of the 475th Weapons Evaluation Group) began from Tyndall AFB from 19 November 1981.

The Sperry QF-100s were followed by 169 QF-100Ds and 41 QF-100Fs converted by FSI at Mojave. The conversion of 14 additional aircraft was cancelled in December 1990, by when the QF-4 drone conversion programme was well underway. Six F-100Fs flew back to the 'boneyard', and five more were returned there by road. FSI bought in spares from other former F-100 operators, and even purchased three ex-Turkish aircraft for possible modification. They were ferried to Mojave in 1989, but remained unconverted.

The take-off of the QF-100 was handled by two controllers sitting in a telemetry van positioned at the end of the runway. One controlled the aircraft in pitch and managed the throttle, while the other controlled the ailerons and rudder. Once airborne, the drone was handed off to a third controller sitting in a fixed-base ground station. A dual redundant system was used to get the drone to the mission area and to select from a variety of pre-programmed manoeuvres. If the drone survived the mission, it was flown back to the handover point, where the two take-off controllers brought the aircraft back in.

QF-100s were usually engaged using missiles with inert warheads, and had a digital Doppler system to measure miss distances. This allowed the aircraft to have a useful average life of six or seven drone target missions, since an attacking fighter could score a theoretical kill without destroying the drone. One drone survived 15 missile shots. The QF-100s could carry ECM, chaff and flares in order to enhance their value as a 'realistic target'. If a drone was too badly damaged to be recovered safely, it could be destroyed by remote control, and the same self-destruct system was used if contact with the drone was lost for more than six minutes. The aircraft could be flown by a pilot, conversion back to manned configuration being achieved by the flick of a switch on later conversions.

The 475th Weapons Evaluation Group expended its last QF-100 in 1992, though a handful remained in use with the 6585th Test Group at Holloman, supporting US Army programmes (including the HAWK SAM) after that date. The last two US military F-100Fs and a single F-100D were finally withdrawn in August 1994.

Target towing

During the 1950s and 1960s, many units used their own front-line aircraft to provide such target-towing support

Above left: Mojave's ramp groans under the weight of freshly modified QF-100s at the height of the FSI drone conversion programme in the late 1980s. Of the 198 aircraft whose fate has been published, 84 were shot down by AIM-9 Sidewinders, 63 by AIM-7 Sparrows, 22 by AIM-120 AMRAAMs and three by AIM-4 Falcons. Twenty crashed – mostly on take-off and landing – and six were destroyed by their operators.

With blacked-out canopy, a QF-100 lands in 'no live operator' mode. Many of the QF-100s were fitted with a Drone Formation Control System (DFCS), which enabled several to be flown together to stage multi-bogey engagements.

This view from September 1963 shows the 7272nd Fighter Training Wing's ramp at Wheelus, populated by target-towing F-100Cs. In the foreground are 335-US gal (1268-litre) supersonic tanks for the Super Sabres. Following a military coup led by Colonel Muammar Ghadaffi, US forces left Libya. Wheelus was subsequently renamed Okba ben Nafi air base, and played host to Soviet Tu-22s and MiG-25s.

Flight Systems Inc. (subsequently Tracor Flight Systems, then BAE Systems) operated five TF-100F target tugs in Europe, initially on a USAFE contract from Hurn in England and then on a Luftwaffe contract. For the latter the aircraft were fitted with Dornier Sk 10 darts, as seen on this aircraft at Wittmund, the main operating base in Germany. After retirement in 2001 their place was taken by ex-Israeli A-4 Skyhawks.

duties as was necessary, usually towing basic banner targets from a simple lug or hook, which was often fitted at base or squadron level. Many units adapted one or two F-100s in just this way. Dedicated F-100C target tugs were used by the 7272nd Fighter Training Wing at Wheelus AFB, Libya, from January 1958 until 1965, to provide target facilities for USAFE units visiting Wheelus for gunnery training. Similarly-equipped F-100Cs and F-100Fs were used by the 4758th Defense Systems Evaluation Squadron at Biggs AFB, Texas, from July 1962, and then from April 1966 until October 1970 at Holloman AFB, New Mexico.

Flight Systems Inc. had been a long-term provider of contract target-towing services to the US armed forces using F-86 Sabres, and when Denmark retired its last F-100s the company took the opportunity to acquire six of them for conversion as target tugs. These promised to be able to tow heavier targets than the F-86s, and more quickly, and spares support for the aircraft (especially in the light of the company's participation in the QF-100 programme) would be easier.

The company therefore purchased six TF-100Fs (N414FS (ex-56-3826, RDAF GT-826), N415FS (ex-56-3844, RDAF GT-844), N416FS (ex-56-3916, RDAF GT-916), N417FS (ex-56-3842, RDAF GT-842), N418FS (ex-56-3996, RDAF GT-996), and N419FS (ex-56-3971, RDAF GT-971)) from the Royal Danish Air Force, and converted them to target-towing configuration. A target-towing winch and its associated controls were installed, and the J57 engine was modernised and modified, gaining the afterburner of the (F-106-type) J75 engine.

One aircraft (N415FS (ex-56-3844, RDAF GT-844) was retained in the USA to service a contract at Holloman AFB, but the remainder were based at Bournemouth's Hurn airport to provide target-towing support for USAFE units. Funding constraints meant that this contract was not renewed in 1988, but FSI (by then Tracor Flight Systems Inc.) was able to gain a replacement contract to provide similar services to the Luftwaffe. The Luftwaffe had previ-

Target tugs at Wheelus

Until the USAF was ejected from Libya in 1965, the 7272nd Fighter Training Wing was based at Wheelus AFB. The location offered large unpopulated areas, excellent year-round flying weather, and the extensive El Uota live bombing range. It was used by USAFE units for armament practice. To provide targets for aerial gunnery the wing used F-100Cs modified to tow banners. They were first allocated to the task in 1958.

ously used its own F-4Fs and F-104Gs to tow targets, but FSI was able to offer a more cost-effective service. The TF-100Fs moved to Wittmund, home of JG 71 'Richthofen', from where they flew sorties over the North Sea ranges. They also supported Luftwaffe deployments to Decimomannu.

The TF-100Fs carried a Dornier DATS-3 target pod below the port inboard station, in place of the original USAF Model 15 A/A37U-15 target system. This consisted of an RMU-10 target winch and a Dornier Sk 10 target dart. The TF-100Fs also usually carried a 200-US gal (757-litre) tank to port, with a huge 335-US gal (1268-litre) tank to starboard. Unlike the original TDU-10 target dart, the new dart was fitted with miss distance recorders and a transponder which allowed it to change its radar signature.

N414FS was lost in a fatal accident at Cuxhaven on 11 July 1994 and the Holloman-based aircraft (N415FS) was sold to a private owner after its contract came to an end, but the four remaining F-100s flew on. They were finally retired from the target-towing role in June 2001, and BAE Flight Systems' final four aircraft flew back to Mojave from Wittmund. At least two of the aircraft remained there, maintained and ready for reactivation if required until 2003.

Warbirds

The only F-100s still flying are a trio of civilian-owned jet warbirds in the USA. Two were among the three ex-Turkish Air Force jets that had been sold to FSI for possible drone conversion and ferried to Mojave in August 1989. They were not needed, however, and were sold on to private owners. The two-seater (F-100F N2011V 56-3498) was delivered to Thomas J. Hickman of Addison, Texas, in August 1992, before being sold to Dean F. 'Cutter' Cutshall (or to his company, American Horizons) of Fort Wayne, Indiana, in July 1996. The aircraft has been beautifully restored and now flies in the striking red and white colours of the 354th TFW.

For several years the only other flying F-100 was another two-seater. TF-100F N26AZ (56-3844) was a former FSI aircraft, N415FS, which was owned and operated by David Tokoph's El Paso, Texas-based Grecoair in New Mexico ANG colours until sold (with an asking price of $US700,000). The aircraft is now decorated in 'Thunderbirds' livery.

Flight Systems Inc. – civilian target-towing

From its Mojave base Flight Systems Inc. operated a number of Super Sabres on various government target facilities contracts through the 1980s. Left is Hilda, an ex-Massachusetts ANG F-100F, while at right is an F-100D (56-3022) which had flown with Connecticut ANG before being acquired by FSI in late 1979.

NACA/NASA 'Huns'

Right: 53-1709 (NASA 703) was designated JF-100C and was flown by NACA/NASA from Moffett (Ames) and Edwards (Dryden). It was fitted with a variable-stability system which allowed it to perform inflight simulations of different control systems, effectively allowing it to mimic other aircraft types.

Below: F-100A 52-5778 was operated by the High-Speed Flight Station (Edwards) from 1954 to 1960, investigating various phenomena which affected supersonic fighters, including pitch-up. Between 1957 and 1961, NACA (which became NASA in July 1958) also flew F-100C 53-1717 on basic research duties.

Alias NASA-200, this F-100A was used to test take-off performance improvements. By a combination of blown boundary layer control across the leading- and trailing-edge flaps, and the deflection of the leading-edge flaps to 60°, a 10 percent improvement was achieved. Note the enlarged intake fitted to the aircraft.

F-100D-50-NH (55-2888) was the second ex-Turkish F-100 flown to Mojave, where it was registered to Global Aerospace of Diamond Bar, California, as N2011U in 1993. The aircraft (by then partly cannibalised for spares) was sold to jet warbird restorer Greg Forbes of Sacramento in 1998, who spent three years restoring the aircraft to flying condition. The F-100D was ferried to the former McClellan AFB near Sacramento for a full restoration on 1 November 2001, which took another 18 months. Test pilot Lee Holcomb of El Dorado Hills, California, flew the aircraft on its first post-restoration flight on an unrecorded day in April 2003, but because the weather was so poor on that day, the event officially took place on 10 May 2003. Forbes reportedly plans to paint his F-100D in the colours of the 'Thunderbirds' team, and it will then join his stable of warbirds, which includes a T-33, six Drakens, and an F-86.

The third of the ex-Turkish FSI jets was an F-100C (54-2091 N2011M), which passed into the hands of Al Hansen and Mojo Jets before being sold (reportedly on eBay) in June 2002. The status of this sale is doubtful, however, and most sources suggest that the aircraft remains in Mr Hansen's hands. The oldest and most 'tired' of the trio, the F-100C was the last of the aircraft to find a new owner. Its restoration to full flying condition would seem unlikely, but cannot be ruled out.

Perhaps the most likely F-100 to return to the air will be one of the BAE Flight Systems TF-100Fs, several of which are stored at Mojave and still have FAA airworthiness certificates. Lynn High, owner of the Fightertown Aviation FBO 60 miles (96 km) north of Dallas, partnered with Mike Menez, a former ANG F-100 pilot and now a commercial pilot with Cathay Pacific, to form Heritage Jets; the sole aim is acquiring an F-100 Super Sabre and getting out on the air show circuit – originally hoping to do so in time for the 50th anniversary of the aircraft in 2003. Menez confirmed in 2002 that Heritage Jets had submitted a bid to BAE and had hired a professional marketing firm to market sponsorship opportunities, but things then went quiet. The four surviving TF-100Fs were recently sold, with the FAA recording a change of ownership on 12 June 2003 to Big Sky Warbirds LLC of Belgrade, Montana

Nevertheless, with three F-100s able to make their own unique and noisy contribution to the US air show circuit, it is unlikely that the Super Sabre will be forgotten just yet.

And the F-100 is an aircraft which should not be forgotten. Unfairly dubbed the 'Not so Super Sabre' by some writers and journalists, the F-100 may not have been much of an air-to-air fighter, and its all-weather fighter variant may have been stillborn, but the aircraft did give useful service as a fighter-bomber. Although the aircraft had never been intended for the air-to-ground role – a role other aircraft performed with greater facility – the F-100 was in the right place at the right time, and its dedicated and highly skilled pilots worked wonders in it, earning a formidable reputation in combat over southeast Asia.

Jon Lake

In 1989 FSI purchased three Turkish air force aircraft for potential QF-100 drone conversions. The trio was sold on to private owners, and two of them are flying in 2003 on the airshow circuit. This is the third aircraft, F-100C N2011M, which may yet be restored to airworthy condition. The other flyer was another ex-FSI aircraft, a Danish TF-100F which had been used by FSI for a contract at Holloman AFB.

F-100 Operators

F-100D, 90th TFS/3rd TFW, Bien Hoa AB, South Vietnam, May 1969

UNITED STATES AIR FORCE

In July 1958 the USAF dispensed with the 'Fighter Day' and 'Fighter Bomber' unit designators in favour of 'Tactical Fighter'. This affected many F-100 units.

3rd TFW
Operated F-100Ds and F-100Fs from England AFB, LA, from June 1964. Deployed to Bien Hoa AB, South Vietnam, on 8 November 1965. The 3rd TFW was unmanned and unequipped, existing only on paper after its withdrawal from southeast Asia on 31 October 1970. The wing moved to Kunsan AB, South Korea, on 15 March 1971, where it re-equipped with F-4 Phantoms. While in Vietnam the wing included squadrons operating other aircraft types such as the B-57.
90th Tactical Fighter Squadron 'Pair o' Dice': assigned 9 June 1964 – 19 November 1965 and 3 February 1966 – 31 October 1970, tailcode 'CB', light blue colours. With 401st TFW in interim
307th Tactical Fighter Squadron: assigned 21 November – 6 December 1965, on TDY
308th Tactical Fighter Squadron: assigned 2 December 1965 – 25 December 1966, on TDY
416th Tact. Fighter Squadron 'Silver Knights': assigned 16 June 1964 – 8 November 1965 and 16 November 1965 – 15 April 1967, blue colours
429th Tactical Fighter Squadron: assigned 21 November – 14 December 1965, on TDY
510th Tactical Fighter Squadron: assigned 16 March 1964 – 15 November 1969, tailcode 'CE', purple colours
531st Tactical Fighter Squadron: assigned to the 3rd TFW 16 June 1964 – 19 November 1965 and 7 December 1965 – 31 July 1970, tailcode 'CP', red colours

4th FDW/TFW
Formed by re-numbering the 83rd FDW. Operated F-100Cs from Seymour Johnson AFB, NC, from early 1958 until conversion to the F-105B began on 16 June 1959. Plain broad bands encircled the nose and the tailfin.
333rd Fighter Day Squadron 'Lancers': assigned December 1957 – 1960, red colours
334th Fighter Day Squadron 'Eagles': assigned December 1957 – 1960, blue colours
335th Fighter Day Squadron 'Chiefs': assigned December 1957 – 1960, green colours
336th Fighter Day Squadron 'Rocketeers': assigned December 1957 – 1960, yellow colours

8th TFW
Operated F-100Ds and Fs from Itazuke AFB, Japan, from late 1956 until conversion to the F-105F began in May 1963. Unit markings consisted of three chevrons on the fin, in squadron colours, with a squadron badge in centre of middle chevron.
35th Tactical Fighter Squadron: assigned 1956 – 1963, blue colours
36th Tactical Fighter Squadron 'Flying Fiends': assigned 1956 – 1963, red colours
80th Tactical Fighter Squadron 'Headhunters': assigned 1956 – 1963, yellow colours

18th TFW
Operated F-100Ds and Fs from Kadena AB, Okinawa, Japan, from 1957 until conversion to the F-105D/F began in October 1962. Unit markings consisted of diagonal stripes around the nose in squadron colours, and a three-pronged red white and blue arrowhead on the fin.
12th Tactical Fighter Squadron 'Bald Eagles': assigned 1957 – 1963, yellow/black colours.
44th Tactical Fighter Squadron 'Vampires': assigned 1957 – 1963, blue/white colours
67th Tactical Fighter Squadron 'Fighting Cocks': assigned to the 18th TFW 1957 – 1963, red/white colours

20th TFW
The 20th TFW converted to the F-100D and F-100F from the F-84F from 16 June 1957, initially based at RAF Wethersfield and RAF Woodbridge (79th FS). The 79th FS moved to Upper Heyford on 15 January 1970, and the Wethersfield units moved on 1 June 1970. The wing converted to the F-111E from September 1970. Unit markings consisted of a lightning flash down the fin and repeated on the fuselage, in squadron colours, with squadron and wing badges superimposed to starboard and port. A three-coloured lightning flash was applied to the tail alone after 1961.
55th Tactical Fighter Squadron 'Fighting Fifty Fifth': assigned June 1957 – 1970, blue colours
77th Tactical Fighter Squadron 'Gamblers': assigned 1957 – 1971, red colours
79th Tactical Fighter Squadron 'Tigers': assigned 1957 – 1971, yellow colours

21st TFW
The 21st TFW converted to the F-100D and F at Misawa from July 1958, but deactivated in June 1960, its squadrons being reassigned to the 3rd TFW and 39th AD. Unit markings consisted of two overlapping chevrons on the fin, with bands encircling the intake. These were in squadron colours. The intake bands were edged with blue bands and white stars.
416th Tactical Fighter Squadron 'Silver Knights': assigned July 1958 – June 1960 colours blue
531st Tactical Fighter Squadron: assigned July 1958 – June 1960, colours red

27th TFW
The 27th TFW became an operator of the F-100D and F at Cannon AFB, NM, in February 1959 by the renumbering of the 312th TFW. The wing maintained a Detachment at Takhli from 13 December 1962 – 1 June 1963, before moving to Da Nang until 6 May 1964. The 27th flew the F-100 until 1972. Unit markings consisted of a huge area of colour on the upper half of the tailfin. This was bisected diagonally, with colour above the diagonal which ran from the fin fillet on the leading edge to the trailing edge above the RWR. This was replaced by a smaller 'Arrowhead' chevron high on the fin.
481st Tactical Fighter Squadron: assigned February 1959 – 1972, tailcode 'CA', green colours
522nd Tactical Fighter Squadron 'Fireballs': assigned February 1959 – 1972, tailcode 'CC', red colours
523rd Tactical Fighter Squadron: assigned February 1959 – November 1965, blue colours, to the 405th TFW, November 1965
524th Tactical Fighter Squadron 'Hounds of Heaven': assigned February 1959 – 1972, tailcode 'CD', yellow colours.

31st TFW
The 31st TFW converted to the F-100D and F in mid-1957 at Turner AFB, with the 31st, 306th, 307th, 308th and 309th TFSs. The wing's assets were transferred to the 354th TFW at Myrtle Beach on 15 March 1959, but the wing and squadron designations were then taken over by the units of the 413th TFW at George AFB. The wing moved to Homestead AFB in mid-1960, then deployed to Tuy Hoa between December 1966 and 15 October 1970. Thereafter, the wing returned to the USA as a paper unit only. In Vietnam, the wing included a number of ANG F-100Cs. Unit markings originally consisted of a broad coloured band going up the tailfin, with a squadron coloured nose. At George AFB the 31st TFW's aircraft wore two broad parallel bands across the tailfin and one around the nose.
136th Tactical Fighter Squadron, NY ANG 'Rocky's Raiders': assigned June 1968 – May 1969, tailcode 'SG'
188th Tactical Fighter Squadron NM ANG 'Enchilada Air Force': assigned June 1968 – May 1969, tailcode 'SK'
306th Tactical Fighter Squadron: assigned 1957 – 1970, tailcode 'SD', red colours
307th Tactical Fighter Squadron: assigned 1957 – April 1966, reassigned to Torrejon. Blue colours
308th Tactical Fighter Squadron 'Emerald Knights': assigned 1957 – 1970, tailcode 'SM', green colours
309th Tactical Fighter Squadron: assigned 1957 – 1970, tailcode 'SS', yellow colours
355th Tactical Fighter Squadron: assigned May 1969 – September 1970, tailcode 'SP'
416th Tactical Fighter Squadron, 'Silver Knights': assigned May 1969 – September 1970, to 4403rd TFW. Tailcode 'SE', blue colours

35th TFW
The 35th TFW formed as an F-100 unit by the exchange of designations with the 366th TFW, which became an F-4 wing at Da Nang. The wing was based at Phan Rang AB, from October 1966 until July 1971.
120th Tactical Fighter Squadron, Colorado ANG: assigned April 1968 – April 1969, tailcode 'VS' when 612th at Phu Cat
352nd Tactical Fighter Squadron: assigned October 1966 – July 1971, tailcode 'VM', yellow colours
612th Tactical Fighter Squadron, Det 1 'Misty FAC': assigned October 1966 – January 1967 and April 1969 – 1971, tailcode 'VS', blue colours
614th Tactical Fighter Squadron 'Lucky Devils': assigned 1966 – 1971, tailcode 'VP', red colours
615th Tactical Fighter Squadron: assigned 1966 – 1971, tailcode 'VZ', green colours

36th FDW/TFW
The 36th FDW converted to the F-100C in 1956, mainly at Bitburg but with single squadrons at Ramstein/Landstuhl, Hahn and Soesterberg. The wing was re-designated as a TFW after 1 July 1958, and its surviving squadrons converted to F-105s from May 1961. Unit markings consisted of three broad diagonal stripes on the tailfin, with a chevron tapering back from the intake.
22nd Tactical Fighter Squadron: assigned May 1956 – 1961. Red colours
23rd Tactical Fighter Squadron: assigned April 1956 – 1961. Blue colours
32nd Tactical Fighter Squadron: assigned July 1956 – August 1960 at Soesterberg. To F-102. Green colours
53rd Tactical Fighter Squadron: assigned June 1956 – 1961, at Ramstein. Yellow colours
461st Tactical Fighter Squadron: assigned 1956 – August 1959, at Hahn. Deactivated. Black colours
'Skyblazers': assigned September 1956 – 1961

37th TFW
The 37th TFW activated as an F-100 unit at Phu Cat AB, Vietnam, on 1 March 1967, and remained there until May 1969.
174th Tactical Fighter Squadron: assigned May 1968 – May 1969, tailcode 'HA'
355th Tactical Fighter Squadron: assigned February 1968 – May 1969, tailcode 'HP'
416th Tactical Fighter Squadron 'Silver Knights': assigned April 1967 – May 1969, tailcode 'HE', blue colours
612th Tactical Fighter Squadron Det 1 'Misty FAC': assigned June 1967 – April 1969, tailcode 'HS'

39th Air Division
In 1958 the 39th AD briefly had the 418th TFS allocated with F-100Cs at Clark AB, Philippines. Later, the 356th TFS was directly assigned to the 39th Air Division at Misawa, but was not part of the 21st TFW. It was responsible for maintaining four F-100s on nuclear alert at Kunsan, Korea (supported by the 6175th Air Base Group) and would have deployed forward to Korea in the event of hostilities. When the 21st TFW deactivated in 1960, the 531st TFS was reassigned to the 39th AD. The 27th TFW provided F-100s from June 1964 – June 1965, the 401st TFW from June 1965 – August 1965, and the 3rd TFW from August 1965 – June 1966, before the 401st TFW took over. The 39th AD at Misawa converted to F-4s in June 1967, though these did not immediately

F-100D, 20th TFW at Nouasseur AB, Morocco, June 1958. 79th TFS – yellow tail, 77th TFS – red tail

F-100D, 492nd TFS/48th TFW, RAF Lakenheath, England

take over the nuclear alert duty at Kunsan.
418th Tactical Fighter Squadron: assigned 1958 at Clark AB
356th Tactical Fighter Squadron 'Green Demons': assigned April 1964 – June 1966. Then to Det 1 475th TFW
531st Tactical Fighter Squadron: assigned 1960 – 1964 at Misawa AB

48th FBW/TFW
Transitioned from the F-86F at Chaumont AB, France, in late 1956. Transferred to RAF Lakenheath as the 48th TFW, by 15 January 1960. Converted to F-4D Phantoms from February 1972. Unit markings consisted of alternating stripes across the tailfin in squadron colours, with a shadowed 'V' shaped chevron on the nose added from 1959. A more subdued chevron on the tail replaced stripes when the wing moved to England.
492nd Tactical Fighter Squadron 'Bolars': assigned September 1956 – April 1972, tailcode 'LR', blue colours
493rd Tactical Fighter Squadron 'Roosters': assigned 1956 – 1972, tailcode 'LS', yellow colours
494th Tactical Fighter Squadron 'Panthers': assigned 1956 – 1972, tailcode 'LT', red colours

49th FBW/TFW
Formed at Etain-Rouvres AB, France on 10 December 1957 by renumbering the 388th FBW. Transferred to Spangdahlem as the 49th TFW during 1960. Converted to F-105 from October 1961. Unit markings included a broad band on the fin and a chevron on intake in squadron colours with white four-angled lightning bolt on fin.
7th Tactical Fighter Squadron: assigned December 1957 – 1962, blue colours
8th Tactical Fighter Squadron 'Black Sheep': assigned December 1957 – 1962, yellow colours
9th Tactical Fighter Squadron 'Iron Knights': assigned December 1957 – 1962, red colours

50th FBW/TFW
Transitioned from the F-86F at Toul-Rosières AB, France, in 1957. Transferred to Hahn as the 50th TFW, by 10 December 1959. Converted to F-4C and F-4D Phantoms from October 1966. Unit markings consisted of starred bands across the fin.
10th Tactical Fighter Squadron: assigned December 1957 – 1967, blue bands with white stars
81st Tactical Fighter Squadron: assigned July 1958 – 1966, yellow bands with black stars
417th Tactical Fighter Squadron: assigned 1958 – 1966, red bands with white stars

57th Fighter Weapons Wing
Formed by re-numbering the 4525th FWW at Nellis AFB in October 1969. Its F-100 element was inactivated 31 December 1969.
65th Fighter Weapons Squadron: the former 4536th FWS was assigned to the 57th FWW October 1969 – 31 December 1969, tailcode 'WB', yellow and black colours

58th Tactical Fighter Training Wing
Formed by re-numbering the 4510th CCTW at Luke AFB in October 1969. Its F-100 elements were inactivated during 1971. 'LA' tailcodes were worn by all squadrons.
310th Tactical Fighter Training Squadron: assigned Oct 1969 – 1971
311th Tactical Fighter Training Squadron: assigned 18 January 1970 – 21 August 1971
426th Tactical Fighter Training Squadron: assigned 18 January 1970 – 13 September 1971
4511th Combat Crew Training Squadron: assigned October 1969 – 18 January 1970, replaced by 311th TFTS
4514th Combat Crew Training Squadron: assigned October 1969 – 15 December 1969 became A-7 training unit
4515th Combat crew Training Squadron: assigned October 1969 – 18 June 1970 became 426th TFTS
4517th Combat Crew Training Squadron: assigned from October 1969

67th Tactical Reconnaissance Wing
The 67th TRW was activated on 25 February 1951 and inactivated on 8 December 1960. It briefly parented the 6021st Reconnaissance Squadron with RF-100As, though at other times they were assigned to the 6000th Operations Wing.
6021st Reconnaissance Squadron: attached to the 67th TRW 1 July – 8 December 1957.

83rd FDW
Perhaps the shortest lived F-100 unit was the 83rd FDW at Seymour Johnson AFB. The unit moved to Seymour Johnson in July 1956, and converted to the F-100 in August 1957. It was re-numbered as the 4th FDW in December 1957.
448th Fighter-Day Squadron: assigned August 1957 – December 1957
532nd Fighter-Day Squadron: assigned 1957 – December 1957
533rd Fighter-Day Squadron: assigned 1957 – December 1957
534th Fighter-Day Squadron: assigned 1957 – December 1957

95th Bombardment Wing
The ADC-assigned 4758th Defense Systems Evaluation Squadron came under the control of the 95th Bombardment Wing while operating from Biggs AFB, TX. The squadron moved to Holloman AFB and was officially inactivated on 26 June 1966. Its aircraft wore a small squadron badge on the tail and often carried dayglo orange fuel tanks.
4758th Defense Systems Evaluation Squadron

113th TFW
The 113th TFW formed at Myrtle Beach in March 1968 to act as an RTU for ANG F-100 aircrew deploying to Vietnam, with two ANG squadrons being called to active duty to man it. It closed in June 1969. The wing tailcode 'XD' was sometimes worn.
119th Tactical Fighter Squadron, NJ ANG: assigned March 1968 – June 1969, 'XA' tailcode
121st Tactical Fighter Squadron, DC ANG: assigned March 1968 – June 1969, 'XB' tailcode

312th FBW/TFW
The 312th FBW was another short-lived F-100 user, converting from the F-86H in late 1956, but then re-designating as the 27th TFW in February 1959. The 312th itself went from FBW to TFW on 1 July 1958. Its aircraft wore a small wing badge on the fin, later coloured fins, leaving the fintip and a triangular portion of the rear, from the trailing edge tapering down to the fin leading edge, unpainted.
386th Tactical Fighter Squadron: assigned late 1956 – February 1959, red colours
387th Tactical Fighter Squadron: assigned late 1956 – February 1959, blue colours
388th Tactical Fighter Squadron: assigned late 1956 – February 1959, yellow colours
477th Tactical Fighter Squadron: assigned October 1957 – February 1959, green colours

F-100D, 81st TFS/50th TFW, at Wheelus AFB, Libya, on weapons practice camp

316th Air Division
The 45th FDS was a conversion training and transition unit for USAFE F-100 units, briefly based at Sidi Slimane AB, French Morocco. The squadron's aircraft had a black edged scalloped yellow nose band and a yellow tailfin, with black/yellow chevrons superimposed.
45th Tactical Fighter Squadron: assigned March 1956 – 8 January 1958

322nd FDW
Though it was TAC's first F-100 operator, the 322nd FDW at Foster AFB, TX, was short-lived, converting from the F-86F in mid-1955 but passing its aircraft on to the 4th and 36th FDWs in late 1957. Its aircraft wore a broad band on the fin with playing cards insignia superimposed.
450th Fighter Day Squadron: assigned 1955 – 1957, red colours
451st Fighter Day Squadron: assigned 1955 – 1957, yellow colours
452nd Fighter Day Squadron: assigned 1955 – 1957, green colours

323rd FBW
The 323rd FBW at Bunker Hill AFB initially used a few F-100As with the 323rd FBG, before equipping the 386th FBG with the F-100D. The unit's aircraft wore a band on the tail, and around the nose, edged with small black checkers.
453rd Fighter Bomber Squadron: assigned March 1956 – May 1957
454th Fighter Bomber Squadron: assigned March 1956 – May 1957
454th Fighter Bomber Squadron: assigned March 1956 – May 1957
552nd Fighter Bomber Squadron: assigned late 1956 – August 1957
553rd Fighter Bomber Squadron: assigned late 1956 – August 1957
553rd Fighter Bomber Squadron: assigned late 1956 – August 1957

325th Fighter Weapons Wing
The 325th Fighter Weapons Wing was reformed on 17 Jun 1981. It included the 82nd TATS with QF-100 drones.
82nd Tactical Aerial Target Squadron: assigned 1 Jul 1981 – 15 Oct 1983

354th FBW/TFW
The 354th FBW/TFW had two incarnations as an F-100 operator. The wing formed at Myrtle Beach AFB, SC, in early 1957, taking over the assets of the deactivated 31st FBW, under the command of WWII and Korean ace Francis S Gabreski. It became the 354th TFW on 1 July 1958. Its squadrons were reassigned to units in Vietnam by June 1968 and it began a new existence as the controlling wing for two ANG squadrons deployed to Kunsan AB, Korea.
127th Tactical Fighter Squadron, KS ANG: assigned at Kunsan, 5 July 1968 – 10 June 1969, 'BO' tailcode, blue colours
166th Tactical Fighter Squadron, OH ANG: assigned at Kunsan, 5 July 1968 – 10 June 1969, 'BP' tailcode, red colours
352nd Tactical Fighter Squadron: assigned September 1957 – August 1966, yellow colours. To 366th TFW
353rd Tactical Fighter Squadron 'Black Panthers': assigned September 1957 – April 1966, red colours. To F-4E, designation to A-7 unit 1970
355th Tactical Fighter Squadron: assigned September 1957 – April 1968, blue colours. To 37th TFW
356th Tactical Fighter Squadron 'Green Demons': assigned September 1957 – November 1965, green colours. To F-4C November 1967

366th TFW
The 366th FBW converted from the F-84F to the F-100D in late 1957, but deactivated at England AFB, LA, in early 1959. The wing subsequently reactivated in Vietnam in April 1966, operating there between April and October 1966, when it swapped identities with the 35th TFW at Da Nang. The wing's early markings consisted of a candy striped nose band and alternating diagonal strips across most of the tailfin.
352nd Tactical Fighter Squadron: assigned at Phan Rang August 1966 – October 1966
389th Tactical Fighter Squadron: assigned at England AFB, LA, September 1957 – January 1959
390th Tactical Fighter Squadron: assigned at England AFB, LA, September 1957 – January 1959
391st Tactical Fighter Squadron: assigned at England AFB, LA, September 1957 – February 1959
480th Fighter Bomber Squadron: assigned at

F-100C, 45th FDS/316th Air Division, based at Sidi Slimane AB, Morocco

153

Warplane Classic

England AFB, LA, September 1957 – March 1959
614th Tactical Fighter Squadron: assigned at Phan Rang July 1966 – October 1966
615th Tactical Fighter Squadron: assigned at Phan Rang July 1966 – October 1966

388th TFW
The 388th FBW converted from the F-86F at Etain-Rouvres in late 1956, but re-numbered as the 49th FBW in December 1957. The wing subsequently reactivated at McConnell AFB on 1 October 1962, with a single F-100 unit but converted to the F-105D in mid-1963. As an F-105-equipped wing at Korat, the 388th subsequently parented the Wild Weasel I F-100 unit.
560th Tactical Fighter Squadron: assigned at McConnell, 1 October 1962 – mid-1963
561st Tactical Fighter Squadron: assigned at Etain-Rouvres late 1956 – December 1957, yellow colours
562nd Tactical Fighter Squadron: assigned at Etain-Rouvres late 1956 – December 1957, blue colours
563rd Tactical Fighter Squadron: assigned at Etain-Rouvres late 1956 – December 1957, red colours
6234th Tactical Fighter Squadron 'Wild Weasel 1': assigned at Korat November 1965 – May 1966

401st TFW
The 401st TFW converted from the F-84F at England AFB, LA, in late 1957. It lost most of its squadrons on TDY to wings in Vietnam, having only one remaining unit when it transferred to Torrejon in Spain in April 1966. The wing converted to the F-4 Phantom in 1970.
90th Tactical Fighter Squadron: assigned December 1965 – February 1966
307th Tactical Fighter Squadron: assigned at Torrejon April 1966 – June 1971
353rd Tactical Fighter Squadron 'Black Panthers': assigned at Torrejon April 1966 – June 1971
531st Tactical Fighter Squadron: assigned November 1965 – December 1965
612th Tactical Fighter Squadron 'Screaming Eagles': assigned September 1957 – November 1965. Blue colours. To 366th TFW
613th Tactical Fighter Squadron 'Squids': assigned September 1957 – 1970. Yellow colours
614th Tactical Fighter Squadron 'Lucky Devils': assigned September 1957 – April 1966. Red colours. To 35th TFW
615th Tactical Fighter Squadron: assigned 1957 – April 1966. Green colours. To 35th TFW

402nd FDW
The 402nd FDW operated F-100Cs for only six months in late 1956. Its aircraft wore two bands high on the tailfin.
320th Fighter Day Squadron: assigned to the 402nd FDG – 1956
442nd Fighter Day Squadron: assigned to the 402nd FDG – 1956
540th Fighter Day Squadron: assigned to the 402nd FDG – 1956

405th FBW/TFW
The 405th FBW was TAC's first F-100D unit, converting from the F-84F in late 1956 at Langley AFB, VA. The wing deactivated on 1 July 1958, but reactivated at Clark AFB on 9 April 1959 with a single F-100 unit. The wing's aircraft wore a broad fin band in the squadron colour, edged in dark blue/black, with a stylised white bird, and there was a checkerboard around the nose intake in the squadron colour and white.
508th Fighter Bomber Squadron: assigned 1956 – March 1958. Yellow colours
509th Tactical Fighter Squadron: assigned April 1959 – August 1964, and November 1965 – 1967. Red colours
510th Tactical Fighter Squadron: assigned at Langley 1956 – April 1958. Assigned at Clark AB 9 April 1959 – March 1964. Purple colours. To 3rd TFW
511th Fighter Bomber Squadron: assigned 1956 – May 1958. Blue colours
522nd Tactical Fighter Squadron: assigned August 1965 – November 1965
523rd Tactical Fighter Squadron: assigned November 1965 – 1967
531st Tactical Fighter Squadron: assigned November 1964 – February 1965
612th Tactical Fighter Squadron: assigned at Langley 1956 – 1958. Assigned February 1964 – June 1964
615th Tactical Fighter Squadron: assigned 4 – 6 June 1964

413th FDW/TFW
The 423th FDW at George AFB converted to the F-100C from the F-86H in late 1957. It transitioned to the F-100D in 1958, becoming the 413th TFW. The wing was redesignated as the 31st TFW on 15 March 1959. The wing's aircraft wore a squadron coloured fin band containing the squadron badge, and a coloured nose band
1st Fighter Bomber Squadron: assigned October 1957 – 15 March 1959, red colours
21st Fighter Bomber Squadron: assigned October 1957 – 15 March 1959, blue colours
34th Fighter Bomber Squadron: assigned October 1957 – 15 March 1959, green colours
474th Fighter Bomber Squadron: October 1957 – March 1959, yellow colours

450th FDW/TFW
The 450th FDW converted from the F-86F to the F-100C at Foster AFB in 1955, but deactivated in December 1958. The wing's aircraft wore an approximation of the stars and stripes, with seven red and six white stripes on the trailing edge, and three stars in white on the blue forward portion of the fin. They also used a coloured, scalloped nose chevron.
720th Fighter-Day Squadron: assigned 1955 – July 1958
721st Fighter-Day Squadron: assigned 1955 – July 1958, red colours
722nd Fighter-Day Squadron: assigned 1955 – July 1958
723rd Fighter-Day Squadron: assigned 1955 – August 1958

474th FBW/TFW
The 474th FBW converted from the F-86H to the F-100D at Cannon AB, NM in late 1957. The unit converted to the F-111 in 1968. The wing's aircraft wore a thick band on the tail containing a double-headed 'Machbusters' shockwave device, and later used alternating chevrons covering most of the fin. Three of the wing's constituent squadrons were sent to Vietnam on TDY.
428th Tactical Fighter Squadron 'Buccaneers': assigned 1957 – September 1965. Blue colours
429th Tactical Fighter Squadron 'Black Falcons': assigned 1957 – December 1965. Yellow and black colours
430th Tactical Fighter Squadron 'Tigers': assigned 1957 – September 1965. Red colours
478th Tactical Fighter Squadron: assigned 1957 – September 1965. Green colours, but later black and white

475th TFW
The 475th TFW activated on 21 December 1967, but was a 'paper' unit which borrowed assets from other wings in SEA. The 475th TFW's Det 1 (rotated between F-100 and F-4 units) took over the Korean nuclear alert commitment from the 39th AD, despite the presence of ANG F-100s at Kunsan. The wing's three squadrons converted to F-4s later during 1968.
67th Tactical Fighter Squadron 'Fighting Cocks': assigned 15 January 1968 – 15 March 1971
356th Tactical Fighter Squadron 'Green Devils': assigned 15 January 1968 – 15 March 1971
391st Tactical Fighter Squadron: assigned 22 July 1968 – 28 February 1971

475th Weapons Evaluation Group
The 475th WEG parented the QF-100s at Tyndall AFB after the re-designation of the 325th FWW. Det 1 of the 475th WEG looked after the QF-100 drones at Holloman AFB.
82nd Tactical Aerial Target Squadron: assigned October 1983-1992

479th FDW/TFW
The 479th FDW at George AFB converted from F-86F to F-100A in 1954. The wing subsequently received F-100Cs, becoming TAC's first F-100C unit, before converting to the F-104A in 1958, when it also became the 479th TFW.
434th Fighter Day Squadron: assigned September 1954 – 1959, red colours
435th Fighter Day Squadron: assigned September 1954 – 1959, green colours
436th Fighter Day Squadron: assigned September 1954 – 1959, yellow colours
476th Fighter Day Squadron: assigned 8 October 1954 – 1959, blue colours

506th TFW
The 506th FBW transitioned from the F-84F at Tinker AFB in September 1957. After a number of deployments to West Germany, the wing deactivated on 1 April 1959. The wing's F-100s wore markings reminiscent of those briefly used by the 450th TFW, with an approximation of the stars and stripes, with four coloured and three contrasting stripes on the trailing edge and two stars on the forward portion of the fin. Coloured, scalloped nose chevron.
457th Fighter Bomber Squadron: assigned 1957 – December 1958, red/white colours
458th Fighter Bomber Squadron: assigned 1957 – December 1958, yellow/black colours
462nd Fighter Bomber Squadron: assigned 1957 – December 1958, blue colours
470th Fighter Bomber Squadron: assigned 1957 – December 1958, green colours

3525th CCTW
The 3525th Combat Crew Training Wing was a pilot training unit based at Williams AFB. When responsibility for advanced training passed from Air Training Command, the 3,000 series CCTWs were redesignated, and new basic training units were formed using the old designations. The 3525th CCTW became the 4530th Combat Crew Training Wing. When it was redesignated in July 1958 it had both F-86 Sabres and F-100s on charge.

3595th CCTW
The 3595th Combat Crew Training Wing was based at Nellis AFB, and used F-86s and F-100s between 1954 – July 1958, when it became the 4520th Combat Crew Training Wing. The 3595th CCTW's most famous F-100 element was the 'Thunderbirds' display team, which moved to Nellis AFB in 1956, becoming the 3595th Air Demonstration Flight.

3600th CCTW
The 3600th Combat Crew Training Wing was based at Luke AFB, and used F-86s and F-100s between December 1957 – July 1958. It was redesignated as the 4510th Combat Crew Training Wing when training passed to TAC. The 'Thunderbirds' display team was originally based at Luke as the 3600th Air Demonstration Flight.

4403rd TFW
The 4403rd TFW was a provisional unit established at Homestead to parent F-100 units returning from SEA before re-equipment. The unit operated F-100Ds and F-100Fs from England AFB, La, from 1970-1972.
68th Tactical Fighter Squadron: assigned 1970 – 1972, 'SD' tailcode, yellow colours
416th Tactical Fighter Squadron: assigned September 1970 – 1972, 'SE' tailcode, blue colours
431st Tactical Fighter Squadron: assigned 1970 – 1972, 'SM' tailcode, red colours

4510th CCTW
When responsibility for advanced training passed from Air Training Command to TAC in July 1958, the 3600th Combat Crew Training Wing at Luke AFB was re-designated as the 4510th CCTW. 'LA' tailcodes were used from July 1968, and the wing was re-designated as the 58th TFTW in October 1969.
4511th Combat Crew Training Squadron: assigned July 1958 – 18 January 1970
4512th Combat Crew Training Squadron: assigned July 1958 – 1970
4514th Combat Crew Training Squadron: assigned July 1958 – 15 December 1969
4515th Combat Crew Training Squadron: assigned 1 September 1966 – 18 June 1970
4517th Combat Crew Training Squadron: assigned 1 September 1966 – October 1969

4520th CCTW
The 3595th Combat Crew Training Wing at Nellis AFB was redesignated as the 4520th CCTW in July 1958 when TAC

F-100D, 355th TFS/354th TFW, refuelling during deployment to Aviano, February 1959

F-100D, 614th TFS/401st TFW, at Langley AFB, Virginia

F-100 Operators

F-100D, 429th TFS/474th TFW, combat deployment to Bien Hoa, July to November 1965

took responsibility for advanced training. The F-100 originally served with two squadrons, though these aircraft transferred to Luke AFB in 1962. The remaining F-100s at Nellis were then concentrated with the 4536th FWS of the 4525th CCTW.
4523rd Combat Crew Training Squadron: assigned July 1958 – 1960/1961
4526th Combat Crew Training Squadron: assigned July 1958 – 1960/1961

4525th CCTW/FWW
The 4525th CCTW was a Nellis-based advanced and weapons training unit. The wing was redesignated as the 57th FWW in October 1969, and then on 15 October, as the 57th TTW.
4536th Fighter Weapons Squadron: assigned to the 4525th FWW September 1966 – October 1969

4530th CCTW
The 4530th CCTW was formed by the re-designation of the 3525th CCTW at Williams AFB in July 1958. The wing ceased F-100 operations in October 1960.

6000th Operations Wing
The 6000th Operations Wing parented the 6021st Reconnaissance Squadron with RF-100As, though at other times they were assigned to the 67th TRW.
6021st Reconnaissance Squadron: attached 2 June 1955 – June 1958

6200th Air Base Wing
The 6200th ABW parented the 72nd Tactical Fighter Squadron (formerly the 418th FBS) at Clark AB, Philippines. The unit subsequently became the 510th FBS with the 405th FBW. The squadron's markings comprised a red fin band, edged in white, with white chevron superimposed, with a white-edged red nose band.
72nd Tactical Fighter Squadron: assigned 1 July 1958 – 9 April 1959
418th Fighter Bomber Squadron: assigned May 1958 – 30 June 1958

6585th Test Group
The 6585th Test Group supported activities at the White Sands missile range, and included Det 1 of the 82nd Tactical Aerial Targets Squadron from Tyndall.

7272nd Flying Training Wing
The 7272nd FTW supported training detachments by USAFE combat wings to use the ranges in Libya. It was based at Wheelus AFB, Libya, from January 1958 until 1965. Its F-100 target tugs wore red and yellow markings originally, but these were replaced by dark blue bands on the tail and around the nose, with white arrows superimposed.
7235th Support Squadron: assigned 1959 – January 1970

7499th Support Group
The 7499th SG parented the 7407th Support Squadron with RF-100As. They were replaced by six RB-57Fs from 9 June 1959, which were detached from the 4080th SRW.
7407th Support Squadron: assigned May 1955 – 1 July 1958

F-100C, 436th FDS/479th FDW, first unit with F-100A and F-100C

AIR NATIONAL GUARD

102nd TFG, Massachusetts ANG
The 101st TFS briefly operated the F-100D from Logan Airport between May 1971 and the spring of 1972, when F-106s were received.
101st Tactical Fighter Squadron: assigned June 1971 – June 1972

103rd TFG, Connecticut ANG
The 118th TFS converted from F-86Hs to F-100As during the summer of 1960, at Bradley Field, re-designating as the 118th FIS on 1 September 1960 and becoming ADC-gained. F-100As gave way to F-102As in January 1966, but F-100Ds were received in June 1971, when the squadron became the TAC-gained 118th TFS again. Early aircraft wore a coloured nose band, edged in white, but markings on the F-100D were limited to a 'CT' tailcode.
118th Tactical Fighter Squadron: assigned to the 103rd TFG October 1959 – September 1960, to the 103rd FIG September 1960 – January 1966, and to the 103rd TFG April 1971 – mid-1979

104th TFG, Massachusetts ANG
The 131st TFS converted from F-84Fs to F-100Ds in June 1971, at Barnes Field, and to the A-10A in July 1979. These aircraft wore a red/white/red fin stripe, and sometimes used 'MA' codes.
131st Tactical Fighter Squadron: assigned June 1971 – 1979

107th TFG, New York ANG
At Niagara Falls MAP the 136th TFS's F-86Hs gave way to F-100Cs in August 1960. The squadron was called to active duty between 1 October 1961 – 24 August 1962 during the Berlin Crisis (but remained at Niagara Falls MAP), and then again between 26 January 1968 – 11 June 1969 during the *Pueblo* Crisis, deploying to Tuy Hoa, where it became known as 'Rocky's Raiders', and wore the tailcode 'SG'.
136th Tactical Fighter Squadron: assigned to the 107th TFG 1960 – 1971, and to the 31st TFW June 1968 – May 1969

113th TFG, DC ANG
The 121st TFS converted from F-86Hs to F-100Cs in mid-1960, becoming TAC-gained in July. The unit was called to active duty 1 October 1961 – 24 August 1962 during the Berlin Crisis, but remained at Andrews AFB. The squadron was called to active duty again between 26 January 1968 – 18 June 1969 during the *Pueblo* Crisis, forming part of the F-100 CCTW at Myrtle Beach using the 'XB' tailcode. F-100Cs gave way to F-105s in July 1971. Early aircraft wore a dark fin stripe containing five white stars on natural metal aircraft.
121st Tactical Fighter Squadron: assigned to the 113th TFG mid-1960 – 1971, and to the 113th TFW March 1968 – June 1969

114th TFG, South Dakota ANG
The 175th TFS converted from the F-102 to the F-100D at Sioux Falls in the spring of 1970. These remained in use until 1977 when they gave way to A-7Ds. The squadron's F-100s often wore a stylised white line drawing of a wolf's head on the fin cap.
175th Tactical Fighter Squadron: assigned to the 114th TFG May 1970 – 1977

116th TFG, Georgia ANG
The 128th Military Airlift Squadron at Dobbins AFB gained a new aircraft type and a new role when it converted from the C-124C Globemaster to the F-100D, becoming the 128th TFS in April 1973. The F-100s gave way to F-105s in late 1979. They wore a yellow fin cap, with a blue fin stripe lower on the fin.
128th Tactical Fighter Squadron: assigned to the 116th TFG April 1973 – mid 1979

121st TFG, Ohio ANG
The 166th TFS traded F-84Fs for F-100Cs at Lockbourne AFB during August 1970. The unit was called to active duty 26 January 1968 – 18 June 1969 during the *Pueblo* Crisis, deploying to Kunsan, where it wore the 'BP' tailcode, and red colours. The squadron gained F-100Ds in November 1971 and converted to the A-7 in December 1974.
166th Tactical Fighter Squadron: assigned to the 121st TFG August 1962 – July 1968 and June 1969 – 1974, and to the 354th TFW at Kunsan, 5 July 1968 – 10 June 1969

122nd TFG, Indiana ANG
F-84Fs gave way to F-100Ds in June 1971. The 122nd TFG disbanded on 9 December 1974, and the F-100s were assigned directly to the 122nd TFW. The squadron converted to the F-4Cs in early 1979. The aircraft wore a yellow fin band, edged in white.
163rd Tactical Fighter Squadron: assigned to the 122nd TFG June 1971 – 1979

127th TFG, Michigan ANG
RF-101As gave way to F-100Ds with the Selfridge-based 107th TFS in the summer of 1972. They were replaced by A-7Ds in the summer of 1978. The squadron's F-100s wore a thin red fin band with 'MICHIGAN' superimposed in white.
107th Tactical Fighter Squadron: assigned to the 127th TFG June 1972 – October 1978

131st TFG, Missouri ANG
The 110th TFS replaced its F-84Fs with F-100Cs at Lambert Field in August 1962, and replaced these with F-100Ds in December 1971. The F-100 gave way to the Phantom in early 1979. The unit's Super Sabres were decorated with a thin red fin band, thinly outlined in white, with 'Missouri' superimposed in white.
110th Tactical Fighter Squadron: assigned to the 131st TFG fall 1962 – 1979

132nd TFG, Iowa ANG
The 124th TFS converted from F-84Fs to F-100Cs in April 1971, at Des Moines MAP, and then gained F-100Ds in 1975, replacing what were the ANG's last C-models. The unit converted to the A-7D in January 1977.
124th Tactical Fighter Squadron: assigned to the 132nd TFG April 1971 – July 1978

138th TFG, Oklahoma ANG
The 125th ATS relinquished its C-124Cs in January 1973, gaining F-100Ds and becoming the 125th TFS. The Tulsa-based unit flew these until July 1978, when it converted to the A-7. The unit's F-100s wore a red fin band, thinly outlined in white, with 'Oklahoma' superimposed in white.
125th Tactical Fighter Squadron: assigned to the 138th TFG January 1973 – July 1978

140th TFG, Colorado ANG
The 120th TFS exchanged F-86Ls for F-100Cs on 1 January 1961, switching from ADC to TAC. The unit was called to active duty on 1 October 1961 during the Berlin Crisis, but remained at Buckley ANGB until stood down on 24 August 1962. The unit was called to active duty again between 26 January 1968 – 30 April 1969 during the *Pueblo* Crisis, deploying to Phan Rang and flying 5,905 combat sorties. The unit returned to state control on 30 April 1969 and converted to F-100Ds in October 1971. These gave way to A-7s in April 1974. Its early aircraft wore a

F-100C, 121st TFS/113th TFG, District of Columbia ANG, Andrews AFB, April 1963

Warplane Classic

F-100D/Fs, 113th TFS/181st TFG, Indiana ANG, Hulman Field – last F-100 operator

F-100D, 122nd TFS/159th TFG, Louisiana ANG – three F-4 'kill' markings

large chevron on the tailfin, while camouflaged aircraft had a red mountain lion's head on nose gear door and sometimes on the tailfin in Vietnam.
120th Tactical Fighter Squadron: assigned to the 140th TFG January 1961 – April 1974, and to the 35th TFW at Phu Cat April 1968 – April 1969, tailcode 'VS'

149th TFG, Texas ANG
The 182nd TFS traded F-84Fs for F-100Ds in the spring of 1971, at Kelly AFB. These were used until spring 1979, when the unit converted to the F-4C. The squadron's Super Sabres wore a thin red fin band, thinly outlined in white, with 'TEXAS' superimposed in white.
182nd Tactical Fighter Squadron: assigned to the 149th TFG April 1971 – mid-1979

150th TFG, New Mexico ANG
Ancient F-80Cs gave way to F-100As with the 188th FIS at Kirtland AFB in April 1958. The unit became TAC-gained in 1960, and converted to F-100Cs in spring 1964. The squadron was called to active duty 26 January 1968 – 4 June 1969 during the *Pueblo* Crisis, deploying to Tuy Hoa as the 'Enchilada Air Force'. Returning to Kirtland, the 188th TFS finally converted to the A-7 in late 1973. Camouflaged aircraft wore a small yellow roadrunner on the fin, together with a Distinguished Unit Citation ribbon. Natural metal and ADC grey aircraft carried yellow-edged black chevrons on the fin, and a yellow-edged black flash on the fuselage. In Vietnam the squadron used the tailcode 'SK'.
188th Tactical Fighter Squadron: assigned to the 150th FIG/TFG April 1958 – June 1968 and May 1969 – 1973, and to the 31st TFW June 1968 – May 1969

159th TFG, Louisiana ANG
At NAS New Orleans, the 122nd TFS converted from F-102s to F-100Ds in late 1970, and used these until F-4Cs arrived in April 1979.
122nd Tactical Fighter Squadron: assigned to the 159th TFG July 1970 – April 1979

162nd TFG, Arizona ANG
The group's 152nd FIS converted from the F-84F to the F-100A in May 1958, and transitioned to the F-102A from February 1966. The F-102 was replaced by the F-100C in 1969, when the 152nd became a training unit. F-100Ds arrived in June 1972. The F-100 was phased out in March 1978. Natural metal aircraft had a large yellow fin chevron and intake band, both thinly outlined in black.
152nd Tactical Fighter Squadron: assigned to the 162nd FIG at Tucson IAP from mid 1958 – 1964, and to the 162nd TFTG September 1969 – 1978, as the 152nd TFTS

177th TFG, New Jersey ANG
At Atlantic City, the 119th TFS converted from F-86Hs to F-100Cs in September 1965. The unit was called to active duty 26 January 1968 – 17 June 1969 during the *Pueblo* Crisis, going to Myrtle Beach to form part of the F-100 CCTW there, using the 'XA' tailcode. The unit began conversion to the F-105 in June 1970.
119th Tactical Fighter Squadron: assigned to the 177th TFG 1964 – June 1970, and to the 113th TFW March 1968 – June 1969

178th TFG, Ohio ANG
The 162nd TFS at Springfield traded F-84Fs for F-100Ds in April 1970. They were replaced by A-7Ds in April 1978. The aircraft wore a red or green fin band, thinly outlined in white, with 'Ohio' superimposed in white.
162nd Tactical Fighter Squadron: assigned to the 178th TFG April 1970 – 1977

179th TFG, Ohio ANG
The F-84F-equipped 164th TFS converted to F-100Ds in February 1972, at Mansfield-Lahm airport. The squadron converted to the C-130B Hercules during the winter of 1975. The unit's F-100s wore a white-edged yellow tail band.
164th Tactical Fighter Squadron: assigned to the 179th TFG February 1972 – 1975

180th TFG, Ohio ANG
The 112th TFS at Toledo Express Airport swapped its F-84Fs for F-100Ds in October 1970, and flew these until late 1979, when it converted to the A-7. The squadron's Super Sabres wore a black and white checkerboard fin band, thinly edged in yellow.
112th Tactical Fighter Squadron: assigned to the 180th TFG October 1970 – 1979

181st TFG, Indiana ANG
The 113th TFS converted from RF-84Fs to F-100Ds in September 1971, at Hulman Field, Terre Haute, but converted to the F-4C in the summer of 1979. Its aircraft had a fin cap divided into red, white and blue horizontal bands, with 'Indiana' superimposed on the white stripe.
113th Tactical Fighter Squadron: assigned to the 181st TFG September 1971 – November 1979

184th TFG, Kansas ANG
The 127th TFS converted from the F-86L to the F-100C in April 1961, at McConnell. The unit was called to active duty 26 January 1968 – 18 June 1969 during the *Pueblo* Crisis, deploying to Kunsan where they used 'BO' codes. The squadron converted to F-100Ds in March 1971, and then to F-4D Phantoms in October 1979.
127th Tactical Fighter Squadron: assigned to the 184th TFG April 1961 – March 1971 and to the 354th TFW at Kunsan, 5 July 1968 – 10 June 1969

185th TFG, Iowa ANG
The 174th converted from the RF-84F to the F-100C during the summer of 1961, at Sioux City MAP. The squadron was called to active duty 26 January 1968 – 28 May 1969 during the *Pueblo* Crisis, and deployed to Phu Cat. The squadron converted to the F-100D in June 1974, and then to the A-7 in December 1976. The squadron's early aircraft wore a yellow chevron on the fin, a curved flash on the fuselage and a yellow nose. In Vietnam the unit used the tailcode 'HA'.
174th Tactical Fighter Squadron: assigned to the 185th TFG June 1961 – 1968, and 1969 – July 1977 and to the 37th TFW May 1968 – May 1969

188th TFG, Arkansas ANG
The 184th TFS converted from the RF-101C to the F-100D in June 1972, and retained the Super Sabre until 1979, when it converted to the F-4 Phantom. The aircraft carried a red fin band, thinly outlined in white, with 'ARKANSAS' superimposed in white.
184th Tactical Fighter Squadron: assigned to the 188th TFG June 1972 – mid-1979

Miscellaneous Units

A number of units used small numbers of F-100s in support, test and training roles. These units included:
Armament Development Test Center (ADTC), previously the Air Proving Ground Command
Air Force Logistics Command (AFLC) at McClellan AFB (Sacramento Air Materiel Area/SMAMA)
Air Force Flight Test Center (AFFTC), Edwards AFB
Air Force Special Weapons Center (AFSWC), Kirtland AFB
Aerospace Medical Division (AMD) at Brooks and then Kelly AFBs
Air Research and Development Command (ARDC)
NACA and NASA – mostly at Ames (Moffett) and Dryden (Edwards)
Wright Air Development Center (WADC)
1708th Ferrying Wing, MATS, which trained ferry pilots for the F-100
4925th OMS at Kirtland AFB
4758th DSES at Holloman AFB, which provided target facilities for ADC units and the White Sands missile range, until replaced in 1970 by an element of the New Mexico ANG
US Army operated QF-100s at Holloman AFB in support of missile activities at nearby White Sands

F-100D, Sacramento Air Material Area, McClellan AFB, California

F-100F, Air Force Special Weapons Center, Kirtland AFB, New Mexico.

QF-100D, US Army/FSI, Holloman AFB, New Mexico for HAWK SAM trials

F-100 Operators

OVERSEAS OPERATORS

DENMARK – KONGELIGE DANSKE FLVEVÅBNET

Danish aircraft wore colourful markings until they were camouflaged, when large areas of colour gave way to small and discreet unit badges. Danish F-100s wore serials consisting of the prefix G- or GT- (Fs) followed by the last three numerals of the original USAF serial.

Esk 725: operated F-100s from Karup from April 1961 to September 1970, when the squadron converted to the F35 Draken.
Esk 727: operated F-100s from Karup from May 1959 to April 1974, then from Skrydstrup until conversion to the F-16 in April 1981. Esk 727's aircraft initially had a red nose chevron, thinly outlined in black, and had a red RWR antenna on the fin trailing edge.
Esk 730: operated F-100s from Skrydstrup from July 1961 to August 1982. Esk 730's aircraft originally had a dark green nose, edged with white and dark blue diagonal bands, and had green/white/green stripes on the RWR antenna.

Denmark's F-100s were delivered from May 1959, and initially retained a natural metal finish, with squadron markings applied on the fin. This aircraft also wears the red nose flash of Eskadrille 727.

FRANCE – ARMÉE DE L'AIR

3ème Escadre de Chasse
EC 1/3 'Navarre': F-100s replaced F-84Fs in January 1959 at Reims. Redeployed to Lahr 1961 – 66, transitioning to the Mirage III. Returned to Nancy-Ochey September 1967. Codes '3-IA' to '3-IZ'
EC 2/3 'Champagne': F-100s replaced F-84Fs in January 1959 at Reims. Redeployed to Lahr 1961 – 66, transitioning to the Mirage III. Returned to Nancy-Ochey September 1967. Codes '3-JA' to '3-JZ'

11ème Escadre de Chasse
EC 1/11 'Roussillon': F-100s replaced F-84Fs and Gs in May 1958 at Luxeuil-St Sauveur. Redeployed to Bremgarten June 1961 – September 1967. Returned to France (Toul-Rosières) September 1967. Converted to Jaguar October 1975. Codes '11-EA' to '11-EZ'
EC 2/11 'Vosges': F-100s replaced F-84Fs and Gs in May 1958 at Luxeuil-St Sauveur. Redeployed to Bremgarten June 1961 – September 1967. Returned to France (Toul-Rosières) September 1967. Converted to Jaguar late 1976. Codes '11-MA' to '11-MZ'
EC 3/11 'Corse': formed at Bremgarten in April 1966, returning to France (Colmar and Toul) in late 1967. Became the F-100 conversion training unit. Converted to Jaguar late 1976. Codes '11-RA' to '11-RX'
EC 4/11 'Jura': activated at Djibouti 1 January 1973. Ended F-100 operations 12 December 1978. Aircraft carried large sharkmouth from March 1978. Codes '11-YA' to '11-YZ'

Esc. de Convoyage EC 070
This escadrille was responsible for ferrying AdA aircraft, and transported the remaining F-100s to RAF Sculthorpe in 1977-78. Codes 'MA' to 'MZ'.

France's Super Sabres had two 'lives': as NATO-controlled nuclear strikers based in Germany, and as nationally-controlled conventional fighter-bombers in France.

TAIWAN – REPUBLIC OF CHINA AIR FORCE

All Taiwanese Super Sabres (except a handful of camouflaged jets) wore the RoCAF's standard blue and white rudder stripes, and most also wore colourful squadron markings. Hsinchu aircraft usually wore the 2nd FBW badge on their tailfins. Taiwanese F-100s wore sequential four-digit serials.

2nd FBW
The 2nd Fighter Bomber Wing at Hsinchu had three F-100 units.
41st Fighter Bomber Squadron: formerly the 17th FBS from Chiayi. Assigned 1959 – 1983. The squadron's F-100s had a red lightning flash on the nose. Transitioned to the F-104G.
42nd Fighter Bomber Squadron: assigned 1959 – 1984. Its aircraft had blue nose markings like those of the 23rd FBS. Transitioned to F-104G 1983-1984
48th Fighter Bomber Squadron: assigned 1959 – 5 September 1984. Red, white and blue 'Skyblazer' type nose marking. Last Taiwanese F-100 operator. Transitioned to the F-104G

4th FBW
The 4th FBW at Chiayi had a maximum strength of three F-100 units, one transferring to Hsinchu after forming.
17th Fighter Bomber Squadron: transferred to Hsinchu and became the 41st FBS
21st Fighter Bomber Squadron: assigned 1960 – 1978. 21st FBS aircraft wore a red fin chevron and nose markings. Transitioned to the F-5E
22nd Fighter Bomber Squadron: assigned 1960 – 1978. Squadron had a yellow fin chevron and nose markings. Transitioned to the F-5E
23rd Fighter Bomber Squadron: assigned 1960 – 1979. Markings comprised a blue fin chevron and nose markings. Transitioned to the F-5E

5th/401st Tactical Combined Wing
The 5th Wing at Taoyuan parented the four RF-100A Slick Chick reconnaissance aircraft of the 4th Squadron.

This F-100A is preserved in 2nd TFW markings

4th Reconnaissance Squadron: assigned 1 January 1959 – December 1960. No squadron markings were carried

TURKEY – TÜRK HAVA KUVVETLERI

1nci Ana Jet Ussu
Between 1958 and 1979 the First Main Jet Air Base at Eskisehir included three F-100 squadrons.
111 Filo 'Panter' (Panther): assigned November 1958 – 1979. 111 Filo aircraft carried a black panther nose badge and coloured bands around the nose
112 Filo 'Seytan' (Devil): assigned 1962 – 1965 and 1969 – 1974.
113 Filo 'Isik' (Light): assigned 1959 – 1972, became 171 Filo. 113 Filo aircraft carried a black skull and crossbones superimposed on a black edged yellow lightning flash on nose, and had a black edged yellow flash on the fin

3ncu Ana Jet Ussu
The 3rd Main Jet Air Base at Konya had two F-100 squadrons.
131 Filo 'Ejder' (Dragon): assigned 1974 – 1978
132 Filo 'Hancer' (Dagger): assigned 1974 – 1987

7nci Ana Jet Ussu
The 7th Main Jet Air Base at Malatya had two F-100 squadrons.
171 Filo: assigned 1972 – 1977, ex 113 Filo
172 Filo: assigned 1972 – 1979, ex 182 Filo.

8nci Ana Jet Ussu
The 8th Main Jet Air Base at Diyarbakir had two F-100 squadrons.
181 Filo 'Pars' (Leopard): assigned 1972 – 1986
112 Filo 'Atmaca' (Sparrowhawk): assigned 1969 – 1986

Turkey's F-100s mostly wore standard SEA camouflage. The national insignia changed from this square to a roundel.

Type Analysis

He 177 Greif
The Luftwaffe's Lighter

The conventional lines of the He 177 (this is the A-02) hide the radical nature of its powerplant and also mask its size – its wingspan being bigger than that of the Avro Lancaster. Unlike the British bomber, the He 177 was limited to a paltry 4,000 lb (1814 kg) of bombs in the bomb-bay. While design started before the beginning of the war, a long and difficult gestation period resulted in the He 177 only entering service in mid-1942. Its service career was not glorious, and the whole story of the He 177 reflected the deepening crisis that the Luftwaffe faced as the war entered its final stages.

During World War II, the Luftwaffe suffered from the lack of an effective long-range bomber force. According to some sources its planners had, foolishly, ignored this weapon during the years leading up to the war. Yet, that was not the case. In November 1939, three months after the outbreak of war, the prototype of the four-engined Heinkel He 177 heavy bomber took off on its maiden flight. This came after the first flight of the British Short Stirling but before that of the Handley Page Halifax; after the American B-17 Flying Fortress but before that of the B-24 Liberator. Yet when the Heinkel bomber finally went into action, it did so long after its contemporaries, and in far smaller numbers. What went wrong?

If there is one point on which historians can agree regarding the Heinkel He 177, it is that the bomber had no measurable effect on the course of World War II. In all, the Luftwaffe accepted delivery of nearly 1,100 of these aircraft, but it is doubtful whether more than 200 of them ever flew operational missions. During the final nine months of the conflict the surviving He 177s – many of them in mint condition – sat on the ground in aircraft parks

The He 177 V8 had a mock-up installation of a gun barbette mounted on the rear upper fuselage, part of the effort devoted to exploring the best methods of defending the He 177 from Allied fighters. Of note in this picture is the open crew entrance under the nose of the aircraft.

Heinkel He 177 Greif

Right: The first prototype He 177 (V1, Werk-Nr. 00001) made its first flight on 19 November 1939 with Dipl.-Ing. Francke, chief of the Rechlin-based Erprobungsstelle E-2, at the controls. The V1 was the only He 177 to be fitted with a nose-mounted flight data boom, the device being mounted under the starboard wing on other He 177s.

Below: The He 177 V2 (Werk-Nr. 00002), the second prototype, is seen at the Heinkel works at Rostock-Marienehe. Note the full-span Fowler flaps, a feature of the He 177. During early diving tests this aircraft suffered control flutter and broke-up in mid-air, killing the crew which included the test pilot Rickert. While the first prototype carried a crew of three, from the V2 onwards a crew of four was carried (being increased again later). Twin bomb bays were also fitted to the V2, the V1 only having the aft bay.

where they swelled the scores of strafing Allied pilots. It was an ignominious ending to the story of the He 177.

The root cause of the He 177's problems was that, in its requirement for a long-range bomber, the Luftwaffe specification had demanded too much. To meet those requirements the Heinkel Company had to introduce several new and untried features into its design. In retrospect, we can see there were too many.

Development

In 1936, the Heinkel Company received details of the new Bomber A requirement from the Reichsluftfahrtministerium (RLM – the German Air Ministry). The specification called for an aircraft with a maximum speed of 335 mph (539 km/h), able to carry 4,400 lb (1995 kg) of bombs to a target 1,000 miles (1609 km) away or, alternatively, 2,200 lb (998 kg) to one 1,800 miles (2897 km) away.

For that time it was a formidable specification, calling for an aircraft able to outrun any fighter then in service and to outperform, by a considerable margin, any bomber. To meet the challenge Heinkel's chief designer, Siegfried Günther, displayed great ingenuity and incorporated several untried features in his P.1041 design study. The drawings depicted an aircraft that was exceptionally clean aerodynamically – but aerodynamic cleanliness comes at a price.

To power the new bomber, Günther chose a pair of Daimler-Benz DB 606 coupled engines, each producing 2,600 hp (1940 kW). Each DB 606 comprised a pair of DB 601 liquid-cooled, 12-cylinder, inverted-vee, inline motors, mounted side-by-side and driving a single four-bladed propeller through a connecting gear train. A clutch arrangement allowed either engine to be shut down in flight, to enable the aircraft to cruise on two engines and extend its endurance. Mounting the engines in a coupled installation gave an aerodynamically neater arrangement than placing the four engines in separate nacelles along the wing.

To reduce the drag from the engine installation further, Günther sought to dispense with the usual system of drag-producing radiators used to lower temperatures in liquid-cooled engines. Instead, he planned to use the evaporative cooling system pioneered by the Heinkel company. In this system the water coolant was pressurised, allowing the temperature to rise to 230° F (110° C) before steam began to form. The super-heated water was then ducted away and depressurised, at which point steam

Above: V4 (CB+RP, Werk-Nr. 00004) was retained at Heinkel's Rostock-Marienehe airfield for stability tests. Heavier than the previous prototypes and featuring increased internal fuel capacity, at some point during its test programme a large aerial-like rig was erected behind the glazed-cockpit area and another device was attached between the port engine and fuselage (above left). V4 was destroyed over the Baltic near Ribnitz while being flown by the Rechlin test pilot Ursinus. The cause of the loss was the malfunction of the airscrew pitch control mechanism.

Left: The fifth prototype (V5, PM+OD, Werk-Nr. 00005) was to be used for armament tests at Rechlin and thus featured the triple bomb bays and hand-operated 7.9-mm MG 15 machine-guns in the extreme nose, the front of the gondola, in a turret aft of the flight deck and in the extreme tail. It was lost in early 1941 when both DB 606 engines burst into flame during a simulated low-level attack, before hitting the ground and exploding.

Type Analysis

Left: V7 (SF+TB, Werk-Nr. 00007) featured a revised reinforced nose section with fewer glass panels, that was also fitted to V6. While V6 was armed with MG 131s, V7 featured 20-mm MG FFs in the gondola and dorsal turret but retained the MG 131 in the nose. It was the first He 177 with the shorter fuselage (from 67 ft 6½ in to 66 ft 11 in) that was incorporated into the pre-production aircraft. V7 (along with V6) was sent to conduct operational trials with IV./KG 40 at Bordeaux-Mérignac from August 1941 (above). The tests highlighted many deficiencies in the design, and led to a lengthy programme of modifications to the heavy bomber.

Left: The He 177 V8 (SF+TC, Werk-Nr. 00008) was the last prototype to be built as such from the outset, with all the other Versuchs He 177s being modified from pre-production or production aircraft. Apart from the tail gun (an MG 131), the V8 was armed with the MG 15 in all positions. For 40 days from September 1941 it was used at Rechlin for engine development testing before returning to Heinkel.

formed. The water was separated and returned to the motor, while the steam was pumped to condensers in each wing that were cooled by the airflow. When the steam condensed, that water was also returned to the engines. This system worked reasonably well when tested on small engines, but even before the design of the new bomber was finalised it became clear that it could not dissipate the amount of heat generated by the big DB 606 engine. Guenther was forced to revert to conventional radiators for the new bomber, with their attendant drag.

For self-defence, the bomber was originally to have had three remotely-controlled gun barbettes positioned above and below the fuselage, and a hand-operated gun position in the tail. Compared with a manned powered gun turret, the remotely-controlled barbette traded technical complexity for greatly reduced drag and an improved field of fire. At the time, the remotely-controlled gun barbette was too novel a concept for the Luftwaffe to accept, however. Günther had to alter his design to use manned gun turrets, instead, and accept the resultant drag penalties.

While those changes were in progress, the Luftwaffe Technical Office issued a requirement that the new heavy bomber be able to deliver attacks in a 60° dive. Without doubt, there were great advantages to be gained if that could be achieved. The contemporary Luftwaffe horizontal bombsight, the Goetz vizier, gave poor results during attacks from medium and high altitude. Dive-bombing was the much more accurate method, especially against a moving target such as a warship. Yet the pullout from the dive would place great strain on the airframe, so it had to undergo considerable strengthening – which incurred much extra weight – to cope with the greatly increased loads.

Those various changes combined to push the P.1041 design into a vicious spiral. Each increase in drag meant the bomber would fly more slowly for a given power setting. The aircraft burned more fuel to cover a given distance, so to meet its range specification its fuel tankage had to be increased. And at each stage the airframe had to be strengthened further to take this extra weight. Almost every change to the design gave another twist to the spiral of more weight or more drag – less speed – less range – more fuel needed – more weight incurred.

Eventually, the increases in weight became too great for the main undercarriage in its original form, which had a single undercarriage leg

The eight prototypes were followed by 35 pre-production He 177A-0s, with 15 being produced both at Heinkel's Rostock-Marienehe and Oranienburg factories, and five by Arado at Warnemünde. The first (A-01 DL+AP, Werk-Nr. 00016) flew in November 1941, while A-02 (DL+AQ, Werk-Nr. 00020, above and left) was used for engine tests from 8 February 1942. A-02 was lost in May 1942 when its engines caught fire, its crew escaping from a crash-landing seconds before the aircraft exploded. All of the A-0 series had a crew of five (as did the V8) and were armed with one 7.9-mm MG 81 in the nose glazing, a 20-mm MG FF cannon pointing forward in the ventral gondola, two MG 81s in the rear of the gondola, a 13-mm MG 131 in the dorsal turret and another in the tail.

Heinkel He 177 Greif

He 177A-03 shows the clean lines of the bomber and the unusual undercarriage, the two legs retracting either side of the DB 606 mounting. This solution was required to counter the increase in the weight of the aircraft from the early design period to operations, and the need to keep the size of the main wheels within the internal dimensions of the wing.

below each engine nacelle. So Günther introduced an unusual double leg system, with each outer leg retracting into the wing outboard of the engine nacelle, and each inner leg retracting into the wing inboard of the nacelle. Of course, that change added yet more weight to the airframe.

Although daunting, such a range of problems was not uncommon during the design stage of a large combat aircraft. The Luftwaffe Technical Office expressed confidence in the revised design, and in November 1938 it awarded Heinkel an order to build six prototypes. The type now received its official designation: Heinkel He 177.

Into the air

Almost exactly a year later, on 19 November 1939, the prototype He 177 V1 made its maiden flight. Dipl. Ing. Franke, chief of the Rechlin flight test centre, flew the bomber for 12 minutes before engine overheating forced an abrupt return to base. Apart from the engine problem Franke thought the bomber handled satisfactorily, though he complained of some aileron flutter, inadequate effect of the rear flying surfaces and undue vibration from the airscrew shafts. Following the initial flight, the tail surfaces of the prototype were increased in size. In terms of performance, the prototype was disappointing. Its maximum speed of 286 mph (460 km/h) was some 50 mph (80 km/h) below specification, while its calculated range, at just over 3,100 miles (4988 km), was about one-quarter less than specified.

The He 177 V2, the second prototype, joined the V1 in the test programme soon afterwards. This aircraft was generally similar to the V1, but when it began its flight trials it had the original tail surfaces. The V2 made the early diving tests, and during one of these it developed severe control flutter and broke up in mid-air, killing the pilot. Following this incident the V3, V4 and V5 prototypes, then nearing completion, were all fitted with the enlarged tail surfaces.

The diving trials resumed, but during one of these the V4 suffered a malfunction of the propeller-pitch mechanism. It failed to pull out of its dive and crashed into the Baltic. Early in 1941 the V5 suffered a double engine fire during a simulated low-level attack, and crashed into the ground and exploded. Thus three of the first five prototypes, with their crews, had been lost during the early test programme. Already, the new bomber was gaining notoriety as a crew killer.

In its efforts to cure the new bomber's failings, the Heinkel Company introduced several design changes. So did Daimler-Benz, but the problem of the DB 606 engines overheating and catching fire proved more intractable. Due to poor lubrication, the connecting rod bearings were liable to seize up, in which case the connecting rod sometimes smashed through the crankcase. If lubricating oil spilled through the hole and fell on the hot engine exhaust, that could lead to a fire. Moreover, if the pilot handled the throttles roughly, the fuel injectors

Below: The cockpit of the He 177 (this is an A-5/R6) had several unusual features including the centre position of the yoke which could be swung to the left or right for use by either the pilot or co-pilot (observer). Each was provided with a set of rudder pedals and engine controls to fly the aircraft. Some of the German instructions have been repainted in English on this aircraft evaluated at Farnborough in late 1944.

Right: The view from the cockpit was excellent in comparison to other bombers of the period, with the glazing being modelled on that used by the glass-nosed versions of the He 111. This He 177 is an A-3/R2 and has the sub-type's MG 151 cannon in place of the MG FF in the ventral gondola, as well as an improved electrical system. Of note is the crew access ladder built into the door in the bottom of the gondola.

Type Analysis

He 177 powerplants

This preserved DB 606 is on display at the Deutches Museum (above). The two DB 601s were inclined towards each so that the inner banks of cylinders were almost vertical (below).

KG 40 maintenance personnel work on the DB 610 powerplant of a He 177A-5. The engine was difficult to work on as the design was tight, with ancillary equipment being packed in close to the engine itself.

The Daimler-Benz DB 606 was an ingenious solution to the problem of quickly developing a high power powerplant without the expense and time needed for a completely new design. By 'bolting' two inverted-V12 DB 601s together, with a common gearbox connected to a single propeller, a powerplant of around 2014 kW (2,700 hp) for take-off was produced. This solution also allowed the manufacturer to concentrate on effectively building one core engine design, the DB 601, instead of two types. DB 606As were fitted to the port wing of the He 177 and Bs on the starboard, the designations indicating 'handed' engines. The DB 610 (coupled DB 605s) rated at 2200 kW (2,950 hp) was flown on the He 177 V15 and V16 before powering the He 177A-5 series (although the small number of He 177A-3/R5s produced also used the powerplant).

It was planned that the He 177's engine was to rely on surface evaporation, but this was changed to orthodox radiators early in the development phase. Even so, engine cooling problems were to plague the He 177, with one of the design's more severe problems being the tendency of its DB 606s to burst into flames. In service the He 177 was known as the Luftwaffenfeuerzeug (Luftwaffe's Lighter).

There were several features of the DB 606 that increases its flammability. The common exhaust manifold on the two inner cylinder blocks would become excessively hot, causing any oil or grease accumulation at the bottom of the engine cowling to burst into flames. The injection pump had a tendency to pump more fuel than was required by the engine when the pilot throttled back. The connections for the injection pumps also leaked.

The engine did not have a firewall, omitted in order to save weight, and the engines were attached next to the main spar. This resulted in a tight squeeze to get the fuel lines, electrical leads and other ancillary equipment in, increasing the likelihood that if a fire started it would soon become critical.

At altitude the oil used to lubricate the engine had a tendency to foam, partly because the pump used to return the oil back to a reservoir was too large. Once in that state, it no longer lubricated the engine, resulting in the disintegration of the connecting rod bearings that would break through the engine crankcase and puncture oil tanks, resulting in a fire.

The loss of He 177A-02 in February 1942 to engine fires resulted in recommendations that the engine mounting be lengthened by some 8 in (20.32 cm); fuel and oil pipes be relocated; the oil tank moved to a less dangerous position; and a redesign of the exhaust system. Only the oil tank was moved, as this would have a minimal impact on production schedules, although the exhaust system was redesigned when operational units required flame dampeners for night sorties. The engine mountings were lengthened when the He 177A-3 entered production.

An investigation into He 177 engine fires highlighted 56 different causes. In January 1943 the Erprobungsstelle at Rechlin modified a He 177 to conform to the recommendations made and it flew safely without any engine problems whatsoever over a protracted period. Unfortunately, by this time production of the aircraft would have been seriously delayed if the improvements had been carried out – the solution had come too late for the He 177. One final way of overcoming the aircraft's problems was investigated. Heinkel re-engined the airframe with four DB 603As as the He 277 (He 177B), but only eight production aircraft were completed before the programme was abandoned.

David Willis

might leak petrol which ran to the bottom of the engine bay. If petrol fumes wafted over a hot exhaust pipe, that, too, could lead to a fire.

The pre-production batch of He 177s – the A-0 series totalling 35 aircraft – was used to explore various aspects of the bomber's performance envelope and its technical features. These aircraft carried a defensive armament of a power-operated dorsal turret to the rear of the cabin, with three flexible machine-gun mountings in the nose and one more at the tail.

To get the He 177 into a dive, the pilot had to reduce speed before slowly applying pressure on the control wheel to get the nose to drop. Almost as soon as the bomber was established in the dive, it was time to initiate a relatively gentle pull-out to avoid overstressing the airframe. Even so, rivets were popped on several occasions, necessitating lengthy repairs. As the He 177 test programme progressed it became clear that, requirement or not, the

Throughout its life the He 177 remained intolerant of rough treatment, and more were lost in accidents than while flying sorties against the enemy. This A-5 (left) crash-landed, and its starboard engine has torn away from its mounting and is lying behind the trailing edge of the wing. Because of the tendency for the engines to catch fire, a large percentage of the accidents ended in the aircraft being consumed by flames (above).

Heinkel He 177 Greif

He 177 tail turrets

Above and below: Two views show the rear gun position fitted to He 177A-3s (A-3/R2 above) and A-5s, with an MG 151 20-mm (0.787-in) cannon. The cone above the weapon housed the illuminated gunsight for night operations. Protecting the gunner was a laminated glass windscreen, an 18-mm (0.7-in) thick steel gun mounting and a slab of 9-mm (0.35-in) armour under the gun. As can be seen, his view to the rear was poor and the sides of the compartment limited the field of fire of the hand-operated weapon.

Above: Several ways of improving the rear turret were tested. This mock-up had two MG 151 20-mm cannons. It offered a much improved view to the rear for the gunner and a better field of fire. Another mock-up (below) was fitted with four MG 131 13-mm machine-guns.

aircraft was far too big to be a dive-bomber.

Meanwhile, progress in another area had lessened the need for the dive-bombing capability. The new Lotfe tachometric bombsight (working on the same principal as the American Norden) was being introduced into the Luftwaffe. This promised bombing accuracies in horizontal attacks similar to those obtained during diving attacks. The Luftwaffe dropped its requirement for the He 177 to dive, and the wing-mounted dive brakes were omit-

Much time and effort was expended developing the defensive systems of the He 177. This diagram shows the remotely-controlled gun barbette fitted to the He 177A-1, with a single MG 131 13-mm (0.51-in) machine-gun. In the later A-3 and A-5 variants the barbette was enlarged to carry a pair of MG 131s. The key to the numbers is: 1 – sighting station, with gunsight and operating handles; 2 – gun barbette, situated about 12 ft (3.65 m) from sighting station; 3 – main power amplifier, with 1.5 hp (1.11 kW) motor; 4 – MG 131 13-mm machine-gun; 5 – ammunition feed.

ted from aircraft built after the initial pre-production batch.

The first production variant, the He 177A-1, featured a redesigned fin and rudder with the rounded top, which would be a feature of all subsequent sub-types. The A-1 carried a

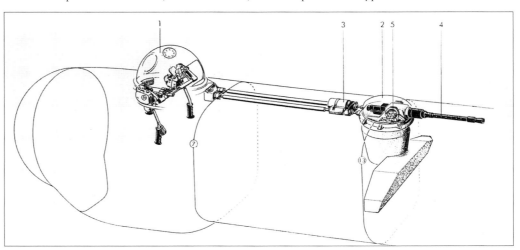

Type Analysis

The He 177 had been unable to meet the requirements that it was designed to meet in several areas, including range. Although not designed specifically for the bomber, He 177A-3 TM+IU (above) was engaged in trials of a rigidly-towed auxiliary fuel tank (right), complete with its own spatted undercarriage and thick aerofoil section (top right). The Luftwaffe tested this system using several different aircraft types, but it was never used operationally. On the He 177 it would have ruled out any use of the tail gun.

remotely-controlled barbette with a 13-mm (0.51-in) MG 131 machine-gun mounted above the fuselage, aimed from a sighting position above the rear of the cabin. The flexible gun positions remained unchanged, as did the rear mounting.

Between March 1942 and June 1943, 130 examples of the A-1 left the production line at Warnemuende. Myriad technical problems remained, however, of which the most serious was that of the disconcerting number of engine fires. Of the 130 A-1s produced, 19 were lost in accidents – and the bomber's combat career had not yet begun.

Initial operations

In the summer of 1942, 34 He 177A-1s were sent to I. Gruppe of Kampfgeschwader 40, based at Bordeaux-Mérignac in France, to test the aircraft under operational conditions. After a few weeks the unit pronounced the type unsuitable for operational use in its current form, and the A-1s were withdrawn. Following that débâcle, the majority of A-1s were relegated to second-line duties.

One oft-repeated canard has it that during August and September 1942, single He 177s flew occasional high-altitude raids on targets in southern England. In fact the 'He 177s' reported by RAF pilots were Junkers Ju 86R stratospheric bombers, two of which operated from Beauvais in northern France.

The next production version, the He 177A-3, had the fuselage lengthened by 5 ft 3 in (1.6 m) and the engine nacelles moved forwards by 8 in (20.3 cm), to improve handling. The armament included an additional manned gun turret in the dorsal position, with a single 13-mm (0.51-in) MG 131 machine-gun. Also, the remotely-controlled barbette farther along the fuselage was enlarged to house two 13-mm guns instead of one. The new variant incorporated several smaller changes aimed at improving handling and reducing the causes of engine fires.

The first 15 A-3s built were fitted with DB 606 engines. The 16th and subsequent aircraft were fitted with the more powerful DB 610, which comprised two DB 605 engines coupled together in the same manner to develop 2,950 hp (2200 kW).

Late in 1942, I. Gruppe of Ergänzungs Kampfgeschwader (Replacement or Reserve Training Bomber Geschwader) 50, based at Brandenburg/Briest, converted to the He 177A-3. Following a period of hurried training, the unit was redesignated as I./KG 50 and sent to Zaporozh'ye in Russia for winter trials. Soon after its arrival the German 6th Army was surrounded at Stalingrad, and the unit was ordered to join the desperate airlift to fly supplies into the besieged city. The new bomber was quite unsuitable for the transport role, however, for its fuselage offered a relatively small volume for the stowage of supplies and little room to carry wounded men. Flying from primitive forward airfields subjected to the full rigours of the Russian winter, the dozen or so He 177s achieved little. After a few resupply missions, the unit reverted to the bombing role. Seven He 177s were lost in the course of 13 missions, none attributable to enemy action. In February 1943 the last of the German troops at Stalingrad surrendered, and I./KG 50 withdrew to Germany with its surviving He 177s.

In total, 170 He 177A-3s were built before production switched to the A-5 in February 1943. Like its predecessor, the new variant was powered by DB 610 engines, although they had further modifications to reduce the risk of

The Starrschlepp ('rigid-tow') method was also tested with a He 177A towing a Go 242A-1 with its undercarriage still attached (the A-1 usually dropping its mainwheels and landing on skids attached to the underside of the glider). The usual tug for the Go 242 was the He 111H, and the He 177A was not used operationally in this role.

engine fires. Other changes included a revised wing flap mechanism and strengthening of the outer wing structure to permit the carriage of heavier external weapons. During 1943, the factories at Oranienburg and Warnemünde turned out 261 examples of the A-5.

In action over the Atlantic

Although the mechanical problems of the He 177 were not all solved, the operational need for the long-range bomber had now become urgent. The crews would have to make the best use they could of the new aircraft when they took it into action.

The He 177 had long been earmarked to replace the Focke-Wulf Fw 200 patrol bombers of the anti-shipping unit KG 40 based at airfields on the French Atlantic coast. After its previous unsuccessful encounter with the new bomber, II. Gruppe began converting to the He 177 in the summer of 1943. Its aircraft were to go into action carrying the latest German

Heinkel He 177 Greif

Kampfgeschwader 40

Major KG 40 units to see action with the He 177:

I. Gruppe of Kampfgeschwader 40.
Flew bombing attacks against Great Britain, operating from bases in France and Germany from January 1944. Later in 1944 part of the Gruppe moved to Norway, where it was disbanded.

II. Gruppe of Kampfgeschwader 40
The longest-serving unit operating the He 177, this Gruppe flew the aircraft in the anti-shipping role from airfields in France from the autumn of 1943 to the time of the Normandy invasion in June 1944, before moving to Norway where it was disbanded. (Identification code F8).

Above: A He 177A-5/R6 of II./KG 40 sits with its engines running and the aircraft crew waiting to board.

Left: Around 14 aircraft are visible in this view of II./KG 40 He 177s with Hs 293 anti-shipping missiles under the wings at Bordeaux-Mérignac in June 1944.

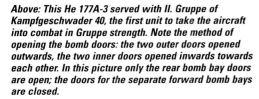

Above: This He 177A-3 served with II. Gruppe of Kampfgeschwader 40, the first unit to take the aircraft into combat in Gruppe strength. Note the method of opening the bomb doors: the two outer doors opened outwards, the two inner doors opened inwards towards each other. In this picture only the rear bomb bay doors are open; the doors for the separate forward bomb bays are closed.

anti-shipping weapon, the Henschel Hs 293 glider-bomb. The He 177A-3 and A-5 variants had provision to carry two glider-bombs, one under each outer wing section, or one of these weapons under the fuselage.

On 21 November 1943, the He 177 flew its first anti-shipping attack operation. Major Rudolf Mons, commander of II./KG 40, led 25 of these bombers into the air from the unit's base at Bordeaux-Mérignac. Their target was a large Allied convoy located some 420 miles (676 km) northeast of Cape Finisterre, Spain, and heading north. Each He 177 carried two glider-bombs.

In the area of the convoy, low cloud hindered the search for the ships. Some bombers concentrated their attentions on two straggling merchantmen, the *Marsa* (4,505 tons) and the *Delius* (6,066 tons). Both ships took hits from glider-bombs. The former sank, but the latter reached port under its own power.

Meanwhile, a remarkable fight had broken out as other He 177s tried to attack the main body of the convoy. Pilot Officer A. Wilson and his crew, flying a Liberator of No. 224 Squadron, RAF Coastal Command, was part of the convoy escort. As Heinkels tried to move into position to launch their glider-bombs, Wilson blocked their path at each opportunity. The Liberator caused several of the missiles to be abandoned after launch, and no ship in the main body of the convoy was hit. The Liberator suffered no damage from the return fire.

It was an undistinguished start to the He 177's operational career. One bomber was lost without trace near the convoy and a second crashed near its base on the return flight, reason unknown. A third He 177 ran out of fuel on the way home and the crew had to bail out. At Mérignac one bomber made a belly-landing and suffered serious damage, and four more returned with battle damage. Three aircraft and two crews lost, and one aircraft seriously damaged, was a poor exchange for one small ship sunk and another damaged.

Five days later, on 26 November, II./KG 40 struck again. On this occasion, 21 Heinkels carrying glider-bombs set out to attack a convoy off the coast of Algeria. Shortly after take-off one bomber suffered a broken crankshaft in one of its engines; it made a crash landing and burned out. Worse was to follow.

Centre: This II./KG 40 He 177A-3's nose is fitted with the antenna of the FuG 200 Hohentwiel ship-search radar. The radar operated on frequencies in the 550 MHz band, and it could detect a large ship or a convoy at a range of up to 50 miles (80 km).

Above: Officers of KG 40 are pictured during an award presentation ceremony early in the war. Second from the left is Rudolf Mons who, with the rank of Major, led the first two attacks by He 177s against Allied shipping in November 1943. Mons was killed leading the disastrous second attack on 26 November.

When the He 177s arrived over the convoy, they had to fly through a vigorous gun barrage from the ships. Then Allied fighters engaged: Spitfires of the French squadron GC 1/7, P-39 Airacobras of the US 350th Fighter Group, and Beaufighters of No. 153 Squadron, RAF. In the mêlée that followed six Heinkels were shot down, including that flown by Major Mons. For their part, the German bombers scored one and possibly two glider-bomb hits on the troopship *Rohna*, which sank rapidly. The approaching darkness impeded rescue work and more than 1,000 US soldiers – more than half of those on board – lost their lives.

The surviving Heinkels returned to find poor weather over their base at Mérignac. Two more He 177s were wrecked in landing, though their crews survived. Taken together, those first two

Above: An He 177A-5/R2 carries an Hs 293A under the fuselage hardpoint located under the position of the first of the bomb bays. The Hs 293 provided the He 177 (and other carriers) with a 10- to 15-km (6- to 9-mile) stand-off ability when attacking shipping. An operator in the He 177 guided the Hs 293 optically, following the flares attached to it, and sent guidance instruction through a Knüppel joystick attached to the FuG 203b Kehl III transmitter to the FuG 230b Strassburg receiver in the Hs 293 to effect course changes. Daylight attacks were dangerous as carrying the missile slowed the He 177 down, so night attacks were practised, using a flare-dropping aircraft to silhouette targets, a requirement for optical guidance. A launch is shown on the right.

attacks with He 177s had cost II./KG 40 12 aircraft destroyed and one seriously damaged, as well as eight crews.

It was obvious that the defences protecting Allied convoys in the Atlantic and the Mediterranean were too powerful for II./KG 40 to attack them by day without risk of prohibitive losses. Hauptmann Dochtermann assumed command of the unit, and from then on the Heinkels restricted their activities to less effective but also less costly night attacks on Allied convoys. The plan was for part of the attack force to drop flares on the far side of the convoy, so the other aircraft could launch their glider-bombs at the ships thus silhouetted. The technique required a high degree of co-ordination between the flare-carrying aircraft and the missile-carriers, however, and it achieved few successes.

On 22 January 1944, Allied forces landed at Anzio south of Rome. For a week or so II./KG 40 flew several sorties against the concentration of shipping off the coast, but here, too, the Allied defences were strong. Again the unit suffered serious losses and achieved little. Thereafter, II./KG 40 resumed its night harassing operations against convoys passing through the Mediterranean.

In January 1944 the Luftwaffe girded itself for another bout of attacks on Britain's cities, code-named Operation Steinbock (Ibex). Two newly converted He 177 units were earmarked to take part in the attacks, I./KG 40 and 3. Staffel of KG 100. Initially with a combined strength of 46 aircraft, the two units operated together from airfields at Rheine near Osnabrück and Chateaudun near Orléans.

The first attack in the new series, against London, took place on the night of 21 January. The total German effort for the night amounted to some 270 sorties. The bombers attacked in two waves, with around eight hours between each wave. About a dozen He 177s flew with each wave, and some aircraft flew twice.

During the attack by the first wave, Flying Officer H. Kemp of No. 151 Squadron, RAF, flying a Mosquito, shot down the first He 177 to fall on British soil. He was heading towards a searchlight cone when his radar operator, Flight Sergeant J. Maidment, picked up a target about 2 miles (3.2 km) away and in front. Kemp closed in and briefly caught sight of the bomber making a violent evasive turn. For a time the Mosquito crew lost contact with the bomber, but then Maidment regained it. Kemp closed in and opened fire with his four 20-mm (0.787-in) cannon, and saw hits on the port wing. There was a violent explosion, then the bomber went into a steep dive and crashed near Haslemere in Surrey. Two of the crew, from I./KG 40, were killed and the other four were taken prisoner. After attacking with the first wave, the He 177s landed at Chateaudun.

During the second attack wave I./KG 40 lost another He 177 to fighter attack, and 3./KG 100 lost one to 'friendly' flak near Dieppe. After attacking, the surviving He 177s landed at Rheine where two of them suffered damage in crash landings.

The He 177 carried the heaviest bomb load of the German aircraft taking part in Steinbock, up to 12,345 lb (5600 kg) being carried to a nearby target like London. The pace of He 177 operations during the remainder of the first month of the operation is shown below:

- Night of 29/30 January: some 285 aircraft took off to attack London. About 20 He 177s of I./KG 40 and 3./KG 100 took part, taking off from Rheine and returning there without loss. Following this action I./KG 40 withdrew to

He 177 colours

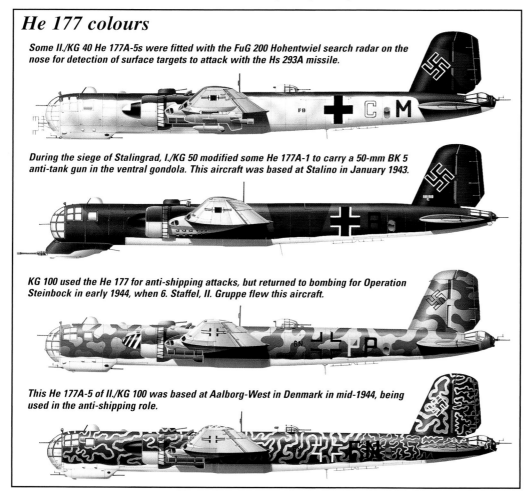

Some II./KG 40 He 177A-5s were fitted with the FuG 200 Hohentwiel search radar on the nose for detection of surface targets to attack with the Hs 293A missile.

During the siege of Stalingrad, I./KG 50 modified some He 177A-1 to carry a 50-mm BK 5 anti-tank gun in the ventral gondola. This aircraft was based at Stalino in January 1943.

KG 100 used the He 177 for anti-shipping attacks, but returned to bombing for Operation Steinbock in early 1944, when 6. Staffel, II. Gruppe flew this aircraft.

This He 177A-5 of II./KG 100 was based at Aalborg-West in Denmark in mid-1944, being used in the anti-shipping role.

Heinkel He 177 Greif

Kampfgeschwader 1

Major KG 1 units to see action with the He 177:

I., II., and III. Gruppen of Kampfgeschwader 1
Based in East Prussia, these units operated over the Eastern Front during the summer of 1944, until the fuel famine halted operations in August.

Above: Oberstleutnant Horst von Riesen commanded KG 1, the only He 177 unit to go into action in full Geschwader strength.

Below: In the late spring of 1944 I./KG 100 was withdrawn from the Western Front, redesignated III./KG 1 and sent into action on the Eastern Front. Although the unit operated by day, several of its aircraft – such as those depicted below and right – retained their night camouflage. Five KG 1 He 177s (right) are seen flying over the Eastern Front during daylight in the late spring or early summer of 1944. Soon after the lack of fuel greatly affected KG 1 and reduced its sortie numbers.

Above: The first two letters of its tactical code of this He 177A-5 of III./KG 1 have been greatly reduce in size. Note the non-standard two-gun rear mounting on this aircraft. The weapons were probably MG 131s.

By the end of 1943 more than 400 He 177A-3s and A-5s had been built, yet the number in front-line service at this time was derisively low. Relatively few were lost during Steinbock, but the heavy bomber's continuing unreliability meant it could play only a relatively minor part in the operation.

Adolf Hitler had scathing words for the heavy bomber. Addressing Luftwaffe Chief of Staff Generaloberst Günter Korten at a conference, the Führer pungently commented: "One gets the impression that once again the Heinkel 177 has suffered 50 per cent breakdowns. They can't even get that far [London]. This rattletrap is obviously the biggest load of rubbish that has ever been built. It is the flying Panther; and the Panther is a crawling Heinkel!" The Panther tank, to which he referred, also suffered numerous mechanical breakdowns at that time.

He 177 tactics

Operation Steinbock continued through the rest of February and March, and petered out in April 1944. The final large-scale attack on London by manned bombers took place on the night of 18/19 April. About 125 bombers were involved, including five He 177s from I./KG 100. The tactics of the latter, described below, were typical of those employed during Steinbock.

For this attack, I./KG 100 operated from Rheine. Its more experienced aircrews flew

Germany and detached one Staffel to Norway to operate in the maritime patrol role.

■ Night of 3/4 February: some 240 aircraft set out for London. About half a dozen He 177s from 3./KG 100 took part in the attack, taking off from Rheine and landing there without loss. After this attack the He 177s of 2. Staffel of KG 100 were declared operational, and joined those of 3. Staffel at Rheine. Henceforth the two units operated together as I./KG 100.

■ Night of 13/14 February: some 200 aircraft set out for London. I./KG 100 put up 17 He 177s from Rheine and landed there without loss.

■ Night of 20/21 February: some 200 aircraft set out to attack London. Fourteen He 177s of I./KG 100 took off from Chateaudun and returned to Rheine. During the transfer flight to Chateaudun one He 177 suffered an engine failure and was wrecked in the subsequent crash landing.

Type Analysis

Kampfgeschwader 100

Major KG 100 units to see action with the He 177:

I. Gruppe of Kampfgeschwader 100
Flew bombing attacks against Great Britain, operating from bases in France and Germany from January to May 1944, when it was redesignated as III./KG 1 (see previous page) (Identification code 6N).

Right, below and bottom: These three pictures, probably taken at Rheine early in 1944, depict the same He 177A-3: 6N+SK belonging to 2. Staffel of KG 100. Painted in night bomber camouflage, this aircraft took part in the Operation Steinbock raids on London and other British cities. Note that although there is a fuselage cross on the starboard side of the aircraft, no such cross is visible on the port side.

aircraft loaded with two 1800-kg (3,968-lb) and two 1000-kg (2,205-lb) bombs; the remaining crews carried four 1000-kg bombs. Before take off, the aircraft lined up at one end of the runway in order of weight. The more heavily laden bombers went first, because they climbed more slowly than the others. Each aircraft took off with its tail light switched on and as it left the ground the light was extinguished, as a signal to the next He 177 in line that the runway was clear.

After take-off the bombers headed for the radio beacon at Noordwijk, northeast of The Hague in Holland, which served as the concentration point for the entire attack force. The bombers then continued over the North Sea, past an on-track fixing point comprising six flame floats laid by 'pathfinder' aircraft. As it neared the coast of England, each bomber began releasing metal foil strips (code-named Düppel in the Luftwaffe) at a rate of two bundles per minute, to confuse the British ground and airborne radars. The stream of bombers crossed the coast just north of Orfordness, flying in a gaggle at altitudes around 17,000 ft (5182 m). The bombers continued on to their next turning point, just east of Newmarket, marked with four red parachute flares. From there they turned on a south-southwesterly heading for their target: the Tower Bridge area of London.

On their way to the capital the bombers made frequent changes of altitude and heading, to make night-fighter interception and anti-aircraft gun engagement more difficult. During this phase of the action one He 177 was intercepted and shot down by a Mosquito of No. 410 (Canadian) Squadron, and crashed near Saffron Walden.

As each bomber neared the target, its pilot throttled back and began a shallow descent to 15,500 ft (4724 m). At the same time the crew increased the rate of Düppel release to one bundle every five seconds, to complicate the prediction problem for defending radar-laid anti-aircraft guns.

At the target, pathfinder crews had laid down clusters of red parachute flares. Each raider levelled off for a minute or so to make its bombing run. Then, immediately after bomb release, each pilot commenced a full throttle descent at about 600 ft (183 m) per minute to build up speed rapidly. During their high-speed withdrawal across Kent, with frequent heading changes to avoid the prowling RAF night-

Above: He 177 A-3s of II./KG 100 are pictured during a training flight. This unit was one of those in the process of converting to the aircraft in the summer of 1944, when the fuel famine brought an end to heavy bomber operations. The same aircraft is seen having its compass aligned (above left).

Left: The 6. Staffel of KG 100 was based at Toulouse-Blagnac during May 1944. This He 177A-5/R2 still wears its factory codes and is finished in a camouflage scheme applied to Greifs used in the anti-shipping role.

Heinkel He 177A-5/R2 Greif
4. Staffel, II. Gruppe of Kampfgeschwader 100 'Viking' Bordeaux-Mérignac, France 1944

Crew accommodation
The crew comprised six men: pilot, observer, wireless operator/dorsal barbette gunner, flight engineer/ventral gunner (all in the main cabin), dorsal turret gunner and tail gunner. The pilot and observer sat side by side, and the centrally-mounted control column could be swung to the left or to the right, for operation by either. The dorsal and rear gun turrets accommodated the remaining two crew members. The rear gunner could either sit or lie prone; although sitting was more comfortable, in that position his field of fire was limited.

Defensive armament
The defensive weapons on this model consisted of the following: one hand-held 20-mm (0.787-in) MG 151 cannon firing forwards from the lower part of the nose; one hand-held 7.9-mm (0.31-in) MG 81 machine-gun firing forwards from the upper part of the nose; a pair of hand-held MG 81s firing rearwards from the lower rear of the main cabin; a pair of 13-mm (0.51-in) MG 131 heavy machine-guns in the remotely-controlled dorsal barbette; a single MG 131 in the powered dorsal turret; a single handheld 20-mm (0.787-in) MG 151 cannon in the rear turret.

Offensive armament
During Operation Steinbock the He 177s flown by the more experienced crews carried two 1800-kg (3,968-lb) and two 1000-kg (2,205-lb) bombs. The other aircraft carried four 1000-kg (2,205-lb) bombs.

Armour protection
The He 177 carried extensive armour protection for its crew. The pilot's seat had 9-mm (0.35-in) thick armour plate at the back and 6-mm (0.24-in) on the seat. The 'chin' gun position had 7-mm (0.28-in) and 6-mm (0.24-in) armour and bullet-proof glass; the rear ventral gun position had 9-mm (0.35-in) armour plate. A slab of 10-mm (0.39-in) thick armour plating protected the gunner in the remote sighting position. In addition, slabs of armour plating were fitted to the fuselage around the forward crew position. The mid-upper gunner's position was protected from the rear by 7-mm (0.28-in) armour, which also protected the dinghy compartment; to the rear of this compartment was a large semi-circular section of armour, which extended down the top two-thirds of the fuselage. The tail gunner was protected by an 18-mm (0.71-in) thick armoured gun mounting, a bullet-proof glass screen, and 9-mm (0.35-in) armour underneath the gun.

Specification – Heinkel He 177A-5
Span: 103 ft 2 in (31.45 m)
Length: 72 ft 2 in (22.0 m)
Height: 21 ft (6.4 m)
Wing area: 1,098 sq ft (102 m²)
Weights: empty 37,038 lb (16800 kg); normal loaded 59,966 lb (27200 kg); maximum loaded 68,343 lb (31000 kg)
Maximum speeds: fully loaded 270 mph (435 km/h) at 20,000 ft (6096 m); lightly loaded 303 mph (488 km/h) at 20,000 ft (6096 m)
Maximum cruising speed: 258 mph (415 km/h) at 20,000 ft (6096 m)
Economical cruising speed: 210 mph (338 km/h) at 20,000 ft (6096 m)
Still-air maximum range (carrying two Hs 293 missiles): 3,420 miles (5504 km) less reserves
Service ceiling: 26,250 ft (8000 m)

Flugzeugführerschule B/16

Above: Flugzeugführerschule B/16 (flight leader school B 16, FFS(B) 16), based at Burg bei Magdeburg, used ex-KG 40 He 177A-1s as the operational conversion school for the type. The Arado Company at Warnemünde built 130 examples of the He 177A-1, the first production variant. After tests these aircraft were judged unsuitable for combat, and they were used only for second-line tasks.

Right: This particular FFS(B) 16 He 177A-1 has two horizontally-mounted balloon cable cutters round the front of the extreme nose, and the MG FF 20-mm (0.787-in) cannon in the lower nose gun position.

fighters, some He 177s reached speeds over 350 mph (563 km/h). The bombers crossed the French coast at Boulogne at altitudes around 2,500 ft (762 m), then headed for Rheine.

After Steinbock, many of the units involved were allowed a brief breathing space to reform. They now had to prepare to meet the long-expected Allied invasion on a point in north-west Europe. An exception was I./KG 100: in May it withdrew to Fassberg, where it was redesignated as III. Gruppe of KG 1. We shall return to this unit later in this account.

Invasion

On 6 June 1944, D-Day, the anticipated Allied invasion of France began. On the evening following the landings, the glider-bomb-carrying He 177s of II./KG 40 were out in force. So, too, were Allied night-fighters. Flying Officer F. Stevens and his radar operator, Flying Officer Kellett, in a Mosquito of No. 456 (Australian) Squadron, were particularly successful. Their radar control ship guided them to an incoming bogie (unidentified aircraft) and Kellett made radar contact. He guided Stevens into visual contact some 300 yd (275 m) behind the contact, which was identified as an He 177.

Stevens later reported: "It was easily recognised by its large fin with a bite out of it and the protruding bulbous nose – also by its extreme wing span, seemingly out of proportion to the length of the fuselage. At this stage, glider-bombs (one under each wing outboard of the engine nacelles) and a large four-pronged aerial array on the nose were seen."

Undetected by the bomber's crew, Stevens dropped back about 100 yd (91 m) behind his prey and moved into a firing position. His initial two-second burst with the four 20-mm cannon caused the bomber's port engine to burst into flames. After two more bursts there was an explosion in the fuselage of the He 177, and it dived into the sea.

About 20 minutes later Stevens and Kellett repeated the performance, and shot down another He 177. That night, No. 456 Squadron claimed the destruction of four of the heavy bombers over the beachhead area.

During the battle for Normandy the Luftwaffe anti-shipping units smashed themselves bravely against the powerful Allied defences, but achieved little. In the 10 days following the invasion, only five ships were sunk by air attack. After that, the Luftwaffe abandoned the idea of direct attacks on ships, and instead concentrated on dropping mines off the coast. By the end of June, airdropped mines had sunk nine warships and 17 auxiliary vessels and merchantmen, and caused much delay and inconvenience. Yet such a scale of losses was never going to achieve a decisive effect. In mid-July, the surviving He 177s of KG 40 withdrew to bases at Trondheim and Gardemoen in Norway.

Operations on the Eastern Front

During the spring of 1944 a full Geschwader of He 177s had been forming at Lechfeld in Bavaria: Kampfgeschwader 1 (KG 1), commanded by Oberstleutnant Horst von Riesen. In May I. and II. Gruppen were ready for action, and they moved to the operational airfields of Prowehren and Seerappen near Königsburg in East Prussia. There they were

Heinkel He 177 variants

Heinkel He 177 V1
The first prototype, the He 177 V1 (V for Versuchs, test) was powered by two Daimler-Benz DB 606 engines rated at 2,600 hp (1940 kW) each. The DB 606 comprised two 12-cylinder, liquid-cooled, inverted-vee engines, coupled together and driving a single airscrew. The prototype He 177 carried no defensive armament, and only the aft bomb bay was fitted. On 18 November 1939 the aircraft made its maiden flight, which was cut to 12 minutes due to engine overheating. That fault would trouble the bomber for most of its career. After its initial flight trials, this aircraft was fitted with larger tail surfaces.

He 177 V2
Generally similar to the V1 with the original tail surfaces, except that it had twin bomb bays fitted. During an early diving test it encountered serious flutter of the control surfaces and the machine broke up in mid-air, killing the pilot.

He 177 V3, V4 and V5
After the V2 crashed, the V3, V4 and V5 prototypes nearing completion had their tail surfaces enlarged as in the case of the V1. The V3 went to Rechlin for engine trials. The V4 featured increased fuel tankage, and was retained at Heinkel for stability trials. During a test dive it failed to pull out and crashed into the Baltic. Attempts to recover the wreckage were only partly successful, but it was believed the cause of the loss was a malfunction of the airscrew pitch control gear.

The He 177 V5 was the first prototype to be fitted with armament. It had triple bomb bays and carried one hand-operated 7.9-mm (0.31-in) MG 15 machine-gun in each of the nose, dorsal, ventral and tail positions. Early in 1941, during a simulated low-altitude attack, both engines caught fire and the aircraft crashed. Thus, of the first five He 177 prototypes, three were lost during the flight test phase. It was not a happy omen for the bomber's future.

He 177 V6 and V7
These were similar to the V5, but were fitted with production DB 606 engines developing 2,700 hp (2014 kW). Also, these aircraft were fitted with a redesigned nose section. The V7 carried an armament of one 13-mm (0.51-in) MG 131 machine-gun in the nose and a 20-mm (0.787-in) MG FF cannon in the lower nose position. In the autumn of 1941, these two aircraft were delivered to Bordeaux-Mérignac in France for operational trials with KG 40 for an initial evaluation in the anti-shipping role. Crews at the unit commented unfavourably on the aircraft.

He 177 V8
This was the last aircraft to be built from the outset as a prototype.

He 177 V9 to V24
These 16 aircraft assigned to the trials role were conversions of pre-production or production aircraft. The He 177A-05, A-06 and A-07 aircraft became the V9, V10 and V11, respectively. The V11, the V15 and V16 were fitted with the more powerful Daimler-Benz

DB 610 engine comprising two DB 605 engines coupled together.

He 177A-0
Fifteen He 177A-0 pre-production aircraft were laid down at Rostock, the first being flown in November 1941. Another 15 A-0s were simultaneously laid down at Heinkel's Oranienburg factory, and five more were produced at Arado's Warnemünde plant. As related above, 16 of the 35 A-0 aircraft were employed as prototypes in a variety of development roles, with much emphasis on engine development. The defensive armament fitted to the A-0 comprised a single power-operated gun turret to the rear of the cabin, three guns on flexible mountings in the nose and one at the extreme tail.

He 177A-1
Between March 1942 and the end of June 1943, the Luftwaffe accepted 130 production He 177A-1 aircraft. This variant carried a remotely-controlled barbette above the centre fuselage mounting a single 13-mm (0.51-in) MG 131 machine-gun, controlled from a sighting blister at the rear of the cabin. The flexible gun positions on the A-1 remained unchanged, however, as did the rear gun mounting.

Several A-1 aircraft were sent to I./KG 40 based at Bordeaux-Mérignac in July 1942, for testing under operational conditions. After two months of spasmodic flying, the aircraft were pronounced unsuitable for operational use and returned to Germany. They were then employed on second-line tasks.

He 177A-3
The He 177A-3, the next major production variant, appeared in the autumn of 1943. This had the fuselage lengthened by 5 ft 3 in (1.6 m) to improve stability, and had modifications to the engines to cure some of the causes of engine fires. The armament was increased by the addition of a manned dorsal turret, mounting a single 13-mm (0.51-in) MG 131 machine-gun. Also, the remotely-controlled dorsal gun barbette was enlarged to house two MG 131s instead of one, and the rear gun position was upgraded to carry an MG 151 20-mm (0.787-in) cannon.

The DB 606 engine powered the first 15 A-3 aircraft. The 16th and subsequent aircraft were fitted with the more powerful DB 610 engine made up of two DB 605 engines coupled together in a similar manner.

This He 177A-3/R5 Stalingradtyp was fitted experimentally with a nose-mounted 75-mm (2.95-in) BK 7.5 cannon, for use in the ground attack role, at Rechlin in the spring of 1943. The trials followed A-1s fitted with BK 5s by I./KG 50 used around Stalingrad.

He 177A-3/R2
This variant featured a redesigned nose gun position, with the 20-mm (0.787-in) MG FF cannon replaced by the more effective MG 151 weapon of the same calibre. The tail gun position was given a domed hood, providing room for the tail gunner to sit upright instead of having to lie prone throughout the time he operated his gun.

He 177A-3/R3 and R4
These were the first variants capable of operating with the Hs 293 guided missile, the latter variant having an enlarged nose section to house the missile control equipment.

He 177A-3/R5
Five aircraft were experimentally fitted with a forward-firing 75-mm (2.95-in) BK 7.5 cannon in a mounting under the nose, for use in the ground attack role. Tests revealed that the installation was unsuitable for operational use and it proceeded no further.

He 177A-3/R7
Three aircraft were modified to carry two L5 aerial torpedoes each. The first machine carried the two weapons externally under the fuselage, while the other two carried one torpedo under each outboard wing section. Tests showed that the Hs 293 guided weapon was a much more effective anti-ship weapon than the torpedo, and neither variant progressed further.

He 177A-5
This, the last major production variant of the bomber, appeared in February 1943. The main difference between it and the late production A-3 was that the Fowler-type trailing-edge flaps were reduced in span, and the ailerons no longer drooped with the flaps. Other changes included a shortening of the main undercarriage legs, and strengthening of the wing to allow the carriage of heavier external loads outboard of the engines.

171

Type Analysis

Heinkel He 274/AAS 01 – high-altitude bomber

Its He 177 roots clearly visible, the AAS 01A is seen at Orléans-Bricy prior to its maiden flight in late 1945.

The DB 610-powered He 177A-4 was intended to be a high-altitude bomber with a pressurised cabin, but otherwise maintaining maximum commonality with the A-3 then being developed. However, as work progressed it became clear the altitude performance sought by the Oberkommando der Luftwaffe could not be met with a variant of the He 177, as an entirely new high aspect ratio wing was required. The adoption of the Heinkel-Hirth 9-2291 exhaust driven turbo-superchargers also resulted in coupled engines having to be abandoned. Redesignated the He 274, work on the aircraft was transferred to the Société Anonyme des Usines Farman (SAUF) at Suresnes near Paris, France, and an order was placed for two prototypes and four pre-production aircraft.

Other changes were also required. The fuselage was lengthened (from 64 ft 3½ in to 73 ft 9 in, and then again with a further 4 ft 3¼ in plug, and a twin tail arrangement was adopted, while the main undercarriage abandoned the twin leg arrangement for a single two-wheel example, retracting rearward into the nacelle. DB 603A-2s with DVL TK 11 turbo-superchargers were selected to power the prototypes, providing 1,900 hp (1417.4 kW) for take-off and 1,450 hp (1081.7 kW) at 36,000 ft (10973 m). The crew had been increased to four (pilot, co-pilot, navigator-bombardier and radio operator).

In late 1943 the pre-production aircraft were cancelled, but work continued (slowly) on the prototypes. He 274 V1 was ready to fly in July 1944, but the approach of the Allied armies necessitated the evacuation of Heinkel's workers from the project, taking all the drawings with them. Explosives were placed on the engines and the airframe was abandoned, but the French restored the damage and replaced the engines. The aircraft was flown for the first time on 30 December 1945 from Orléans-Bricy as the AAS 01A, SAUF having been nationalised as the Ateliers Aéronautiques de Suresnes (AAS) prior to the event. Tests continued at Brétigny-sur-Orge with the Centre d'Essais en Vol (CEV), mainly into cabin pressurisation but also carrying scale models of the SO 4000 and NC270 for test flights.

The He 274 V2 was completed as the AAS 01B, making its first flight on 22 December 1947. It also served with the CEV until a lack of spares grounded it in 1952. Both aircraft were scrapped in 1953, the AAS 01A at Marseilles-Istres.

David Willis

---

joined by III. Gruppe (previously designated I./KG 100). With some 90 He 177s, the Geschwader represented the most powerful long-range striking force possessed by either side on the Eastern Front.

KG 1 had serious problems, however. According to von Riesen, "The airfields were not well equipped, there were no hangars or revetments for the aircraft. Also the logistic support was poor. Fuel was always critically short. There was no reserve at all, it went straight from the tanker trucks into the aircraft. On one occasion we had the aircraft all ready for an operation, we were told that the necessary fuel was on its way. But enemy fighter-bombers caught the train and the fuel never arrived. The operation had to be cancelled."

KG 1 went into action soon after arriving at its new bases, hitting Soviet Army supply centres and troop assembly areas in order to disrupt the offensive expected to commence at any time. During one of the largest raids, von Riesen led his entire Geschwader in an 87-aircraft attack on the important rail centre at Velikye Luki, some 250 miles (402 km) west of Moscow. The force must have made an impressive sight as it attacked in formation in three V-shaped waves, each comprising about 25 to 30 heavy bombers. Von Riesen later recalled: "The formation was carefully worked out beforehand, so that it was as wide as the target. Each bomber carried a full load of 250-kg [550-lb] bombs. I broadcast the release order over the R/T, and all the bombers released their bombs simultaneously in sticks spaced to achieve an even pattern on the ground."

Although the He 177s attacked by day from high altitude, losses were light. The Soviet Air Force, equipped mainly for the low-level interception and ground-attack roles, could do little to hinder the high-flying raiders. A few Soviet pilots tried to engage the heavy bomber formations, but the heavy return fire disconcerted them. Also, unaccustomed to encountering such large aircraft, the Soviet pilots usually opened fire at too great a range to score hits.

During these operations KG 1 crews suffered relatively few instances of engines overheating. The various modifications had reduced the risks of this happening, and all crews had been carefully briefed to avoid the main causes – over-rough use of throttles, and holding high power settings for too long.

On 23 June 1944, the Soviet Army launched its long-expected offensive on the central front. The German forward positions were soon overwhelmed and the Soviet tank spearheads penetrated deeply into the rear areas. Every available fighter and fighter-bomber was thrown into action to slow the Soviet advance.

Then von Riesen was surprised to receive a telephone call from Reichsmarschall Hermann Göring in person. "It was rather a strange procedure, the top man in the Luftwaffe telling a Geschwader commander what he was to do. He said, 'Tomorrow morning you are to attack the Russian tanks from low altitude.' I replied that the 177 was unsuitable for low-altitude

---

Heinkel He 177 variants (continued)

He 177A-5/R2
This variant carried the same revised defensive armament as the He 177A-3 R2 (see above).

He 177A-5/R5
This variant was fitted with a remotely-controlled ventral barbette for two MG 131 machine-guns, fitted aft of the bomb bay. It did not go into production.

He 177 Bomber Destroyer
During the late spring of 1944, five He 177A-5s were converted into bomber destroyers, intended to engage US heavy bomber formations from long range. These aircraft had their bomb bay doors sealed, and in that space were 33 upwards-pointing launching tubes for rockets. The launchers, in three rows of 11, were inclined forward at an angle of 60° and pointed slightly to starboard. The battery was divided into two groups, one of 18 and one of 15 rockets. Either group could be ripple fired, or the full complement of rockets could be ripple fired in one go. The intended tactic was for the He 177 to manoeuvre into position about 6,000 ft (1829 m) below and slightly to port of a US heavy bomber formation, and launch its rockets into the formation.

In June 1944 the modified He 177s arrived at Erprobungskommando 25 based at Tarnewitz, the Luftwaffe weapons testing establishment. By that stage of the war US escort fighters ranged far and wide over Germany, and if they had encountered a Heinkel its chances of survival would have been poor. The modified He 177s flew a few operational trials sorties, but they failed to make contact with a US bomber formation on any of them. The idea was then dropped.

He 177A-6
High-altitude bomber version with a redesigned nose housing a pressurised cabin. Only six aircraft were produced, constructed mainly from A-5 components. This variant was to have carried a powered rear turret with four MG 81 7.9-mm (0.31-in) machine-guns.

He 177A-7
Essentially similar to the A-5, this variant featured a 15-ft (4.6-m) extension to the wings to improve high-altitude performance. Six examples were built. Later aircraft in the series were to have been powered by two Daimler-Benz DB 613 coupled engines, each comprising two DB 603G engines and developing 3,600 hp (2686 kW) for take-off.

He 277, alias the He 177B
To overcome the problems inherent with the coupled engine arrangement, the He 277, also referred to as the He 177B, was fitted with four separately-mounted Daimler-Benz DB 603 engines each developing 1,750 hp (1306 kW) for take off. Eight He 277s were built, but only two or three of them flew before the programme ended in the summer of 1944.

The He 277 V1 (NN+QQ) was tested at Vienna-Schwechat in late-1943 as the He 177B-0, the subterfuge being undertaken to delude Göring and the RLM, who were opposed to its development, wanting Heinkel to concentrate on fixing the He 177 instead.

Captured Greifs

He 177A-5/R6 F8+AP (Werk-Nr. 550062), late of II./KG40, was captured by the French resistance at Toulouse-Blagnac in September 1944, being flown to Farnborough by Wg Cdr R J Falk on the 10th of that month in French markings and carrying PRISE DE GUERRE titles on the fuselage (right). Allocated the serial TS439 (above) and RAF roundels, the aircraft made 20 flights with the RAE before damage to one of its DB 610 engines ended its evaluation. It was dismantled and handed over to the USAAF as FE-2100, replacing He 177A-7 Werk-Nr. 550256 (below) that was damaged beyond repair in a ground loop on 28 February 1945. FE-2100's final fate is unknown.

operations, but he insisted. Such a use would have been all right for the Fw 190, but it was a ridiculous way to use the 177. But what could I say? He was the head of the Luftwaffe and I was just an Oberstleutnant. Later I received a confirming order from headquarters. We were to patrol certain roads, and attack any Russian tanks we saw."

Von Riesen sent his He 177s out to patrol in pairs, so that they could support each other with covering fire. Several were picked off by the hordes of Soviet fighters operating over the battle area. On the first day the Geschwader lost about 10 aircraft of 40 committed. Von Riesen commented, "The operation was a complete failure. It achieved no useful result, because the heavy bombers were not manoeuvrable enough to hit small tanks on the move.

The operation was never repeated, but Göring had had his way and I heard no more from him."

During the summer of 1944 the He 177 at last became available in useful numbers on the main battlefronts. Then fate dealt it a final blow. In the late spring of 1944, the Allied bomber forces concentrated their efforts against the German oil industry. The attacks were a spectacular success. Compared with 195,000 tons in May, production of aviation fuel fell to a meagre 16,000 tons in August.

Lack of fuel forced the Luftwaffe to make drastic cuts to its flying. The heavy bomber force, with its voracious appetite for fuel, was the first to suffer. An attack by 80 He 177s on a distant target used about 500 tons of aviation fuel – close to the average daily output of the entire German synthetic fuel industry in August 1944.

There could be no arguing with the simple arithmetic: the heavy bomber operations had to cease. KG 1 returned its bombers to aircraft parks in Germany, then the unit was disbanded. Many He 177s spent the remainder of the war parked in the open on fields, where they added to the scores of ground-strafing Allied pilots.

He 177 production ended in August 1944, after 565 examples of the A-5 version had been completed. That brought total production of the heavy bomber to 1,094 aircraft of all variants.

The He 177 in retrospect

The He 177 was intended to equip the long-range bomber arm of the Luftwaffe, yet that service's demands for a high-performance machine, and one capable of dive-bombing, proved too much for the available technology to deliver. The heavy bomber made its maiden flight in November 1939, but the next four years of its development career were characterised by frequent losses in accidents.

In the late spring of 1944, production of the He 177 reached just over 100 aircraft per month. By then, four heavy bomber Gruppen were operational with the type and two more were about to convert to it. Then the Allied heavy bombers shifted their attack to the German synthetic fuel industry, stifling the heavy bomber force shortly after its birth.

Yet, even had there been no fuel famine, it is difficult to see how the He 177 could have achieved much in combat. This was not a great wartime aircraft to compare with the Avro Lancaster or the Boeing Flying Fortress: they could be operated and maintained by men with comparatively little training, they were relatively easy to fly, they forgave mishandling and they could absorb considerable punishment. The He 177, in contrast, was over-complex, unforgiving of mishandling and vulnerable to battle damage. Soon after its maiden flight it gained notoriety as a killer of its crews, and it never shed that image.

Dr Alfred Price

A German strategic bombing campaign?

The inability of the Luftwaffe to assemble a strategic bomber force to compare with those of the Western Allies was an embarrassment to Hitler. He dreamed of being able to retaliate for the crushing Allied air attacks on the German homeland. Yet, even had the He 177 performed to specification and its production moved at the rate originally planned, could the Luftwaffe have mounted a bomber offensive to compare with those of the Western Allies? There is reason to believe it could not.

If the He 177 had suffered no problems, 500 of these heavy bombers might have been operational by the summer of 1943. Yet, by that time the Luftwaffe already faced a growing shortage of aviation fuel, as supply failed to meet demand during periods of high air activity. Allowing 6 tons of fuel for a medium-range penetration to a target, a force of 500 bombers required about 3,000 tons of fuel per raid, or 18,000 tons if the force mounted six such attacks during a month. Adding 50 per cent extra for training and essential non-operational flying would bring the force's monthly fuel requirement to about 27,000 tons. In 1943 that figure represented about one-sixth of the average German monthly production. As the conflict progressed, the fuel situation got considerably worse. In short, from the summer of 1943, the Luftwaffe lacked sufficient aviation fuel to sustain a large-scale bomber offensive.

INDEX

Page numbers in **bold** refer to illustration captions.

50 KingBird (see Fokker)
340AEW&C (see Saab)
407 (see Bell)
580 (see Convair)
707 (see Boeing)
767 Tanker Transport (see Boeing)
767-200ER (see Boeing)
1900 Airliner (see Beech)
2000 (see Saab)

A

A-1 (see AMX International)
A-1H Skyraider (see Douglas)
A-6 Intruder II concept (see Grumman)
A-6E Intruder (see Grumman)
A-7 Corsair (see Vought)
A-12 Avenger II (see Lockheed Martin/McDonnell Douglas)
A-29 Super Tucano (see EMBRAER)
A-50 'Mainstay' (see Ilyushin/Beriev)
A109 Light Observation Helicopter (see AgustaWestland)
A-X programme: 40
AAS 01A (see Ateliers Aéronautiques de Suresnes)
Abraham Lincoln, USS: **47**, **48**, 51, **52**, **53**, **55**, **56**, 63, 71
AC-130U (see Lockheed Martin)
ACJ (see Airbus Industrie)
AD Skyraider (see Douglas)
'Advanced F-86D/E' proposals (see North American)
Advanced Hawkeye, E-2 (see Northrop Grumman)
Aero
 L-39 Albatros: 11, 12
 L-39C: **12**
Aérospatiale SA 330H Puma: 7
Aerostar/Mikoyan Lancer: 11
A/F-X programme: 40
Agusta MH-68A Stingray: 13
AgustaWestland
 A109 Light Observation Helicopter (LOH): 10
 CH-149 Cormorant: 8
AH-1W Super Cobra (see Bell)
AH-1Z (see Bell)
AH-64D Apache (see Boeing Helicopters)
Airacobra (see Bell)
Airbus Industrie ACJ (Airbus Corporate Jet): 10
Air Combat – Marine Corsairs in Korea: 104-119, **104-119**
Airliner, 1900 (see Beech)
Air Power Intelligence: 4-15, **4-15**
Ajax, HMS: 94
Albatros, L-39 (see Aero)
Alligator, Ka 52 (see Kamov)
Almirante Irizar, ARA: 102
ALX (see EMBRAER A/AT-29 Super Tucano)
AMX International A-1: 6
An-2 *et seq* (see Antonov)
Antonov
 An-2 'Colt': **11**
 An-26 'Curl': 10
 An-72 'Coaler': **11**
 An-124 Ruslan: 9, 16
'ANZAC'-class frigate: 8
Apache, AH-64D (see Boeing Helicopters)
Argentina
 Comando de Aviación Naval (COAN – naval aviation command): 97, 98, **98**, **99**, 100, **100**, **101**
  Escuadrilla Aeronaval de Exploración: 94-103, **94-103**
  Flotilla de Exploración: 94, 95, 98
  Primera Escuadrilla Aeronaval de Sostén Logístico Móvil (EA51): 98, **98**, 100
 Prefectura Nacional Marítima (Coast Guard)
  Servicio de Aviación: 94
Argus, S 100B (see Saab)
ARL-M, RC-7B (see de Havilland Canada)
AS 332 Super Puma (see Eurocopter)
AS 532 Cougar (see Eurocopter)
AS 532A2 Cougar Mk 2 (see Eurocopter)
AS 565 Panther (see Eurocopter)
AT-29 Super Tucano (see EMBRAER)
Ateliers Aéronautiques de Suresnes AAS 01A (He 274): 172, **172**
Atlantic I (see Breguet)
AU-1 Corsair (see Vought)
Austria
 Österreichische Luftstreitkäfte (OLk – air force): 16-17, **16**, **17**
  Überwachungsgeschwader, Fliegerregiment 2: 32-35, **32-35**
Avenger II, A-12 (see Lockheed Martin/McDonnell Douglas)
Avro
 Lancaster: 173
 Lincoln B.Mk 1: 95
 Shackleton: 94

B

B-2 (see Northrop Grumman)
B-17 Flying Fortress (see Boeing)
B-26 Invader (see Douglas)
B-52 Stratofortress (see Boeing)
B80 Queen Air (see Beech)
B200 Super King Air (see Beech)
Badoeng Strait, USS: 106, 107, **107**, 108, **108**, 109-110, **110**, **113**, 114
BAE Systems
 Hawk Mk 115: 9
 Hawk Mk 120: **11**
 Nimrod: 24
 Sea Harrier: 98
Bahia Paraiso, ARA: 98
Bairoko, USS: **113**, **114**
Bataan, USS: **113**, 114, **114**
Beaufighter (see Bristol)
Beech
 1900 Airliner: 79
 1900C: 79
 1900D: 83, 87
 B80 Queen Air: 99, 101, **101**, 103
 B200 Super King Air: **76**, 77, 97, **100**, 101, **101**, 103
 C-12A Huron: 77-78, 81, 83, **84**
 C-12C: **77**, 78, 79, 81, 82, 83, 84, 85, 86, **86**, 87, **87**
 C-12D Super King Air: 80, 81, 82, **82**, 83, 84, 87
 C-12D1: 78, 79
 C-12D2: 78, 79
 C-12F: 78, 79, 81, 83, **83**, 84
 C-12F1: 78
 C-12F2: **77**, 78
 C-12F3: 78, **84**
 C-12J: 79, 81, 83, **83**, 84, **84**
 C-12L Huron: **76**, 77, 79, 93
 C-12M: 87
 C-12R: 78, 81, 83
 C-12R1: 81
 C-12T: 81
 C-12U: 81, 82
 C-21A: 82
 C-21C: 79
 EU-21: 75
 EU-21A: 89, **89**
 JRC-12G: 83
 JU-21A Left Jab: 75, **75**, 82, 84, **84**, 89, **89**
 JU-21H: 93, **93**
 King Air: 74, 75
 King Air B200: 9, 87, **87**
 King Air B200C: 78
 King Air C90: 87
 King Air F90: 87
 NC-12B: 85
 NU-8F: 74, 88
 RC-12: 82
 RC-12D Improved Guardrail V: **79**, 80, **80**, 82, **82**, 83
 RC-12F: 78, **78**, 85
 RC-12G: 81
 RC-12H: 80, 82
 RC-12K: **79**, 80, 82, 87
 RC-12M: **78**, 79, 81, 85
 RC-12N: 80, **80**, **81**, 82
 RC-12P: 80, 82
 RC-12Q: 80-81, **81**, 82
 RU-21 Ute: 75
 RU-21A Crazy Dog (Cefirm Leader): 76, 89-90, **89**, **90**
 RU-21B Cefirm Leader: 76, 90, **90**
 RU-21C Cefirm Leader: 76, **76**, 90, **90**
 RU-21D Laffing Eagle: 75, **75**, 77, 82, 84, 90-91, **90**, **91**, 93
 RU-21E: 76, 77, 82
 RU-21E Guardrail II: 91, **91**
 RU-21E Guardrail IIA: 76, 91
 RU-21E Guardrail IV: 76, 91, **91**
 RU-21E Left Foot: **74**, 75-76, 90, 91, 93
 RU-21G Guardrail I: 77, 92, **92**, 93
 RU-21H Guardrail V: **76**, 77, 80, 82, 91, 93, **93**
 RU-21J Cefly Lancer: 93, **93**
 RU-21J Huron: **76**, 77, **79**
 Super King Air 200: **76**, 77, 97, **100**, 101, **101**, 103
 T-44A Pegasus: 77, 85, **85**, 86
 TC-12B: **78**, 79, 81, 85, 86, **86**
 U-8D: 91, **91**
 U-12F: 75
 U-21A Ute: 74-75, **75**, 79, 82, 88, **88**
 U-21D: 91, **91**
 U-21F: 79, 87, 92, **92**
 U-21G: 75, 76, 77, 92, **92**, 93
 U-21H: 93, **93**
 UC-12B Huron: **74-75**, 78, **78**, 79, 81, 85, **85**, 86, **86**, 87, **87**
 UC-12F: 78, 79, 81, 85, 86
 UC-12M: 78-79, 81, 85
 VC-6A: **76**
 YU-21: 88, **88**
Bell
 407: 10
 AH-1W Super Cobra: 5
 AH-1Z: 5, **10**
 Airacobra: 164
 NAH-1Z: 5
 UH-1N Huey: 5
 UH-1Y: 5, **10**
 X-1: 130
Bell-Boeing MV-22B Osprey: 13
Bell Canada CH-146 Griffon: 18, 19
Black Hawk (see Sikorsky)
BO105 (see MBB)
Boeing
 707: 126
 767 Tanker Transport: 4
 767-200ER: 4
 B-17 Flying Fortress: 173
 B-52 Stratofortress: 126
 C-40B: 14
 E-3 Sentry: 24, 30
 E-3A: 7
 KC-135 Stratotanker: 5, 7, 126
 KC-135A: **146**
 KC-135E: 14
 KC-135R: 14-15, **56**
 KC-767A: 5
 Sentry AEW.Mk 1: 11
 VC-137C: 13
 X-20 Dyna-Soar: 134
Boeing F/A-18E/F Super Hornet: 36-73, **36-73**
Boeing Helicopters AH-64D Apache: 8
Boeing/McDonnell Douglas
 C-17A: 15
 CF-188 Hornet: 18
 EA-18G 'Growler': 38, 39, 41, 57, 71-73, **72**, **73**
 F-15 Eagle: 14
 F-15B: **6**
 F-15C: 58
 F-15E Strike Eagle: 42
 F-15J: **13**
 F-15T Strike Eagle: 10
 F/A-18 Hornet: 6, 39
 F/A-18A: 12, 39, 43
 F/A-18B: 43
 F/A-18C: 12, **38**, 43-45, 46, 50, 52, 54, 56, **56**, 58, 59, 60, 66, **71**
 F/A-18D: 43-45, 46, 47, 50, 52, 54, 56, 58, 59, 73
 F/A-18E Super Hornet: **36-39**, 40, 41, **41-43**, 47, **47**, **48**, 50, **50**, 51, **51-53**, **55**, **56**, **60**, **61**, 62-66, **64-68**
 F/A-18E1 to F/A-18E4: 40, **40**, 47
 F/A-18E5: **39**, 40, **40**, 47, 48
 F/A-18E/F Super Hornet: **36**, 37-73, **37-73**
 F/A-18E/F Super Hornet Block I upgrade: 73
 F/A-18E/F Super Hornet Block II upgrade: 56-57, 61, **70**, 73
 F/A-18E/F Super Hornet Lot 26: **63**, **70**
 F/A-18F Super Hornet: 13, 40, 41, **41-46**, 47-49, **49**, 50, **50**, **54**, **56-60**, 62, **62**, 63, **70**, 72
 F/A-18F1: 40, **40**, 47-49, **73**
 F/A-18F2: **39**, 40, **40**, 47, 48
 Hornet 2000: 39, 40
 Hornet HUG 2.1: 7, **7**
 KC-10A: **52**
Boeing Vertol CH-113 Labrador: 8
Bombardier (see also Gates)
 Learjet 45: 9
Bomber Destroyer, He 177 (see Heinkel)
Boxer, USS: 110
Brasilia, EMB-120EW (see EMBRAER)
Breguet Atlantic I: **94**, 96
Bristol Beaufighter: 164
Bronco, OV-10C (see Rockwell)

C

C-5A (see Lockheed Martin)
C-9A Nightingale (see McDonnell Douglas)
C-12 (see Beech)
C-17A (see Boeing/McDonnell Douglas)
C-26D Metro (see Fairchild)
C-37A (see Gulfstream Aerospace)
C-40B (see Boeing)
C-47 (see Douglas)
C-54 (see Douglas)
C-119F (see Fairchild)
C-123 (see Fairchild)
C-130 Hercules (see Lockheed Martin)
C-141B/C (see Lockheed Martin)
C.295 (see EADS-CASA)
Canada
 Aerospace Engineering Test Establishment (AETE): 18-20, **18**, **19**, **20**
Canadair
 CT-114 Tutor: 18, 19, 20
 CT-133 Silver Star: 18-20, **18**, **19**, **20**
Carl Vinson, USS: **46**
Catalina, PBY-5A (see Consolidated)
CC-130J (see Lockheed Martin)
Cefirm Leader, RU-21A/B/C (see Beech)
Cefly Lancer, RU-21J (see Beech)
Cervantes, ARA: 95
Cessna
 O-1: 140
 O-2: 140
 T-37B: 14
CF-5B Freedom Fighter (see Northrop)
CF-188 Hornet (see Boeing/McDonnell Douglas)
CH-113 Labrador (see Boeing Vertol)
CH-146 Griffon (see Bell Canada)
CH-149 Cormorant (see AgustaWestland)
Cheetah (see HAL)
Chengdu
 JF-17 Thunder (FC-1 Super 7): 4
CN-235 (see EADS-CASA)
'Coaler' (see Antonov An-72)
'Colt' (see Antonov An-2)
Combat Colours – Escuadrilla Aeronaval de Exploración – Argentina's Lockheed patrollers: 94-103, **94-103**
Combat Talon, MC-130E (see Lockheed Martin)
Combat Talon II, MC-130H (see Lockheed Martin)
Combat Shadow, MC-130P (see Lockheed Martin)
Conqueror, HMS: 97
Consolidated
 Liberator: 164
 P2Y-3A Ranger: 94
 PBY-5A Catalina: 94, 95
Constellation, USS: 63
Cormorant, CH-149 (see AgustaWestland)
Convair
 580: 28
 EC-131K: 28
 F-102 Delta Dagger: 123
 PQM-102A/B Delta Dagger: 149
 QF-102A Delta Dagger: 149
Convair 5800 (see Kelowna Flightcraft)
Corsair (see Vought)
Cougar, AS 532 (see Eurocopter)
Cougar Mk 2, AS 532A2 (see Eurocopter)
Crazy Dog, RU-21A (see Beech)
CT-114 Tutor (see Canadair)
CT-133 Silver Star (see Canadair)
'Curl' (see Antonov An-26)
Curtiss FSL: 94

D

Dagger (see IAI)
Dassault
 Mirage 2000: 11, 31
 Mirage F1CH: **15**
 Mirage F1EH: **15**
 Super Etendard: 97, **97**, 98
 Super Etendard Modernisé Standard 4 (SEM 4): 14, **14**
DC-8 (see Douglas)
Debrief: 16-21, **16-21**
de Havilland Mosquito: 166, 168, 170
de Havilland Canada RC-7B Airborne Reconnaissance Low-Multimission (ARL-M): 82
Delius: 164
Delta Dagger (see Convair)
Denmark
 Kongelige Danske Flyvevåbnet (Royal Danish Air Force): 137-138, 148, **148**, 150, 157
  Esk 725: 137, 138, 157
  Esk 727: 137-138, **137**, 157, **157**
  Esk 730: 137, **137**, 157
Denver, USS: 4
Dornier G-52 Wal: 94
Douglas (see also McDonnell Douglas)
 A-1H Skyraider: 138
 AD Skyraider: **109**, 112
 B-26 Invader: 107, 117
 C-47: **105**, 116
 C-54: 98
 C-54D: 133
 DC-8: 126
 F3D Skynight: **115**, 118
 F4D Skyray: 124
Draken, J 35 (see Saab)
Dyna-Soar, X-20 (see Boeing)

E

E-2 (see Northrop Grumman)
E-3 Sentry (see Boeing)
EA-6B Prowler (see Northrop Grumman)
EA-18G 'Growler' (see Boeing/McDonnell Douglas)
EADS-CASA
 C.295: 10
 C.295M: 10
 CN-235: 8
Eagle, F-15 (see Boeing/McDonnell Douglas)
EC-131K (see Convair)
EF-100A (see North American)
EF-2000 Typhoon (see Eurofighter)
Electra (see Lockheed)
Electra Explorador (see Lockheed)
Electra Wave (see Lockheed)
EMB-120EW Brasilia (see EMBRAER)
EMB-145 (see EMBRAER)
EMBRAER
 A/AT-29 Super Tucano (ALX): 28
 EMB-120EW Brasilia: 28, **28**, 29, 30
 EMB-145AEW&C: **23**, **26**, 27, **27**, 30-31, **30**
 EMB-145AEW&C (R 99A): **22**, **25**, 27, 28, **28**, 29-30, **31**
 EMB-145MP: 31
 EMB-145RS (R 99B): 29, 30
 ERJ-145LR: 29
 ERJ-145XR: 29
 Legacy: 9
 R 99A (EMB-145AEW&C): **22**, **25**, 27, 28, **28**, 29-30, **31**
 R 99B (EMB-145RS): 29, 30
Enterprise, USS: 15, 50, 51
EP-2E Neptune (see Lockheed)
EP-3 Orion (see Lockheed)
Erieye – Sweden's Eyes on the World: 22-30, **22-31**
Erieye concept, 2000 (see Saab)
ERJ-145LR/XR (see EMBRAER)
Escuadrilla Aeronaval de Exploración – Argentina's Lockheed patrollers: 94-103, **94-103**
'Espora'-class frigate: 102
EU-21 (see Beech)
Eurocopter
 AS 332 Super Puma: 102
 AS 532 Cougar: 16
 AS 532A2 Cougar Mk 2: **8**
 AS 565 Panther: 10
 Tiger HAD: 10
 Tiger HAP: 10
Eurofighter
 EF-2000 Typhoon: 4, **4**
 Typhoon DA.1: **4**
 Typhoon Tranche 1: 33, **35**
Exercise
 Composite Training Unit (COMPTUEX): 15
 Co-operative Key 2003: 11, **11**
 Goldfire: 145
 Maple Flag: **19**
 Opera: **15**
 Red Flag: 52
 Reforger: 76

F

F1CH/F1EH, Mirage (see Dassault)
F3D Skynight (see Douglas)
F-4 Phantom II (see McDonnell Douglas)
F4D Skyray (see Douglas)
F4U Corsair (see Corsair)
F7F Tigercat (see Grumman)
F-14 Tomcat (see Northrop Grumman)
F-15 (see Boeing/McDonnell Douglas)
F-15 Reporter (see Northrop)
F-16 (see Lockheed Martin)
F-22 (see Lockheed Martin)
F27 Friendship (see Fokker)
F28 (see Fokker)
F-35 Joint Strike Fighter (see Lockheed Martin)
F-51 Mustang (see North American)
F-80 Shooting Star (see Lockheed)
F-84F Thunderstreak (see Republic)
F-84G Thunderjet (see Republic)
F-86 Sabre (see North American)
F-100 Operators: 152-157, **152-157**
F-100 Super Sabre (see North American)
F-102 Delta Dagger (see Convair)
F-104 Starfighter (see Lockheed)
F-105 Thunderchief (see Republic)
F-105F Wild Weasel III (see Republic)
F-107A (see North American)
F-111 (see General Dynamics)
F-117 (see Lockheed Martin)
F/A-18 (see Boeing/McDonnell Douglas)
F/A-22 Raptor (see Lockheed Martin)
Fairchild
 C-26D Metro: 78
 C-119F: 133
 C-123: 133
 Metro III (Tp 88): **24**, 25
 OA-10A: 14
FC-1 Super 7 (see Chengdu JF-17 Thunder)
'Fencer-D' (see Sukhoi Su-24MK)
Fighting Falcon, F-16 (see Lockheed Martin)
Fire Scout VTUAV, RQ-8A (see Northrop Grumman)
Flying Fortress, B-17 (see Boeing)
Focke-Wulf Fw 200: 164
Focus Aircraft – Boeing F/A-18E/F Super Hornet: 36-73, **36-73**
Fokker
 50 KingBird: 28
 F27 Friendship: 28
 F28: **98**
Formosa: 97
France
 Armée de l'Air (air force): 136-137, 138, 147, 148, **148**, 157, **157**
  Escadre de Chasse (EC) 3: 136-137, 157
  Escadre de Chasse 1/3 'Navarre': 136, 137, 157
  Escadre de Chasse 2/3 'Champagne': 136, 157
  Escadre de Chasse 11: 136-137, 148, 157
  Escadre de Chasse 1/11 'Roussillon': 136, **136**, 157
  Escadre de Chasse 2/11 'Vosges': 136, 157
  Escadre de Chasse 3/11 'Corse': 136, **136**, 157
  Escadre de Chasse 4/11 'Jura': 136, **136**, 148, 157
  Escadrille de Convoyage EC 070: 157
Freedom Fighter, CF-5B (see Northrop)
Friendship, F27 (see Fokker)
FSL (see Curtiss)
Fuji T-1B: **13**
'Fulcrum' (see Mikoyan MiG-29)
Fw 200 (see Focke-Wulf)

G

G-21A Goose (see Grumman)
G-52 Wal (see Dornier)
Gates (see also Bombardier)
 Learjet 35: 78
'Gearing'-class destroyer: 146
General Belgrano, ARA: 97
General Dynamics
 F-111: 15
 F-111C: **15**
George H.W. Bush, USS: 15
Germany
 Luftwaffe (air force): 150
  Flugzeugführerschule B/16 (FFS(B) 16): **170-171**
  KG 1: 167, 170, 172-173
  I./KG 1: 170, 172
  II./KG 1: 170, 172
  III./KG 1: 168, 170, 172
  I./KG 40: 165, **165**, 166-167, 170, 171
  II./KG 40: 164-166, **165**, **166**, **168**, 170, **173**
  I./KG 50: 164, **166**
  I./KG 100: 167-168, **167**, 170, 172
  II./KG 100: **166**, 169, **169**
  2./KG 100: 167, **168**
  3./KG 100: 166, 167
  4./KG 100: 169, **169**
  6./KG 100: **166**, **168**
Global Hawk UAV, RQ-4A (see Northrop Grumman)
Go 242A-1 (see Gotha)
Golden Eagle LIFT, T-50 (see KAI)
Goose, G-21A (see Grumman)
Gotha Go 242A-1: **164**
Greif, He 177 (see Heinkel)
Griffon, CH-146 (see Bell Canada)
Gripen (see Saab)
'Growler' (see Boeing/McDonnell Douglas EA-18G)
Grumman (seen also Northrop Grumman)
 A-6 Intruder II concept: 39
 A-6E Intruder: 39, 40, 60
 F7F Tigercat: **105**, 109, **109**, 112, 116
 F7F-3N Tigercat: 107, 108, 117, 118
 G-21A Goose: 94
 JRF-6B: 94
 S-2E Tracker: **100**
Guardrail I, RU-21G (see Beech)
Guardrail II/IIA/IV, RU-21E (see Beech)
Guardrail V, RU-21H (see Beech)
Gulfstream IV, U-4 (see Gulfstream Aerospace)
Gulfstream 450/500 (see Gulfstream Aerospace)
Gulfstream Aerospace
 C-37A: 14
 Gulfstream 450/500: 6
 U-4 Gulfstream IV: 12

H

HAL Cheetah: 6
Harrier, TAV-8B (see McDonnell Douglas/BAe)
Harry S. Truman, USS: **40**, 48-49
Hawk Mks 115/120 (see BAE Systems)
Hawkeye, E-2 (see Northrop Grumman)
Hawkeye 2000, E-2C (see Northrop Grumman)
HC-130P Hercules (see Lockheed Martin)
He 177 *et seq* (see Heinkel)
He 177 Greif – The Luftwaffe's Lighter: 158-173, **158-173**
Heinkel
 He 177 Bomber Destroyer: 172
 He 177 Greif: 158-173, **158-173**

He 177 V1: **159**, 161, 171
He 177 V2: **159**, 161, 171
He 177 V3: 161, 171
He 177 V4: **159**, 161, 171
He 177 V5: **159**, 161, 171
He 177 V6: **160**, 171
He 177 V7: **160**, 171
He 177 V8: **158**, **160**, 171
He 177 V9 to V24: 171
He 177 V15: 162, 171
He 177 V16: 162, 171
He 177A: **164**
He 177A-0: **160**, 162, 171
He 177A-02: **158**, **160**, 162
He 177A-03: **161**
He 177A-05: 171
He 177A-06: 171
He 177A-07: 171
He 177A-1: 163-164, **163**, **166**, **170-171**, 171
He 177A-3: 162, **163**, 164, **164**, 165, **165**, 167, **168**, 171
He 177A-3/R2: **161**, **163**, 171
He 177A-3/R3: 171
He 177A-3/R4: 171
He 177A-3/R5: 162, 171, **171**
He 177A-3/R7: 171
He 177A-4: 172
He 177A-5: 162, **162**, **163**, 164, 165, **166**, 167, **167**, 169, 171, 172, 173
He 177A-5/R2: **166**, **168**, 169, **169**, 172
He 177A-5/R5: 172
He 177A-5/R6: **161**, **165**, **173**
He 177A-6: 172
He 177A-7: 172, **173**
He 177B (He 277): 162, 172
He 177B-0: **172**
He 274 (AAS 01A): 172, **172**
He 277 (He 177B): 162, 172
He 277 V1: **172**
P.1041 design study: 159-161
Henschel Hs 293: 164, **165**, **166**, 171
Hercules (see Lockheed Martin)
HH-60G (see Sikorsky)
'Hind' (see Mil Mi-35P)
Hornet (see Boeing/McDonnell Douglas)
Hornet 2000 (see Boeing/McDonnell Douglas)
Hornet HUG 2.1 (see Boeing/ McDonnell Douglas)
Hs 293 (see Henschel)
Huey, UH-1N (see Bell)
Huron (see Beech)

I

IAI Dagger: 98
Ilyushin Il-76: 12, 27
Ilyushin/Beriev A-50 'Mainstay': 9
Improved Guardrail V, RC-12D (see Beech)
Intruder, A-6E (see Grumman)
Intruder II concept, A-6 (see Grumman)
Invader, B-26 (see Douglas)
Israel
Defence Force/Air Force (IDF/AF): 5

J

J 35 Draken (see Saab)
JAS 39A Gripen (see Saab)
JF-17 Thunder (see Chengdu)
JF-100C/F (see North American)
John C. Stennis, USS: 47, 48, 52, **59**, **70**
Joint Strike Fighter, F-35 (see Lockheed Martin)
Joint unmanned combat air system (J-UCAS) program (see Northrop Grumman/Lockheed Martin)
JRC-12G (see Beech)
JRF-6B (see Grumman)
JSF, F-35 (see Lockheed Martin)
JU-21A/H (see Beech)
Ju 86R (see Junkers)
J-UCAS program (see Northrop Grumman/Lockheed Martin)
Junkers Ju 86R: 164

K

Ka-52 Alligator (see Kamov)
KAI T-50 Golden Eagle LIFT (Lead-In Fighter Trainer): 4
Kaman SH-2G(A) Seasprite: 8
Kamov Ka-52 Alligator: 10
KC-10A (see Boeing/McDonnell Douglas)
KC-130H/J Hercules (see Lockheed Martin)
KC-135 Stratotanker (see Boeing)
KC-767A (see Boeing)
Kelowna Flightcraft Convair 5800: 28
King, ARA: 95
King Air (see Beech)
KingBird, 50 (see Fokker)
Kitty Hawk, USS: 58

L

L-39 (see Aero)
L-188A/PF Electra (see Lockheed)
Labrador, CH-113 (see Boeing Vertol)
Laffing Eagle, RU-21D (see Beech)
Lancaster (see Avro)
Lancer (see Aerostar/Mikoyan)
Last of the Dragons: 32-35, **32-35**
Learjet 35 (see Gates)
Learjet 45 (see Bombardier)
Left Foot, RU-21E (see Beech)
Left Jab, JU-21A (see Beech)
Legacy (see EMBRAER)
Liberator (see Consolidated)
Light Observation Helicopter, A109 (see AgustaWestland)
Lincoln B.Mk 1 (see Avro)
Lockheed
Electra: **94**, **100**
Electra Explorador: **98**, **99**, 100, 101, 102
Electra Wave: **94**, 100, **100**
EP-2E (P2V-5FE) Neptune: 96, 103
EP-3 Orion: 13
F-80 Shooting Star: 105
F-104 Starfighter: 148
L-188A Electra: 98, **98**, 99-101, **99**, 103
L-188PF Electra: 97, **97**, 98-99, **98**, **99**, 100-101, 103
P-2E (P2V-5) Neptune (MR.Mk 1): 94-96, **95**, **96**, 103
P2V-5FE (EP-2E) Neptune: 96, 103
P2V-5FS (SP-2E) Neptune: 96, **96**, 103
P2V-7S (SP-2H) Neptune: 96-98, **96**, **97**, 103
P-3 Orion: 13
P-3 Orion CUP: 8
P-3B: 96
P-3B Orion TACNAVMOD: **94**, 101-103, **102**, **103**
P-3C Orion: 13
P-3C Orion Update II/II.5: 7
SP-2E (P2V-5FS) Neptune: 96, **96**, 103
SP-2H (P2V-7S) Neptune: 96-98, **96**, **97**, 103
T-33A: 133
U-2: 128, 145
Lockheed Martin
AC-130U: 14
C-5A: 14, 15
C-130 AEW concept: 28, **28**, 30-31
C-130 Hercules: 102, 133
C-130A: 14
C-130E: 14, 15, 28
C-130H: 8, 12, 14, 15, 28
C-130H2: 15
C-130J: 28
C-130J-30: 8
C-141B: 15
C-141C: 14, 15
CC-130J: 14
F/A-22 Raptor: 42
F/A-22A Raptor: **6**, 14
F-16 Fighting Falcon: 9, 11, 14, 58
F-16 Block 40: 20-21
F-16 Block 50/52: 21
F-16 VISTA: **12**
F-16A/B: 6-7
F-16A/B Block 15: 9-10
F-16B: **12**
F-16C: 52
F-16C Block 30: 17, **17**
F-16C/D: 12
F-16C/D Block 30: 17
F-16C/D Block 40: 17
F-16C/D Block 42: 20-21, **20**, **21**
F-16D Block 52+: **10**
F-16E/F Block 60: 4-5
F-16F Block 60: **4**
F-16I Soufa (Storm): 5, **5**
F-16XL: 4
F-22: 40
F-22 Naval Advanced Tactical Fighter (NATF): 39, 40
F-35 Joint Strike Fighter (JSF): 38, 40, 41
F-35C Joint Strike Fighter: **71**
F-117: 42
F-117A: **14**
HC-130P Hercules: 14
KC-130H Hercules: 97
KC-130J: 5
MC-130E Combat Talon: 7-8
MC-130H Combat Talon II: 7-8
MC-130P Combat Shadow: 8
NC-130A: 14
NC-130H: 14
S-3B Viking: 13, 70
'Super-blimp': 5-6
Lockheed Martin/McDonnell Douglas A-12 Avenger II: 39, 40, 46, 67
LOH, A109 (see AgustaWestland)

M

'Mainstay', A-50 (see Ilyushin/Beriev A-50)
Makin Island, USS: 15
Marine Corsairs in Korea: 104-119, **104-119**
Mariner, PBM-5A (see Martin)
Marsa: 164
Martin PBM-5A Mariner: 94, 95
MBB BO105: 10
MC-130E/H/P (see Lockheed Martin)
McDonnell RF-101 Voodoo: 127
McDonnell Douglas (see also Boeing/ McDonnell Douglas)
C-9A Nightingale: 12, 14, 15
F-4 Phantom II: 11, 43, 101
F-4C: 145
TA-4J: 13
McDonnell Douglas/BAe TAV-8B Harrier: 8
Meko 140-type frigate: 102
Metro, C-26D (see Fairchild)
Metro III (see Fairchild)
MH-60R Seahawk (see Sikorsky)
MH-68A Stingray (see Agusta)
Mi-24V (see Mil)
Mi-35P 'Hind' (see Mil)
MiG-15 *et seq* (see Mikoyan)
Mikoyan
MiG-15: 108, 115-116, 123
MiG-17: 138-139
MiG-19: 136
MiG-21: 11
MiG-21bis: **11**
MiG-21UM: 9, **9**
MiG-23: 15
MiG-29 'Fulcrum': **8**, 10
SM-9: 124
Mil
Mi-24V: **9**
Mi-35P 'Hind': 9
Mirage 2000 (see Dassault)
Mirage F1CH/F1EH (see Dassault)
Mosquito (see de Havilland)
Mosquito, T-6 (see North American)
Murature, ARA: 95
Mustang (see North American)
MV-22B Osprey (see Bell-Boeing)

N

N-9MB (see Northrop)
NA-180 *et seq* (see North American)
NAH-1Z (see Bell)
NATF (Naval Advanced Tactical Fighter), F-22 (see Lockheed Martin)
NC-12B (see Beech)
NC-130A/H (see Lockheed Martin)
Neptune (see Lockheed)
NF-100F (see North American)
NH Industries NH 90: 8
Nightingale, C-9A (see McDonnell Douglas)
Nimitz, USS: 12, 15, 50, 51, **54**, **55**, **56**, 59, 60, **66**
Nimrod (see BAE Systems)
North American
'Advanced F-86D/E' proposals: 122
EF-100A: **125**
F-51 Mustang: 105, 106, **109**, **113**, 115
F-51D: **111**
F-86 Sabre: 49, 122, 123, **124**, 126, 128, 130, 135, 136
F-86D: 124
F-100 Super Sabre: 120-157, **122**, **130**, **132**, **138**, **139**, **140**, **143-144**, **148**, **157**
F-100A: **124**, 125-127, **125**, **126**, 128, 130, 132, 136, **136**, 148, **151**, 153, 155, 156, **157**
F-100A 'Rehab': 136
F-100B (F-107A/NA-212): 135
F-100BI: 135
F-100C (NA-214): 126, **127**, **128**, 129, **129**, 130-137, **132**, 140, **147**, 150, **150**, 151, **151**, 152, **153**, 154, 155, **155**, 156
F-100D: **120-121**, **122**, 127, **128**, 129, 130-134, **130**, **131**, **132**, 136-144, **136-139**, **143-147**, 147, 148, 149, **150**, 152-156, **152-156**
F-100D-20-NA: 133, **133**
F-100D-50-NH: 151
F-100D-75-NA: 141-142, **141-142**
F-100F (NA-243): 127, **128**, 133-140, **135**, **136**, **137**, 145, 147-150, **147**, **148**, **150**, 152, 154, **156**
F-100F Wild Weasel I: 145-146, **145**, **146**
F-100F-20-NA (NA-255): 136
F-100I: 135
F-100J: 146-147
F-100K/L/N: 147
F-100S proposal: 147
F-107A (F-100B/NA-212): 135
JF-100C: **151**
JF-100F: 130, **130**
NA-180 Sabre 45 (YF-100A): 122-125, **123**
NA-192: 123
NA-212 (F-100B/F-107A): 135
NA-214 (F-100C): **128**, 129, 130-132, **132**
NA-222: 132
NA-230 (TF-100C): **134**, 135
NA-243 (F-100F): 135-136, **135**
NA-255 (F-100F-20-NA): 136
NF-100F: 134, **134**
P-51 Mustang: 122, 138
QF-100: **149**, 153, 154, 156
QF-100D: **126**, 149, **156**
QF-100F: 149
RF-100A Slick Chick: 127, **127**, 128-129, 136, 138, 153, 155, 157
T-6 Mosquito: **109**, **118**, 137
TF-100C (NA-230): **134**, 135
TF-100F: 137, **137**, 138, 148, 150, **150**, 151, **151**
X-15: **131**, 134
YF-86D: **123**, 124
YF-100A (NA-180 Sabre 45): 122-125, **123**
YF-100C: 130
YF-107A: **135**
YQF-100D: 149
North American F-100 Super Sabre: 120-157, **120-157**
Northrop
CF-5B Freedom Fighter: 8-9
F-15 (RF-61C) Reporter: 109
N-9MB: **12**
Northrop Grumman
B-2: 42
B-2A Spirit: 8
E-2 Advanced Hawkeye: 7
E-2 Hawkeye: 24, 30
E-2C Hawkeye: **7**, 38
E-2C Hawkeye 2000: 7
EA-6B Prowler: 41, 42, 71-72
F-14 Tomcat: 39, 59
F-14A: 13
F-14B: 58
F-14D: 39, **41**, **47**, **62**, 63
RQ-4A Global Hawk UAV: 5, 14
RQ-8A Fire Scout VTUAV: 4-5
Northrop Grumman/Lockheed Martin joint unmanned combat air system (J-UCAS) program: 5
NU-8F (see Beech)

O

O-1 (see Cessna)
O-2 (see Cessna)
OA-10A (see Fairchild)
Operation
Allied Force: 42, 44, 50
Bell Tone: 138-139
Desert Storm: 50, 51, 58, 102
Enduring Freedom: 50, 51, **52**, 70
Falconer: **7**
Flaming Dart: 139
Héraclès: 14
Iraqi Freedom: **7**, 42, **53**, **54**, **55**, **56**, **64-66**, 70
Mobile Zebra: 134
Red Richard: 132
Rolling Thunder: 138, 139
Sawbuck: **139**
Southern Watch: 15, **52**, 70
Steinbock (Ibex): 166-168, **166**, **168**, 169, 170
Orion (see Lockheed)
Osprey, MV-22B (see Bell-Boeing)
OV-10C Bronco (see Rockwell)

P

P-2E Neptune (see Lockheed)
P2V-5 Neptune (see Lockheed)
P2V-7S Neptune (see Lockheed)
P2Y-3A Ranger (see Consolidated)
P-3 Orion (see Lockheed)
P-51 Mustang (see North American)
P.1041 design study (see Heinkel)
Panavia
Tornado: 101
Tornado F.Mk 3: **9**, **102**
Tornado GR.Mk 1: **9**
Panther, AS 565 (see Eurocopter)
PBM-5A Mariner (see Martin)
PBY-5A Catalina (see Consolidated)
PC-6 Turbo-Porter (see Pilatus)
Pegasus, T-44A (see Beech)
Phantom II, F-4 (see McDonnell Douglas)
Philippine Sea, USS: 110
Photo Feature – Last of the Dragons: 32-35, **32-35**
Piedrabuena, ARA: 97
Pilatus
PC-6 Turbo-Porter: **15**
PC-6B/H2 Turbo-Porter: 99, 101, **101**, 103
Point Cruz, USS: **113**, **119**
PQM-102A/B Delta Dagger (see Convair)
Prowler, EA-6B (see Northrop Grumman)
Pueblo, USS: 140, 155, 156
Puma, SA 330H (see Aérospatiale)

Q

QF-100 (see North American)
QF-102A Delta Dagger (see Convair)
Queen Air, B80 (see Beech)

R

R 99A/B (see EMBRAER)
Ranger, P2Y-3A (see Consolidated)
Raptor, F/A-22 (see Lockheed Martin)
Raytheon (see also Beech)
T-6A Texan II: 13-14
RC-7B Airborne Reconnaissance Low-Multimission (see de Havilland Canada)
RC-12 (see Beech)
'Rehab' (see North American F-100A)
Rendova, USS: **108**, **113**
Reporter, F-15C (see Northrop)
Republic
F-84F Thunderstreak: 130, 133
F-84G Thunderjet: 137
F-105 Thunderchief: 138, 139
F-105B Thunderchief: 131, 133, 134
F-105D Thunderchief: 138, 146
F-105F Wild Weasel III: 146
RF-61C Reporter (see Northrop)
RF-100A Slick Chick (see North American)
RF-101 Voodoo (see McDonnell)
Rio Carcarañá: 97
Rockwell OV-10C Bronco: 10
Rohna: 164
RQ-4A Global Hawk UAV (see Northrop Grumman)
RQ-8A Fire Scout VTUAV (see Northrop Grumman)
RU-21 (see Beech)
Ruslan, An-124 (see Antonov)

S

S-2E Tracker (see Grumman)
S-3B Viking (see Lockheed Martin)
S-70A Black Hawk (see Sikorsky)
S 100B Argus (see Saab)
S340 (see Saab)
SA 330H Puma (see Aérospatiale)
Saab
340AEW&C: 25-26, 30
2000: 26, 31
2000 Erieye concept: 26, 31, **31**
Gripen: 11, 27, 31
J 35B/D Draken: 23
J 35F: 33
J 35OE Draken: 32-35, **32-35**
JAS 39A Gripen: **11**
S 100B Argus: **22**, **23**, **24**, 26-27, **26**, **28**, **29**, 30, **30**, 31
S340 (Tp 100A): 24-25, 26
S340B: 25, 26
Sabre (see North American)
Saudi Arabia
Prince Sultan Air Base: 15
Sea Harrier (see BAE Systems)
Seahawk, MH-60R (see Sikorsky)
Seasprite, SH-2G(A) (see Kaman)
Sentry (see Boeing)
SF.260M (see SIAI-Marchetti)
SH-2G(A) Seasprite (see Kaman)
Shackleton (see Avro)
Sheffield, HMS: **97**, 98
Shooting Star, F-80 (see Lockheed)
SIAI-Marchetti SF.260M: **13**
Sicily, USS: 106, **106**, 107, **107**, 109-110, **110**, **113**, 114, 115, **118**
Sikorsky
HH-60G: 14
MH-60R Seahawk: **10**, 38
S-70A Black Hawk: 10
S-70A-42 Black hawk: 16-17, **16**, **17**
UH-3H: 13
UH-60L Black Hawk: 10
UH-60M: **8**
Silver Star, CT-133 (see Canadair)
Skynight, F3D (see Douglas)
Skyraider (see Douglas)
Skyray, F4D (see Douglas)
Slick Chick, RF-100A (see North American)
SM-9 (see Mikoyan)
Soufa, F-16I (see Lockheed Martin)
Southampton (see Supermarine)
SP-2E/H Neptune (see Lockheed)
Spain
Air Force
Ala 11: 4, **4**
Spirit, B-2A (see Northrop Grumman)
Spitfire (see Supermarine)
Starfighter, F-104 (see Lockheed)
Stingray, MH-68A (see Agusta)
Stratofortress, B-52 (see Boeing)
Stratotanker, KC-135 (see Boeing)
Strike Eagle, F-15E/T (see Boeing/ McDonnell Douglas)
Su-24MK 'Fencer-D' (see Sukhoi)
Su-25 *et seq* (see Sukhoi)
Sukhoi
Su-24MK 'Fencer-D': 7
Su-25: 11, 12
Su-27: 9, 12
Su-30MKM: 10
'Super-blimp' (see Lockheed Martin)
Super Cobra, AH-1W (see Bell)
Super Etendard (see Dassault)
Super Etendard Modernisé Standard 4 (see Dassault)
Super Hornet, F/A-18E/F (see Boeing/ McDonnell Douglas)
Super King Air (see Beech)
Supermarine
Southampton: 94
Spitfire: 164
Super Puma, AS 332 (see Eurocopter)
Super Sabre, F-100 (see North American)
Super Tucano, A/AT-29 (see EMBRAER)

T

T-1B (see Fuji)
T-6 Mosquito (see North American)
T-6A Texan II (see Raytheon)
T-33A (see Lockheed)
T-37B (see Cessna)
T-44A Pegasus (see Beech)
T-50 Golden Eagle LIFT (see KAI)
TA-4J (see McDonnell Douglas)
Taiwan
Republic of China Air Force (RoCAF): 126, 136, **136**, 138, 148, 157
2nd TFW/FBW: **136**, 157, **157**
4th FBW: 157
4th Reconnaissance Squadron: 129, 157
5th/401st Tactical Combined Wing: 157
TAV-8B Harrier (see McDonnell Douglas/BAe)
TC-12B (see Beech)
Technical Briefing – Erieye – Sweden's Eyes on the World: 22-31, **22-31**
Texan II, T-6A (see Raytheon)
TF-100C/F (see North American)
Theodore Roosevelt, USS: 58, **61**
Thunder, JF-17 (see Chengdu)
Thunderchief, F-105 (see Republic)
Thunderjet, F-84G (see Republic)
Thunderstreak, F-84F (see Republic)
Tiger HAD/HAP (see Eurocopter)
Tigercat, F7F (see Grumman)
Tomcat, F-14 (see Northrop Grumman)
Tornado (see Panavia)
Tp 88 (see Fairchild Metro III)
Tp 100A (see Saab S340)
Tracker, S-2E (see Grumman)
Tripoli, USS: 129
Turbo-Porter, PC-6 (see Pilatus)
Turkey
Türk Hava Kuvvetleri (THK – air force): 137, **137**, 145, 148, 157, **157**
1nci Ana Jet Ussu: 157
3nci Ana Jet Ussu: 157
7nci Ana Jet Ussu: 157
8nci Ana Jet Ussu: 157
111 Filo: 137, 146, 157
112 Filo: 137, 146, 157
113 Filo: 137, 157
131 Filo: 137, 148, 157
132 Filo: 137, 148, 157
171 Filo: 137, 157
172 Filo: 137, 148, 157
181 Filo: 137, 146, 157
182 Filo: 137, 148, 157
Tutor, CT-114 (see Canadair)
Type Analysis – He 177 Greif – The Luftwaffe's Lighter: 158-173, **158-173**
Typhoon (see Eurofighter)

U

U-2 (see Lockheed)
U-4 Gulfstream IV (see Gulfstream Aerospace)
U-8D (see Beech)
U-12F (see Beech)
U-21 Ute (see Beech)
UC-12B/F/M (see Beech)
UH-1N Huey (see Bell)
UH-1Y (see Bell)
UH-3H (see Sikorsky)
UH-60L/M Black Hawk (see Sikorsky)
United Arab Emirates
air force: 4
United States
Air Force: 84, 152-155
3rd TFW: 140, 152, **152**
3rd Wing: 84
4th FDW/TFW: **128**, 152
8th TFW: **132**, 152
'9th AF Firepower Team': **128**
15th Airlift Wing: 14
15th Expeditionary Mobility Task Force (EMTF): 12
18th TFW: **132**, 152
20th FBW/TFW: **132**, 152, **152**
21st Expeditionary Mobility Task Force (EMTF): 12

21st TFW: **132**, 152
22nd ARW: **84**
27th TFW: 140, 147, **148**, 152
31st TFW: **120**, **122**, 140, **146**, 152
35th TFW: 140, **145**, 152
36th Fighter-Day Wing (FDW)/TFW: **128**, **129**, **132**, **132**
37th TFW: **140**, **141**–142, 152
39th Air Division: 152–**153**
45th FDS: **153**
46th Test Group: 84
48th FBW/TFW: **132**, 147, **148**, 153, **153**
49th FBW/TFW: **132**, 153
50th FBW/TFW: **132**, 153, **153**
51st Fighter Wing: 84, **84**
53rd Fighter-Day Squadron: **128**, 152
57th Fighter Weapons Wing 153
58th Tactical Fighter Training Wing: 153
64th Aggressor Squadron: 12
67th Tactical Reconnaissance Wing: 153
67th TFS: **132**, 152, 154
77th TFS: 152, **152**
79th TFS: 152, **152**
80th FS: 17, **17**
81st TFS: **153**, 154
82nd Tactical Aerial Targets Squadron: 149, 154
83rd FDW: 152, 153
85th Group/85th Operations Squadron: 14
89th MAW: **76**
90th TFS: 152, **152**, 154
95th Bombardment Wing: 153
113th TFW: 153
306th TFS: **122**, 152
308th TFS: **120**–121, 152
309th TFS: **146**, 152
312th TFW/FBW: 152, 153
316th Air Division: 153, **153**
322nd FDW: 153
323rd FBW: 153
325th FWW: 153
333rd TFS: **128**, 152
336th TFS: **128**, 152
352nd TFS: **145**, 152, 153
354th FBW/TFW: **131**, **135**, 150, 153, **154**
355th TFS: 141, 152, 153, **154**
356th TFS: **131**, **135**, 153, 154
366th TFW: 153–154
375th Aeromedical Airlift Wing (AAW): 78, 78, 83, 84
388th TFW: 154
401st TFS: **154**, 154
402nd FDW: 154
405th FBW/TFW: **138**, 154
412th Test Wing: **77**, 84
413th FDW/TFW: 154
416th TFS 'Silver Knights': 138, **139**, **140**, **141**–142, **145**, 152, 154
428th TFS: **130**, 154
429th TFS: 152, 154, **155**
436th FDS: 154, **155**
450th FDS: 131, 153
450th FDW/TFW: 154
474th FBW/TFW: **130**, 154, **155**
475th TFW: 154
475th Weapons Evaluation Group: 149, 154
479th Fighter Day Wing (FDW)/TFW: 125–126, **126**, 154, **155**
492nd TFS: 153, **153**
493rd TFS: **148**, 153
506th TFW: 154
510th TFS: 138–139, **138**, 152, 154
524th TFS: **147**, **148**, 152
531st TFS: **132**, 152, 153, 154
612th TFS, Det 1: 140, 141, 145, 152, **154**
614th TFS: 152, 154, **154**
615th TFS: 139, 152, 154
1708th Ferrying Wing: **126**, 156
3525th CCTW: 154
3595th CCTW: 154
3600th CCTW: 154
4403rd TFW: 154
4510th Combat Crew Training Wing (CCTW): 154
4520th CCTW: 154–155
4525th CCTW/FWW: 155
4530th CCTW: 155
4758th Defense Systems Evaluation Squadron: **147**, 150, 153, 156
4925th OMS: 156
6000th Operations Wing: 155
6021st Reconnaissance Squadron: 129, 155
6200th Air Base Wing: 155
6234th TFS: 146, 156
6585th Test Group: 149, 155
7272nd Flying Training Wing: 150, **150**
7407th Support Squadron: 129, 155
7499th Support Group: 155
Aerospace Medical Division: 156
Air Combat Command: 14
Air Education and Training Command: 14
Air Force Flight Test Center (AFFTC): **12**, **124**, **125**, 156
Air Force Logistics Command (Sacramento Air Materiel Area): 156, **156**
Air Force Security Assistance Center: 84
Air Force Space Command: 14
Air Force Special Operations Command: 14
Air Force Special Weapons Center: 156, **156**
Air Mobility Command: 12, 14–15
Air Mobility Command Air Operations Squadron (AMCAOS): 84
Air Research and Development Command: 156
Armament Development Test Center: 156
Eighteenth Air Force: 12

Pacific Air Forces (PACAF): 15, **132**, 136
'Thunderbirds' aerobatic display team: 122, **124**, **129**, 131–132, **133**, 134, **135**, **147**, 150, 154
Wright Air Development Center: **125**, 130, **130**, 134, 156
Air Forces Europe (USAFE): 15, **129**, 132, 147
'Skyblazers' aerobatic display team: **129**, 132, 152
Air National Guard: 78, 129, 131, 140, 147, 155–156
Arizona, 152nd FIS/TFTS: 126, **126**, **129**
Arizona, 162nd TFG: **147**, 156
Arkansas, 188th TFG: 155
Colorado, 120th TFS: **140**, 155–156
Colorado, 140th TFS: **147**, 155–156
Connecticut, 103rd TFG: 155
Connecticut, 118th TFS: 126, **126**, **147**, 155
District of Columbia (DC), 113th TFG: 155, **155**
District of Colombia, 121st TFS: **140**, 155, **155**
Georgia, 116th TFG: **147**, 155
Indiana, 122nd TFG: 155
Indiana, 113th TFS: 141, 147, 156, **156**
Indiana, 181st TFG: 156, **156**
Iowa, 124th TFS: 129, 155
Iowa, 132nd TFG: 155
Iowa, 174th TFS: **140**, 156
Iowa, 185th TFG: 156
Kansas, 184th TFG: **129**, 155
Louisiana, 122nd TFS: 156, **156**
Louisiana, 159th TFG: 156, **156**
Massachusetts, 102nd TFG: 155
Massachusetts, 104th TFG: 155
Michigan, 127th TFG: 155
Missouri, 131st TFG: 155
New Jersey, 119th TFS: 129, 155
New Jersey, 177th TFG: 156
New Mexico, 150th TFG: 155
New Mexico, 188th FIS/TFS: 126, **126**, **147**, 156
New York, 107th TFG: 155
Ohio, 121st TFG: 155
Ohio, 162nd TFS: 141, **147**, 156
Ohio, 166th TFS: **129**, 155
Ohio, 178th TFG: 155
Ohio, 179th TFS: 156
Ohio, 180th TFG: 156
Oklahoma, 125th FS 'Tulsa Vipers': 20–21, **20**, **21**
Oklahoma, 138th FW: 20–21, **20**, **21**
Oklahoma, 138th TFG: 155
South Dakota, 114th TFG: 155
South Dakota, 175th TFS: **147**, 155
Texas, 149th TFG: 156
Army: 82–84, 156, **156**
1st Military Intelligence Battalion (Aerial Exploitation) (MIB(AE)): 80, 82
2nd MIB(AE): 80
3rd MIB(AE): 80, 82
7th Radio Research Field Station: 84
15th MIB(AE): 80, 82
6-52nd Aviation Regiment (Theater Aviation): 83
138th Aviation Company (Radio Research): 75
146th ASA Company: 76
1-158 AVN: **88**
1-185 AVN: **93**
204th MIB(AE): 82
1-214th Aviation Regiment (Theater Aviation): 83
1-223rd Aviation Regiment (Training): 82
224th AVN BN (Radio Research): 84
224th MIB(AE): 80, **81**, 82
2-228th Aviation Regiment (Theater Aviation): 83, **92**
305th MIB (Training): 82
330th ASA Company: 82
Aeromedical Center Air Ambulance Detachment: 83
Airborne Engineering Evaluation Support Branch: **93**
Air Traffic Control Activity (USAAT-CA): **88**, **91**
Aviation Technical Test Center: 83, 93
Communications-Electronics Command (CECOM) Flight Activity: **82**, 82
Davison Aviation Command: 92, **92**
Echelon Above Corps Aviation Intelligence Company: 81
ECOM: **90**
EW Co: **90**
Operational Support Airlift Command (OSACOM): 78, 79, 83, **83**, **88**, 92
OSACOM Det 42: **77**, 83
Security Agency (USASA) Group: 75
Soldier, Biological and Chemical Command Aviation Det: 79, 83
White Sands Missile Range: 83
Army Reserve Command: **75**
Department of Homeland Security: 87, **87**
Department of Transportation (FAA): 87, **87**
Flight Systems Inc.: 150, **150**
Marine Corps: 86–87
Marine Aircraft Group 12 (MAG-12): 108, **110**, 115
Marine Aircraft Group 33 (MAG-33): 105, 106–107, **107**, **110**
Marine Force Atlantic: 87
Marine Force Pacific: 86
Marine Forces Reserve: 87
VMA-212: **105**, **115**, 116
VMA-312 'Checkerboards': **104**, 113, **114**, 118
VMA-323 'Death Rattlers': 112, **115**, 118, **118**
VMA-332 'Polka Dots': 118, **119**

VMF-214 'Black Sheep': 105–107, **106**, 108, 109, **109**, 110, 111, **111**, 112–113, 114, 115, **115**, 118
VMF-312 'Checkerboards': 108, 111, 112, **112**, 113, 114, **114**, 115–116, 117–119, **118**
VMF-323 'Death Rattlers': **104**–**105**, 105–107, **107**, 108, 109, 110, **110**, 111, **111**, 112–113, 114, **115**, 118
VMF(N)-513 'Flying Nightmares': 105, 106–107, 108, 113, 114, 116, **116**, **117**, 118
VMF(N)-542: 108
National Advisory Committee for Aeronautics (NACA): **151**, 156
National Air & Space Administration (NASA): 87, **87**, **151**, 156
Navy 85–86
Chief Naval Air Training Command: 86
Commander US Naval Air Systems Command: 86
Commander US Naval Atlantic Fleet: 85
Commander US Naval Force Reserve: 85
Commander US Naval Forces Central: 85
Commander US Naval Forces Europe: 85
Commander US Naval Forces Southern Command: 85
Commander US Naval Pacific Fleet: 85
CVW-11: 50, **50**, 51, **56**, **66**
Naval Test Pilot School: **75**, 92
Patrol and Reconnaissance Wing One: 13
Patrol and Reconnaissance Wing Two: 13
VAQ-128 'Fighting Phoenix': 12
VAQ-129 'New Vikings': **73**
VAQ-133 'Wizards': 12–13
VAQ-209 'Star Warriors': **73**
VC-8 'Redtails': 13
VF-154 'Black Knights': 13, 71
VFA-2 'Bounty Hunters': 62, **62**, 63, **63**
VFA-14 'Tophatters': **36**–**37**, 51, **51**, **55**, **56**, 59, 62, **64**–**66**
VFA-22 'Fighting Redcocks': 71
VFA-41 'Black Aces': **49**, 50, **50**, **54**, **56**, 57, 59, 60, 62
VFA-97 'Warhawks': 12
VFA-102 'Diamondbacks': 13, 58, **58**, 59, **59**, 62
VFA-115 'Eagles': **41**, **47**, 48, **48**, **52**, **53**, **54**, 57, 62, 70–71, **70**
VFA-122 'Flying Eagles': **38**, 41, 43, **43**, **44**, **46**, 61–62, **62**, 63
VFA-137 'Kestrels': 62, **71**, **71**
VFA-154 'Black Knights': 13, 62
VMX-22: 13
VRC-30: 78
VT-31: 77
VX-9 'Vampires': **41**, **41**, **42**, **47**, 50, 51, 52–53, 55, **63**, **64**
VX-23 'Salty Dogs': 61, **61**
VX-31 'Dust Devils': 60, **60**, 70
US King Air Operators: 82–87, **82**–**87**
US Military King Airs: 74–93, **74**–**93**
Ute (see Beech)

V

Valley Forge, USS: 110
Variant File – US Military King Airs: 74–93, **74**–**93**
VC-6A (see Beech)
VC10 (see Vickers)
VC-137C (see Boeing)
Vickers VC10: 102, **102**
Vietnam, North
Vietnamese People's Air Force: 138–139
Viking, S-3B (see Lockheed Martin)
Voodoo, RF-101 (see McDonnell)
Vought
A-7 Corsair: 11, 122
AU-1 Corsair: 112, **115**, **116**
F4U Corsair: 104–119, **104**–**115**, **118**, 119
F4U-4: **112**, 113
F4U-4B: **104**–**105**, **106**, 107, **107**, 110, **111**, **114**, 119
F4U-5N Corsair: 95, 105, 107, 116, **116**, 117, **117**, 118
F4U-5P: 107

W

Wal, G-52 (see Dornier)
Warplane Classic – North American F-100 Super Sabre: 120–157, **120**–**157**
Wild Weasel I, F-100 (see North American)
Wild Weasel III, F-105F (see Republic)

X

X-1 (see Bell)
X-15 (see North American)
X-20 Dyna-Soar (see Boeing)

Y

Yakovlev Yak-130: 4
YF-86D (see North American)
YF-100A/C (see North American)
YF-107A (see North American)
YQF-100D (see North American)
YU-21 (see Beech)

Picture acknowledgments

Front cover: Boeing, Warren Thompson, Larry Davis Collection. **4:** Lockheed Martin, Roberto Yañez. **5:** Lockheed Martin, Boeing. **6:** Peter R. March (three), USAF via Tom Kaminski. **7:** RAAF via Nigel Pittaway, Nigel Pittaway, Jamie Hunter. **8:** Alexander Mladenov, Tom Kaminski, Konstantinos Dimitropoulos. **9:** Peter R. Foster (three), Bronco Aviation (two). **10:** Konstantinos Dimitropoulos, Jamie Hunter. **11:** Holger Stüben (two), Jens Schymura (two), BAE Systems. **12:** Peter R. March (three), AFFTC via Tom Kaminski. **13:** Jos Schoofs, Yaso Niwa (two). **14:** Jos Schoofs (two), Daniel J. March. **15:** Didier Lagutere, Gert Kromhout, RAAF via Nigel Pittaway. **16:** Erich Strobl (three). **17:** Erich Strobl, via Warren Thompson. **18:** Stefan Degraef/Edwin Borremans (two). **19:** Stefan Degraef/Edwin Borremans (two), AETE via Stefan Degraef/Edwin Borremans. **20:** Stefan Degraef/Edwin Borremans (two), AETE via Stefan Degraef/Edwin Borremans, Andy Wolfe. **21:** Andy Wolfe. **22:** Embraer via Robert Hewson, Embraer via Robert Hewson, Peter Liander. **23:** Embraer via Robert Hewson (two). **24:** Peter Liander, via Robert Hewson, Saab. **25:** Embraer. **26:** Konstantinos Dimitropoulos (four). **27:** Peter Liander. **28:** Embraer via Robert Hewson (two). **29:** Peter Liander (four). **30:** Embraer via Robert Hewson, Ericsson. **31:** Embraer, David Willis. **32–35:** Luigino Caliaro. **36–37:** Ted Carlson/Fotodynamics. **38:** Boeing via Terry Panopalis. **39:** Boeing via Terry Panopalis (two). **40:** Tom Kaminski (five), Boeing via Terry Panopalis (two). **41:** Boeing via Terry Panopalis (two), Ted Carlson/Fotodynamics, Boeing. **42:** Ted Carlson/Fotodynamics, Boeing. **43:** Rick Llinares (three). **44:** Ted Carlson/Fotodynamics (two). **45:** Ted Carlson/Fotodynamics. **46:** Ted Carlson/Fotodynamics (two). **47:** Boeing, Jonathan Chuck, Eric Katerburg. **48:** Eric Katerburg (three). **49:** Boeing (two). **50:** Boeing (two), Ted Carlson/Fotodynamics. **51:** Ted Carlson/Fotodynamics (three). **53:** VFA-115 via Tony Holmes (three), US Navy. **54:** VFA-115 via Tony Holmes (three), US Navy. **55:** US Navy (two), VFA-14 via Tony Holmes. **56:** US Navy (two), VFA-41 via Tony Holmes. **57:** VFA-41 via Tony Holmes (two). **58:** Tony Holmes (three). **59:** US Navy (four). **60:** Eric Katerburg (two), via Jim Winchester. **61:** US Navy, Eric Katerburg (two). **62:** Tony Holmes, VFA-2 via Tony Holmes. **63:** Tony Holmes (three), Eric Katerburg, Boeing. **64:** VFA-115 via Tony Holmes (five), Eric Katerburg, Boeing, US Navy. **67:** Ted Carlson/Fotodynamics, General Electric. **69:** Ted Carlson/Fotodynamics (six), VFA-115 via Tony Holmes (two). **70:** Boeing, US Navy. **71:** Tony Holmes (four). **72:** Boeing, Boeing via Brad Elward. **73:** Paul Bird via Brad Elward (two), Boeing, Ted Carlson/Fotodynamics. **74:** Ted Carlson/Fotodynamics, Raytheon. **75:** Tom Kaminski, Mike Anselmo, via Dennis Buley. **76:** Frank MacSorley via Tom Kaminski, Stuart Freer, Tom Kaminski, Peter R. Foster, Richard Sullivan via Stephen Miller. **77:** Ted Carlson/Fotodynamics. **78:** Jonathan Chuck, Ted Carlson/Fotodynamics (two). **79:** Raytheon, via Dennis Buley. **80:** Don Spering/AIR, Carey Mavor (two). **81:** Carey Mavor, Ted Carlson/Fotodynamics. **82:** US Army CECOM via Tom Kaminski, Don Spering/AIR, Robert Tourville via Mike Wilson. **83:** Pete Becker, via Tom Kaminski. **84:** via Dennis Bailey, via Tom Kaminski, via Don Spering/AIR, Keith Riddle. **85:** via Robert F. Dorr, Pete Becker via Tom Kaminski. **86:** Tom Kaminski, Pete Lakatosh via Tom Kaminski, Ted Carlson/Fotodynamics (two). **87:** Robert F. Dorr, Tom Kaminski, Ted Carlson/Fotodynamics, Carey Mavor. **88:** Raytheon, T.H. Brewer via Tom Kaminski, via Tom Kaminski (two). **89:** Tom Kaminski, via Dennis Bailey (two). **90:** Raytheon, via Tom Kaminski, Frank MacSorley via Tom Kaminski, via Dennis Bailey. **91:** Tom Kaminski, Brian Rogers via Tom Kaminski, via Dennis Bailey, via Tom Kaminski. **92:** Frank MacSorley via Tom Kaminski, Norris Graser via Tom Kaminski, US Army via Tom Kaminski, Eugene Zom via Stephen Miller. **93:** via Tom Kaminski, via Dennis Bailey, via Tom Kaminski, Keith Riddle, Tom Kaminski, Eugene Zom via Stephen Miller. **94–99:** Santiago Rivas/Juan Carlos Cicalesi. **100:** Santiago Rivas/Juan Carlos Cicalesi (two), Cees-Jan van der Ende. **101:** Santiago Rivas/Juan Carlos Cicalesi (two), Cees-Jan van der Ende. **102:** Santiago Rivas/Juan Carlos Cicalesi (five), Cees-Jan van der Ende. **103:** Cees-Jan van der Ende. **104:** Ed Godfrey via Warren Thompson, J.J. Geuss via WT, USMC via WT. **105:** Doug Canning via WT. **106:** John Perrin via WT (two), via WT, Bill Mitchell via WT (two). **107:** Conrad Buschmann via WT (two), George Glauser via WT, John Perrin via WT, Al Grasselli via WT, Gerald Haddock via WT, Ed Godfrey via WT. **108:** Ray Stewart via WT, Al Grasselli via WT. **109:** TRH, Richard Albert via WT, Al Grasselli via WT, John Perrin via WT. **110:** John Corrigan via WT, via WT, Bill Rockwell via WT, Vance Yount via WT. **111:** via WT, Bill Rockwell via WT, Richard Albert via WT (two), Clarence Chick via WT, John Perrin via WT. **112:** Ed Godfrey via WT, J.J. Geuss via WT, George Kubal via WT (two). **113:** J.J. Geuss via WT, George Kubal via WT, E.H. Yeager via WT, Jim Bailey via WT, Tom Archer via WT. **114:** Robert Howard via WT, Leo Reigel via WT, Stu Nelson via WT, Tom Archer via WT. **115:** Ben Sowaske via WT, Jack Dunn via WT, TRH, USMC via WT. **116:** John Corrigan via WT, Russ Janson via WT, Doug Carter via WT. **117:** Marvin Wallace via WT, via WT, Gene Derrickson. **118:** John Hughes via WT, Ernie Banks via WT, Vance Yount via WT, Ohman Collection via WT, Harold Taylor via WT. **119:** Art Beasley via WT (two), Jim Hallet via WT, H.A. Gamblin via WT. **120–121:** Robert F. Dorr. **122:** Robert F. Dorr, Larry Davis Collection. **123:** Larry Davis Collection, USAF via Terry Panopalis, NAA. **124:** AFFTC via Terry Panopalis, Larry Davis Collection (two), Terry Panopalis. **125:** NASA Dryden via Terry Panopalis, Terry Panopalis. **126:** Larry Davis Collection (three), Terry Panopalis, Peter R. Foster. **127:** NAA, Larry Davis Collection (two). **128:** Larry Davis Collection, Curt Burns via Warren Thompson, Don Scott via Warren Thompson. **129:** Larry Davis Collection (three), Robert F. Dorr. **130:** Terry Panopalis (three), Larry Davis Collection. **131:** Larry Davis Collection, AFFTC via Terry Panopalis, NAA, Terry Panopalis. **132:** Larry Davis Collection (three), Don Scott via Warren Thompson, Bill Myers via Warren Thompson, Terry Panopalis, Chris Ryan. **134:** Larry Davis Collection, Terry Panopalis, USAF via David W. Menard via Terry Panopalis. **135:** Larry Davis Collection, Boeing via Warren Thompson, Terry Panopalis, Bob Terbet via Warren Thompson. **136:** Aerospace (two), Peter R. Foster, Larry Davis Collection (two). **137:** Larry Davis Collection, Aerospace (two), Terry Panopalis. **138:** Bob Terbet via Warren Thompson, Robert F. Dorr. **139:** Robert F. Dorr (three). **140:** Robert F. Dorr, USAF, USAF via Terry Panopalis, Terry Panopalis, Larry Davis Collection. **143:** Robert F. Dorr. **144:** Larry Davis Collection. **145:** Robert F. Dorr, USAF, Larry Davis Collection. **146:** Larry Davis Collection (three). **147:** Terry Panopalis (four), Larry Davis Collection (two). **148:** Larry Davis Collection, Terry Panopalis (two), Peter R. Foster. **149:** Don Spering via Terry Panopalis, David Donald, Larry Davis Collection. **150:** Charles Hutchinson via Warren Thompson, Chris Ryan, EADS, Terry Panopalis (two). **151:** NASA Dryden via Terry Panopalis (three), Terry Panopalis. **152–155:** Larry Davis Collection. **156:** Terry Panopalis (four), AFFTC via Terry Panopalis. **157:** Larry Davis Collection (three), Peter R. Foster. **158:** Dr Alfred Price (three). **159:** Dr Alfred Price. **160:** Dr Alfred Price (two), Aerospace (three). **161:** Aerospace (three). **162:** David Willis (two), Aerospace, Dr Alfred Price (two). **163:** Aerospace, Dr Alfred Price (four). **164:** Dr Alfred Price (two), TRH Pictures (two). **165:** Aerospace (two), Dr Alfred Price (three). **166:** Aerospace, TRH Pictures. **167:** Dr Alfred Price (three), TRH Pictures. **168:** Dr Alfred Price (four), Aerospace (two). **170:** Aerospace, Dr Alfred Price. **171:** Dr Alfred Price. **172–173:** Aerospace.